Praise for *Growing and Marketing Ginseng, Goldenseal*

This book is the complete resource for ginseng. I recommend this book to our members and visitors who are interested in learning and growing their own green gold. Thank you for bringing so much helpful and useful information.

—Michael S. Lee, President of WildGrown.com

The definitive guide to growing our shade-loving native medicinal plants for fun and profit. As an ecologist and conservation biologist, I particularly appreciate the fact that woodlot owners can help take some pressure off wild populations through careful cultivation of medicinals on appropriate plots in their forested landscape. By passing on lessons from their vast hard-won experience in this enterprise, Davis and Persons have done a great service.

—James B. McGraw, Eberly Professor of Biology, Department of Biology, West Virginia University

My office is like a shell midden, with the oldest and most decomposed material at the bottom, and more recent and vital material located somewhere near the surface. I am delighted to report that my copy of Davis and Persons' book has always stayed right at the top! wUpgrades to the information found in this book will help maintain this tradition of growing valuable plants close to home, and I believe the plants will join me in thanking the authors for a job well done. May we all go out to the woodlands, drop to our knees in the cool, soft earth, and cultivate, for the love of life, a rare plant.

—Richo Cech, herbal author and gardener at Horizon Herbs, LLC in Williams, Oregon.

The recent popular interest in wild American ginseng spurred by high prices in Asian markets means that now more than ever it's important to create cultivated woods-simulated supplies of American ginseng and other woodland medicinal plants. Scott Persons and Jeannine Davis have combined decades of experience and expertise to create the most significant, must-have reference on growing ginseng, goldenseal, and other woodland medicinal plants. Anyone interested in understanding any aspect of wild American ginseng, it's biology, history, economics, and the practical details of production needs this book.

—Steven Foster, Senior author, *Peterson Field Guide to Medicinal Plants.*

Important revised work on how we can encourage conservation through cultivation of two medicinal and economically important plants that have been on United Plant Savers At-Risk list since UpS created this list.

—Susan Leopold, Executive Director of United Plant Savers.

This work is a plant lover's treasure. What Jeanine and Scott have accomplished with this book will be revealed for years to come, as the layers of wisdom and knowledge are deep. Scientists, herbalists, growers, conservationists, native plant enthusiasts, 'plantophiles' in general will thrill at the research and clear, user–friendly information that is in these pages. You can easily tell that these authors have made this a life long passion and profession

—Kathleen Maier, RH (AHG), Sacred Plant Traditions, LLC

This book is required reading for anyone interested in growing ginseng and other woodland botanicals in a shady site. Persons and Davis have captured the wisdom of a generation of ginseng growers in this comprehensive book, now updated to include practical information for home gardeners who want to enrich a patch of woods with native medicinal plants.

—Barbara Pleasant, award-winning garden writer and contributing editor to *Mother Earth News*

The first edition of this book became an instant classic in the fields of medicinal plant horticulture, sustainable agriculture, and agroforestry. With this latest edition, Scott and Jeanine have remarkably managed to expand, improve and update this classic so that it is now even more useful and full of up-to-date information. Their combined knowledge, experience and wisdom is abundant throughout this book. I heartily recommended this updated edition to anyone interested in native woodland plants and their culture.

—Eric P. Burkhart, PhD., Program Director, Plant Science, Shaver's Creek Environmental Center, The Pennsylvania State University

Scott Persons and Jeanine Davis have written the only accurate and comprehensive grower's guide to woodland cultivation of American Ginseng and other forest medicinal and culinary herbs. This new revision of their original book is a significant improvement over the first edition, with updated and expanded information. In addition to being full of practical "how to" data, based on both University peer reviewed research, as well as first-hand knowledge and experience, it is a delightful and easy to read textbook. I consider this book as truly a "must read" for anyone who is seriously interested in pursuing this form of Agroforestry.

—Bob Beyfuss, American Ginseng Specialist, Cornell University Cooperative Extension (retired)

This unique book is a comprehensive guide on the history, production and marketing of medicinal plants native to the forests of eastern North America. Practical experiences are included from both an American and Canadian perspective. It is a valuable, easy to read resource for both the beginner and experienced grower.

—Dr. Sean Westerveld, Ginseng and Medicinal Herbs Specialist, Ontario Ministry of Agriculture and Food and Ministry of Rural Affairs

GROWING AND MARKETING

Ginseng, Goldenseal and other Woodland Medicinals

REVISED AND UPDATED

Jeanine Davis and W. Scott Persons

Cover design by Diane McIntosh.

Printed in Canada. First printing June 2014.

New Society Publishers acknowledges the financial support of the Government of Canada through the Canada Book Fund (CBF) for our publishing activities.

Inquiries regarding requests to reprint all or part of *Growing and Marketing Ginseng, Goldenseal & Other Woodland Medicinals* should be addressed to New Society Publishers at the address below.

To order directly from the publishers, please call toll-free (North America) 1-800-567-6772, or order online at www.newsociety.com

Any other inquiries can be directed by mail to:

New Society Publishers
P.O. Box 189, Gabriola Island, BC V0R 1X0, Canada
(250) 247-9737

LIBRARY AND ARCHIVES CANADA CATALOGUING IN PUBLICATION

Persons, W. Scott, 1945–,

[Growing & marketing ginseng, goldenseal & other woodland medicinals]
Growing and marketing ginseng, goldenseal and other woodland medicinals / Jeanine Davis and W. Scott Persons.—Revised and updated 2nd edition.

Revision of: Growing & marketing ginseng, goldenseal & other woodland medicinals / W. Scott Persons, Jeanine M. Davis.—Fairview, N.C. : Bright Mountain Books, c2005.
Includes bibliographical references and index.
Issued in print and electronic formats.
ISBN 978-0-86571-766-4 (pbk.).—ISBN 978-1-55092-563-0 (ebook)

1. American ginseng. 2. Goldenseal. 3. Medicinal plants. I. Davis, J. M. (Jeanine Marie), 1955–, author II. Title. III. Title: Growing & marketing ginseng, goldenseal & other woodland medicinals

SB295.G5P48 2014 633.8'8 C2014-901910-6
C2014-901911-4

New Society Publishers' mission is to publish books that contribute in fundamental ways to building an ecologically sustainable and just society, and to do so with the least possible impact on the environment, in a manner that models this vision. We are committed to doing this not just through education, but through action. The interior pages of our bound books are printed on Forest Stewardship Council®-registered acid-free paper that is 100% post-consumer recycled (100% old growth forest-free), processed chlorine-free, and printed with vegetable-based, low-VOC inks, with covers produced using FSC®-registered stock. New Society also works to reduce its carbon footprint, and purchases carbon offsets based on an annual audit to ensure a carbon neutral footprint. For further information, or to browse our full list of books and purchase securely, visit our website at: www.newsociety.com

Contents

List of Tables ix

Preface I (*Scott Persons*) xi

Preface II (*Jeanine Davis*) xiii

Author Biographies xv

Abbreviations and Definitions (*Jeanine Davis*) xvii

General Introduction (*Jeanine Davis and Scott Persons*) 1

Part One: American Ginseng (*Scott Persons*) 3

1. American Ginseng: Its Life Cycle, Range, Related Species, and Government Regulation (*Scott Persons*) 7
2. History of the Ginseng Trade: Ancient China to the New Millennium (*Scott Persons*) 19

Part Two: A Ginseng Grower's Manual (*Scott Persons*) 43

3. Under Artificial Shade (*Scott Persons*) 45
4. Wild-Simulated Planting (*Scott Persons*) 59
5. Woods Cultivation (*Scott Persons*) 89
6. The Harvest: Picking Berries and Stratifying Seeds, Digging and Drying Roots (*Scott Persons*) 127
7. Business Decisions and the Future Market Outlook (*Scott Persons*) 147
8. A Grower Tells His Own Story: Oscar Wood (*Scott Persons*) 165
9. Ginseng Resources 173
10. Ginseng References 177

Part Three: Other Species of Green Gold: Goldenseal and Ramps (*Jeanine Davis*) . . . 183

11. Goldenseal: Its History, Range, Description, Uses, and Government Regulation (*Jeanine Davis*) . . . 185
12. Goldenseal Growing Instructions: Methods, Care, Protection, Harvesting, and Marketing (*Jeanine Davis*). . . . 193
13. Goldenseal Growers' Stories (*Jeanine Davis*) . . . 229
14. Ramps: History, Description, and Uses (*Jeanine Davis*). . . . 249
15. Ramps Growing Instructions: Methods, Care, Protection, Harvesting, and Marketing (*Jeanine Davis*). . . . 255
16. Ramps Growers' Stories (*Jeanine Davis*) . . . 269

Part Four: There Are Many Other Woodland Medicinals You Can Grow (*Jeanine Davis*) . . . 283

17. Bethroot (*Jeanine Davis*). . . . 285
18. Black Cohosh (*Jeanine Davis*) . . . 291
19. Bloodroot (*Jeanine Davis*) . . . 303
20. Blue Cohosh (*Jeanine Davis*). . . . 315
21. False Unicorn (*Jeanine Davis*) . . . 323
22. Galax (*Jeanine Davis*) . . . 329
23. Mayapple (*Jeanine Davis*) . . . 335
24. Pinkroot (*Jeanine Davis*). . . . 341
25. Spikenard (*Jeanine Davis*) . . . 345
26. Wild Ginger (*Jeanine Davis*). . . . 349
27. Wild Indigo (*Jeanine Davis*) . . . 353
28. Other Forest Botanicals Growers' Stories (*Jeanine Davis*). . . . 357

Part Five: Growing Woodland Medicinals in the Home Garden (*Jeanine Davis*) . . . 371

29. Making the Perfect Woodland Garden Site (*Jeanine Davis*). . . . 373
30. Choosing the Plants to Grow in Your Garden (*Jeanine Davis*) . . . 385
31. How to Grow a Garden in the Woods (*Jeanine Davis*) . . . 393
32. Ginseng—A Horticultural Challenge (*Scott Persons*) . . . 403

33. Making Some Simple Products from Your Woodland Medicinals (*Jeanine Davis*) . . . 407

34. Home Gardeners' Stories (*Jeanine Davis*). . . . 409

Part Six: Sustainable Wild-harvesting (*Jeanine Davis*) . . . 431

35. What Is Wild-harvesting? (*Jeanine Davis*) . . . 433

36. Why There Will Always Be a Place for Wild-harvesting (*Jeanine Davis*). . . 437

37. Rules and Regulations for Wild-harvesters (*Jeanine Davis*) . . . 439

Part Seven: Supplemental Information (*Jeanine Davis*) . . . 441

Appendix 1: Forest Botanicals Bought and Sold in the United States and Canada (*Jeanine Davis*) . . . 443

Appendix 2: Prices Paid for Forest Botanicals from 2004 through 2013 (*Jeanine Davis*) . . . 451

Appendix 3: Interesting and Helpful Calculations, Tables, and Miscellaneous Information (*Jeanine Davis*) . . . 473

Appendix 4: Good Agricultural, Collection, and Manufacturing Practices (*Jeanine Davis*) . . . 475

Appendix 5: Joe-Ann McCoy's Disease List (*Joe-Ann McCoy*) . . . 477

References and Resources for Parts Three through Six . . . 483

Comprehensive Resource Directory . . . 493

Index . . . 499

Tables

1. Recent Prices for Wild and for Field-cultivated Ginseng 41
2. Profit and Loss Statement for One-tenth Acre of Ginseng Grown under Artificial Shade 57
3. Comparison of Average Calcium Levels Found in Soils with Wild Ginseng Stands of Varying Quality 64
4. Projected Nine-year Budget for One-half Acre of Wild-simulated Ginseng . 82
5. Visual Site Assessment and Grading Criteria for Potential Woodland Ginseng Growing Operation for a Northern Forest. 93
6. Ginseng Root Weight Response to Four Plant Spacings at Ages 2 through 6 Years 101
7. Suggested Ginseng Pest Control Guide. 119
8. Projected Seven-year Budget for One-half Acre of Woods-cultivated Ginseng. 122
9. Ginseng Grower's Calendar for Intensive Woods Cultivation 125
10. Estimated Woods-grown Ginseng Farming by State in 2000 160
11. Estimated U.S. Woodland Ginseng Farming in 1994 and 2000 161
12. Four-year Hypothetical Farm Enterprise Budget for One-tenth Acre of Certified Organic, Woods-cultivated Goldenseal Using the Lowest Yields Reported by Growers or Obtained from Research and Demonstration Plots 225
13. Four-year Hypothetical Farm Enterprise Budget for One-tenth Acre of Certified Organic, Woods-cultivated Goldenseal Using the Highest Yields Reported by Growers or Obtained from Research and Demonstration Plots 226

14. Four-year Hypothetical Farm Enterprise Budget for One Acre Certified Organic Goldenseal Grown under Artificial Shade 227
15. Vestal's Estimated Five-year Budget for One-half Acre of Forest-grown, Wild-simulated Goldenseal from Planting to Harvest in 2003 and 2013 . . 235
16. John's Budget for One Acre of Certified Organic Goldenseal Grown under Artificial Shade for Three Years in 2003 and 2013 247
17. Farm Enterprise Budget for One-tenth Acre of Ramps Grown in the Forest. 268
18. Farm Enterprise Budget for One Acre of Black Cohosh Grown under Artificial Shade 300
19. Farm Enterprise Budget for One Acre of Bloodroot Grown under Artificial Shade 313
20. Forest Botanicals Price Information 453
21. Estimated Plants per Acre 474
22. Commonly Used Fahrenheit-to-Celsius Conversions. 474
23. Other Handy Conversions. 474

Preface I

When I realized that my old book, *American Ginseng: Green Gold,* was rapidly becoming outdated and that a new book was needed, I thought that many of my potential readers would be interested in practical, detailed information and instruction on growing other valuable native woodland medicinal herbs—other species of green gold—as well as ginseng. I asked Dr. Jeanine Davis to be a co-author and cover the additional material. Dr. Davis and I have interacted professionally for many years. I grow American ginseng and a little goldenseal on wooded hillsides in western North Carolina at the edge of the Great Smoky Mountains. Dr. Davis is a professor at North Carolina State University's Mountain Horticultural Crops Research and Extension Center, where she conducts research on a wide variety of native woodland botanicals. Dr. Davis works only about an hour's drive northeast of me, and we often share information, and sometimes we find ourselves speaking at the same conferences—I on woodland ginseng production and she on the cultivation of goldenseal, ramps, and many other native herbs. Our approaches to small-scale farming and our advice to prospective growers are similar and compatible.

There is a great deal of material available, both in print and on the Internet that discusses growing woodland botanicals. Some of the information is excellent, but a significant chunk is partial disinformation. It is often not based on sound research—or even on more than one grower's experience—and profitability is not forthrightly assessed. Cultivating native woodland medicinal herbs in a sustainable manner is often advocated primarily as an enjoyable, even noble, activity. Of course, it *is* a noble and enjoyable activity (or it can be), but Dr. Davis and I have a more hard-core point of view: We are interested in using best management practices and in turning a profit.

Many people helped along the way as Dr. Davis and I researched, wrote, and prepared this text for publication. Plant scientists, agriculture extension specialists, herb growers, and herb buyers freely shared their expertise, and many are acknowledged by name within the text. However, we wish to express special thanks to some of those who are not mentioned by name: Dr. William G. Bailey (deceased), professor and ginseng researcher at Simon Fraser University in British Columbia; Claude Deyton, agricultural technician in Yancey County, North Carolina; Ed Fletcher, chief operating officer of Strategic Sourcing, Inc.; Tony Hayes, president of Ridge Runner Trading Co.; Michael McGuffin, president of the American Herbal Products Association; Al Oliver, ginseng specialist for the British Columbia Ministry of Agriculture and Fisheries (retired); John T. A. Proctor, ginseng researcher in the Department of Horticultural Sciences at Guelph University in Ontario (retired); Jan Schooley, ginseng and medicinal herb specialist with the Ontario Ministry of Agriculture at the Simcoe Research Station (retired); and Robin Suggs, executive director of the Yellow Creek Botanical Institute. We also wish to express our great appreciation to Karen Hardy and Jackie Greenfield who researched references and helped in many other ways to provide information for the manuscript.

Because we have limited photographic skills, we are indebted to a broad spectrum of friends, colleagues, growers, and even professional photographers for supplying the photographs that illustrate the text. We thank them all here and give credit next to their pictures. Lastly, we are particularly grateful for the generous contributions of two horticultural experts with experience writing about the propagation of woodland herbs. Robert Beyfuss, ginseng grower, researcher, and Cornell cooperative extension agent (retired) for Greene County, New York, reviewed the original manuscript for the ginseng section of the book, suggested needed improvements, and even contributed photographs. Richo Cech of Horizon Herbs reviewed the second section and made suggestions covering all the other native forest botanicals.

— W. Scott Persons, 2005, 2007, 2013

Preface II

When it came time to update our book for the second time, Scott and I agreed we should make a few changes. Since the book was first published in 2005, I have received hundreds, if not thousands, of requests from home gardeners wanting to grow woodland botanicals. Every year I offer propagation workshops and speak at herb conferences, botanical gardens, and to Master Gardeners about how to grow one's own forest medicine. Many of the people who attend have already purchased our book, and they tell us how much they appreciate it. But we wrote the book for commercial growers, and it definitely has that angle to it. When I speak to home gardeners and hobbyists, I don't talk to them the same way that I speak to commercial growers. So, for this edition of the book, we added a section specifically for the home gardener.

Over the past eight years, smartphones and tablets (the electronic kind) have become ubiquitous. People who did not have reliable or fast Internet connections in 2005 now have instant access. In light of this, we changed some of the reference sections; now there are fewer snail mail addresses and phone numbers, and more website URLs and email addresses. We still included names and contact information for some of the companies offering plants, seeds, and supplies, but there are many more out there. Just use a search engine to find them, and of course, check out their quality and reliability. Finally, I expanded the table that contained raw material prices to provide a historical perspective on more than 60 forest medicinals bought and sold in North America. I also added sections on wild-harvesting and the federal regulations on dietary supplements that impact growers. And lastly, we have made this book available in ebook format so you can carry it with you wherever you go.

Once again, many people have helped us make this new book a reality. We want to thank Cynthia Bright, our first publisher, for all her help and support in transitioning this book over to a new publisher. This was her last project before she retired from the publishing business. I especially want to thank Bob Beyfuss, Eric Burkhart, Joe-Ann McCoy, Randy Beavers, Ed Fletcher,

Tony Hayes, Jackie Greenfield, and David Cozzo for many stimulating discussions about these fascinating plants. And finally, I want to remember Andy Hankins, extension specialist with Virginia State University. He passed away suddenly in November 2012. Andy was a very special person who dedicated his life to helping others. He knew a great deal about growing ginseng and shared his information freely in publications and presentations for over 25 years.

— Jeanine Davis, 2013

Author Biographies

Jeanine M. Davis was born in Oak Park, Illinois, and has lived in many states east and west of the Mississippi. After acquiring an A.A. degree in Fine Arts, she changed majors and earned a B.S. degree in Horticulture from Delaware Valley College in Doylestown, Pennsylvania. She then moved across the country where she earned her M.S. and Ph.D. degrees in Horticulture from Washington State University. In 1988, she moved back east to join the faculty in the Department of Horticultural Science at North Carolina State University. There she is an associate professor and extension specialist at the Mountain Horticultural Crops Research and Extension Center near Asheville, North Carolina.

Jeanine's research and educational programs are diverse, covering commercial production of vegetables, new crops, medicinal herbs, organics, and most recently, biodynamics. A constant in her program over the past 25 years has been a devotion to the conservation of native medicinal herbs through cultivation. She has published numerous papers, reports, and webpages on research and activities demonstrating that cultivated herbs can provide a more reliable and consistent product for the market than wild-harvested material while at the same time helping to conserve precious native populations. Jeanine participates in numerous professional and non-profit organizations and speaks at conferences and workshops across North America. She also runs a small farm with her family, raising steer, horses, donkeys, chickens, bees, and vegetable and herb gardens.

W. SCOTT PERSONS has successfully grown American ginseng for 34 years. Born in Durham, North Carolina, he graduated from Duke University in 1967 with a B.A. in Philosophy, and then went on to earn M.A. and Ph.D. degrees in Educational Psychology from Emory University. When marriage led him to settle in western North Carolina, Dr. Persons became fascinated with the valuable woodland herb that flourished there on the heavily forested hillsides of his new wife's homestead.

Established in 1979, his woodland ginseng farm has not only supported his family but also supplied planting stock (and often advice and counsel as well) to ginseng farmers all over the country—indeed, all over the world. His first book, *American Ginseng: Green Gold* has helped woods growers from Alberta to Tasmania.

Scott represented American woods growers at the International Ginseng Conference (IGC) in Vancouver in 1994, again at IGC 1999 in Hong Kong, and at IGC 2003 in Melbourne; he also speaks frequently about ginseng at conferences in the United States. As owner/operator of Green Gold Enterprises, Inc. he provides consultative advice on all matters related to forest ginseng farming, and he brokers both domestic and international ginseng root sales. Thus, he has both extensive hands-on growing experience and a broad knowledge of the ginseng trade and the international world of ginseng.

Abbreviations and Definitions

CITES: Convention on International Trade in Endangered Species of Wild Fauna and Flora
EPA: United States Environmental Protection Agency
GAPs: Good Agricultural Practices
HRT: Hormone Replacement Therapy
SARE: Sustainable Agriculture Research and Education Program
SMNPA: Smoky Mountain Native Plants Association
USDA: United States Department of Agriculture
USFWS: United States Fish and Wildlife Service

Definitions

(written in context for this book and in easy-to-understand language):

Annual: a plant that completes its life cycle, including producing seed, in one year, and then dies.

Biennial: a plant that lives two years. It blooms only in the second year, and then dies.

Deciduous: a plant, usually referring to a tree or shrub, that sheds its leaves every year.

Dormancy (seed): a period during which a mature seed "waits", until certain conditions (often a cold period) are met before germinating.

Dormancy (bud): a period during which a bud (on a branch or an underground rhizome) cannot open until it has been exposed to a set number of hours below a certain temperature.

Herbaceous: a plant with stems and leaves that die down each fall and regrow from a perennial rhizome and/or roots each spring.

Mycorrhiza: a symbiotic (mutually beneficial) relationship between a fungus and the roots of a plant.

Perennial: a plant that lives for more than two years.

Scarification: the method of scratching, nicking, or cracking a seed coat so the seed can take in water and start the germination process.

Stratification: a treatment in which seeds are exposed to cold (often moist cold), warmth, or alternating cold/warmth to break dormancy so the seed can germinate.

General Introduction

In our complex world of cell phones, virtual shopping malls, processed foods, and managed health care, many people desire to simplify their lives and make use of what Nature has provided us. For a rapidly expanding segment of the population, this return to a more natural life includes the use of medicinal herbs. A growing number of us take herbs as a natural source of medicine, while others use them because they are often less expensive than prescription drugs. Some people want control over what they consume, so they gather or grow their own medicines and food. The forests of the United States and Canada provide habitats for many of the most popular medicinal herbs. These plants have a special mystique that spans cultures and generations.

For some time, we have noted that there is increasing interest in growing native, perennial, woodland medicinal herbs and that many people wish to gain at least some supplemental income from their production. Small landowners, if they go about it wisely, can grow many of these native medicinals profitably while preserving and even enhancing their woodlands. This book provides guidance not only in the cultivation of native forest herbs but also in the economics of their production and sale.

Aspiring herb growers are often attracted first to American ginseng, because it is the most valuable medicinal botanical and has a broad, well-established market, which has existed almost continuously for over 300 years. Indeed, in the southern part of its range, people often refer to ginseng as "green gold." The first part of this book is devoted entirely to this one native plant.

While little information exists on the production of the other species covered in this book, a good deal has already been written on growing ginseng as a commercial venture, including *American Ginseng: Green Gold* by W. Scott Persons. In writing the 2005 version of this book, *Growing and Marketing Ginseng, Goldenseal, and Other Woodland Medicinals*, we borrowed much from the by then out-of-print 1994 edition of *Green Gold*. The many North American woodland ginseng farmers who read and used that first edition will find portions of the first part of the current book to be generally familiar; however, the content has been extensively revised and rewritten to update the material and provide the most comprehensive, detailed, practical, and reliable information available on the woodland production of ginseng.

One complete chapter of *American Ginseng: Green Gold* is included in this revision. That is the interview with Oscar Wood. Oscar has passed on, but his story remains engaging and instructive to a beginning ginseng farmer; moreover, reprinting it again preserves the memory of a good and gracious man a little longer.

The second part of this book provides practical guidance in the production and marketing of other native woodland herbs that also have the potential to yield "green gold." Goldenseal and ramps are covered in detail, because their economic potential is well established and reliable information on their propagation is available. Black cohosh, bloodroot, and nine other lesser-known native botanicals are discussed as thoroughly as present knowledge allows, with emphasis on their potential and the uncertainties associated with each. There is not nearly as much information available on growing and marketing any of these herbs as there is for ginseng. Research studies, the experiences of many growers (including the authors), and the knowledge of several long-time buyers were the basis for the advice provided here. The production budgets are best estimates using all available information.

For the 2014 revision of this book, we completely updated the entire book and added a section for the growing number of gardeners, herbalists, and herb enthusiasts who want to grow these amazing plants for their own enjoyment and use. There is also some information about wild-harvesting and some of the new federal regulations concerning dietary supplements.

One of our hopes in publishing this expanded version is that it will encourage the herb grower to diversify as a means of reducing risk and increasing long-term potential.

PART ONE

American Ginseng

For 33 years now, I have grown American ginseng (*Panax quinquefolius*) in the woods not 30 yards from my front door. It allows me a healthy, comfortable, low-stress life that is a treasure to find in our hectic culture. An individual can cultivate a forest garden of this revered herb just to have the fascinating plant around or for his (or her) own consumption, but ginseng also has great potential as a small-scale cash crop with a ready market. With little capital investment, the small farmer can net a greater profit growing ginseng on a rugged, otherwise idle, woodlot than he can net raising just about any other legal crop on an equal area of cleared land. Of course, you have to be willing to bend your back and get your hands dirty, and to take a risk and persevere when the payoff is years in the future. [Author's note: A non-commercial home gardening approach to growing ginseng is discussed in chapter 32, but the home gardener will certainly learn from the material covered in the first half of this book.]

To guide the reader in growing ginseng, I have drawn from my own hands-on experience, from discussions with other experienced growers and agriculture professionals, and from my observations of ginseng operations throughout the United States, Canada, and Australia. Chapters 1 and 2 provide background information, much of it essential knowledge for a grower. The plant's botany, life cycle, habitat requirements, range, and related species are all covered; the regulation of commerce in ginseng is explained at the international, national, and state levels; and the long history of the ginseng trade, including recent changes in the complex ginseng market, is reviewed.

Chapters 3 through 7, in Part 2, "A Ginseng Grower's Manual," cover the three basic methods of growing ginseng (including rough production budgets for each), the harvesting and processing of seeds and roots, and the important business decisions you will need to make. Among other things, you will learn how to select and prepare a planting site; how to acquire your planting stock; what problems you are likely to encounter and how to prevent or deal with them; what has to be done when throughout the year to care for your crop;

what costs and how much labor to anticipate; and who to sell to and how to get the best price for your roots.

Then, in chapter 8, I have supplemented my own thoughts by interviewing a gentleman who was successful growing ginseng with his own individual methods. That interview personalizes the growing experience, which may help you decide whether ginseng farming is for you.

Finally, the Ginseng Resources section in chapter 9 lists root buyers, sources of planting stock, consultants, ginseng-related organizations, etc.; and the Ginseng References section in chapter 10 provides a listing of selected ginseng literature and websites.

While the References includes a few studies and accounts of ginseng's therapeutic benefits, I certainly claim no expertise in either traditional Chinese medicine or modern pharmacology, and a thorough discussion of ginseng's medicinal properties does not fall within the purview of this book. However, most ginseng growers would surely like to believe (as I do) that they are producing a commodity with real potential for human benefit. So I think the subject is worth a moment's attention before proceeding.

Although ginseng (referring loosely to all species of the *Panax* genus) has an exceptionally long and continuous history of medicinal use with an associated high market value, there remains considerable doubt (especially among many Western scientists) as to its real potency. There is compelling evidence that ginseng contains biologically active compounds (primarily steroidal saponins and polysaccharides), but the evidence for their impact on human physiological functions is less certain. Until quite recently, studies on ginseng's medicinal properties were often undertaken without employing strict experimental controls or standardized doses of ginseng. Consequently, a consistency in scientific results has been lacking, resulting in skepticism as to ginseng's genuine benefits.

But more scientists are studying ginseng than ever before, and their new research findings (many published in respected Western journals), are consistently indicating a potential use for American ginseng, *Panax quinquefolius*, in medical therapy. Studies have shown, for example: that an extract of the ginseng berry has potent antidiabetic effects in laboratory mice; that ginseng root enhances copulatory behavior in male rodents (yes, ginseng really is a consistent and dramatically effective sexual stimulant—at least for male rats!); that regular consumption of ginseng by mice stimulates their immune system response in tissues throughout the body; and that ginseng inhibits the growth of most types of human cancer cells—including lung, skin, liver, GI, prostate, colon, and breast—when they are growing in petri dishes or have been implanted into rodents.

Perhaps the most promising research on the anticancer effects of American ginseng was done by Dr. Laura Murphy at the Southern Illinois University School of Medicine's Department of Physiology, whose entry into ginseng research was initiated by the repeated urging of her younger brother, a woodland ginseng farmer. One line of Dr. Murphy's research focused on American ginseng as a complementary therapy, along with standard chemotherapy, for treatment of breast cancer. When cultured human breast cancer cells are implanted into mice, the mice are regularly injected with a chemotherapy drug, and some of them are also fed American ginseng, tumor shrinkage is much greater in those mice who received the ginseng together with the traditional chemotherapy drug. Thus, ginseng actually appears to help the chemotherapy drug work more effectively, and that suggests the dosage of the toxic drug could be significantly reduced.

Scientists do not yet know how their findings in laboratory animals are clinically relevant to humans, but ginseng, particularly its polysaccharides, may stimulate immune cells located in our digestive tract to produce more potent immune cell stimulators that ramp up the immune system throughout our body. Dr. Murphy investigated one of these immune cell products, called TNF, or tumor necrosis factor, which is a compound known to kill cancer cells. Mice fed whole-ginseng extract for four weeks have four times more TNF in their blood stream. Having obtained these results in mice, Dr. Murphy "fed" ginseng extract to human gut immune cells in petri dishes. After the gut immune cells had time to secrete TNF (and many other compounds), she introduced some of those secretions into petri dishes with human breast cancer or colon cancer cells. Consistently, within 24 hours, the human cancer cells were all dead!

Knowing of Dr. Murphy's work and other recent scientific evidence of ginseng's beneficial properties adds a small sense of satisfaction to the daily chores of my ginseng business (as well as to the writing of this book). I believe it is a good business that I am engaged in and that you are considering.

As this revised edition of *Growing & Marketing Ginseng, Goldenseal & Other Woodland Medicinals* is about to go to press, the prices being paid for wild ginseng are higher than ever before. While this certainly makes woodland ginseng growing even more attractive, should roots continue to bring such high value in the future, wild populations could be threatened by overharvesting, and the United States Fish and Wildlife Service might well feel compelled to prohibit the export of wild ginseng in order to protect the plant. Growers are therefore advised to proactively document their purchases of planting stock and their growing operation in order to be able to prove that their roots were not foraged from wild populations. Increased production of high-grade roots by woodland growers is the best way to keep supply in balance with demand, thereby keeping prices down and protecting the still widespread populations of wild ginseng.

1

American Ginseng: Its Life Cycle, Range, Related Species, and Government Regulation

Though it is one of the world's most valuable herbs, American ginseng, *Panax quinquefolius* (Linnaeus, 1753), is a rather ordinary-looking little plant—about 20 inches high—that grows inconspicuously on the floor of hardwood forests throughout eastern North America. Ginseng produces a new stem and leaf top each year, but its value lies buried in its slow-growing tuberous rootstock. The great demand for its root has led to the regulation of American ginseng's harvest and export.

Life Cycle

The First-year Seedling

When it sprouts between late April and early June, a ginseng seedling has a small, short stem supporting three tiny furled leaflets. Within four or five weeks of sprouting, the herb is about three inches tall and leaflets are unfurled and fully developed. At this point, the seedling looks something like a wild strawberry plant. No further foliar growth occurs after midsummer, even if leaflets are damaged or lost. This is true in subsequent growing seasons as well. In autumn, the foliage turns a rich yellow ocher and soon dies off, often hastened by frost.

When the ginseng seed germinates in the spring, it is the young root, or radicle, that first emerges through the seed husk. However, the root does not develop to any appreciable extent until mid-summer, after the leaflets have unfurled and completed their season's growth. The small skinny root then grows from midsummer through the fall and develops a solitary bud at its top, below the ground. The root survives the winter, freezing as the ground freezes. It is from the bud that the single stem and leaves will grow and unfurl the following spring. Interestingly, examination of the bud under magnification reveals the configuration of the next year's foliar top (that is, the number of prongs and leaflets).

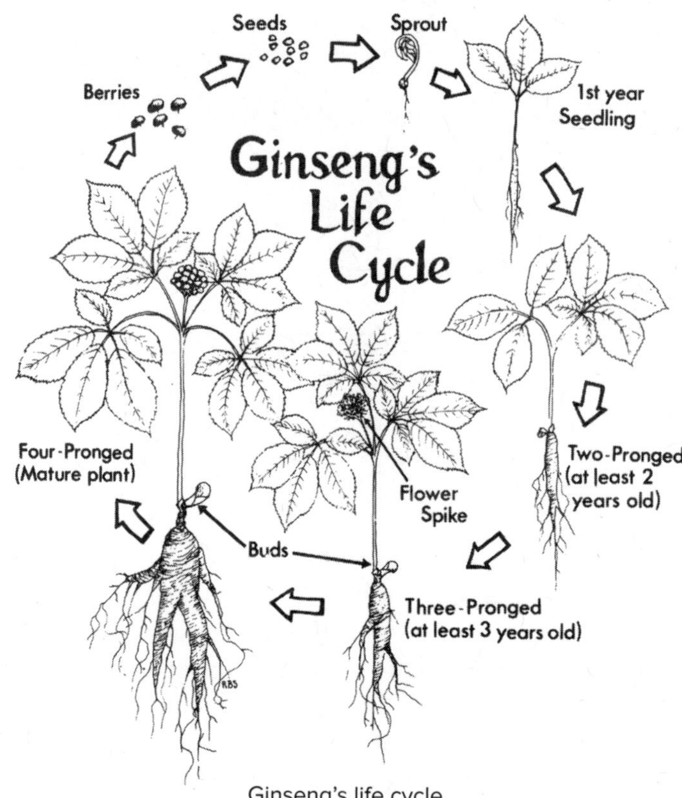

Ginseng's life cycle.

Foliage and Berries

In its second year, under optimal growing conditions, the plant can reach five or more inches in height and produce two prongs branching from the central stem, each prong being a single leaf composed of three to five leaflets. If conditions are friendly and fertile, the number of prongs will increase with age, and the plant may eventually reach a height exceeding two feet. In cultivated shade gardens, ginseng typically produces three prongs in its third growing season and often four prongs in its fourth. However, in the wild, plants are usually five to nine years old before they add a third prong and begin to produce berries (with seeds) in any quantity. In later years, particularly healthy and vigorous specimens can have as many as five prongs radiating from the top of the stem, with each prong typically having five leaflets (occasionally, as many as eight).

The species name, *quinquefolius*, means five-leafed. The two smallest leaflets on a prong are less than two inches long and the other three larger leaflets are three or four inches in length. The shape of the leaflets is *lanceolate*, with saw-toothed edges ending in a sharp point.

From the center of the whorl of prongs, a delicate cluster of small, nondescript blossoms arises in early summer, usually on

American Ginseng (*Panax quinquefolius*) during its first summer of growth. Photo by James W. Wallace, Jr.

Second-year plant with typical two-pronged leaf development. Photo by James W. Wallace, Jr.

Third-year plant with typical three-pronged leaf development. Photo by James W. Wallace, Jr.

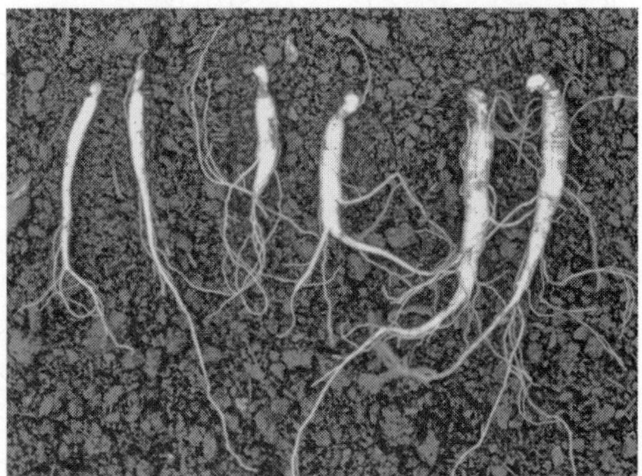

Pairs of roots from one-, two-, and three-year-old woods-cultivated plants. Note bud for next season's growth at top of roots. Photo by Kim Fadiman.

Flower spike beginning to blossom in early summer. Photo by James W. Wallace, Jr.

plants that are at least three years old. Each blossom has five greenish-white petals only a few millimeters in width. A ginseng plant is capable of self-pollination, but reproductive success is greater when sweat bees and other insects cross-pollinate the flower clusters. By July or August, as few as two or three green berries or (on large, older plants) as many as 50 berries follow the blossoms. These kidney-shaped berries about the size of bloated black-eyed peas turn a beautiful bright crimson color as they ripen. Each ripe berry usually contains two slightly wrinkled, hard whitish seeds about the size and shape of a children's aspirin tablet. Young plants sometimes produce berries containing only one seed, and vigorous older plants often have berries with three seeds in them. Under normal conditions, the seeds do not germinate and sprout until 18 to 20 months after they fall from the plant in August or September.

The Root

The root continues to develop each growing season. Young roots are long, slender, and generally light in color. As the root matures, its color often darkens, and the root may become forked with tendrils extending from the main body. Occasionally, the mature root grows into a form suggesting human arms, legs, and torso. The name ginseng means "man root " or "man essence" in Chinese. First-year roots are usually between ⅛ and ¼ inches in diameter, while the main trunk root of four-pronged plants may thicken to an inch or more in diameter and often exceed four inches in length. Under ideal growing conditions, roots can double or triple their size during each of the first few seasons. During harsh conditions such as prolonged drought or if fertilization of otherwise poor soil is stopped, roots can actually decrease in size with commensurate reduction in the size of the foliar top. Of course, malnourished plants eventually die when there is no energy left in the root to support a top. Even under optimal conditions, once the plant begins fruiting heavily, its growth rate gradually slows until increases in root weight are only about 20 percent each year.

When the foliage dies in the fall, the base of the stem breaks off just below ground level, leaving a scar at the top of the root. The next year's bud will have developed on the opposite side of and just above that scar. This yearly scarring produces a root "neck," technically called a rhizome, which bears a series of alternating and ascending marks that indicate the age of the ginseng. Under harsh conditions, plants will lie dormant for one, or even several, growing seasons, and no stem and hence no scar will form. Twenty-year-old plants are not rare, and one venerable survivor over 132 years of age has been documented. (See photo in color section.)

American Ginseng's Wild and Cultivated Range

Ginseng occurs naturally throughout the eastern half of North America as part of the forest flora under hardwood timber. Its range runs from southern Ontario and Quebec to central Alabama, and from the

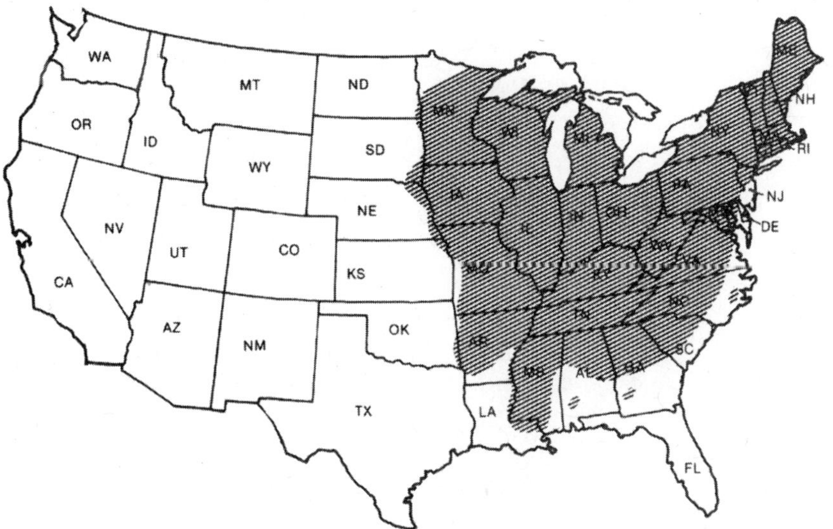

Wild range of American ginseng in the continental United States

East Coast to just west of the Mississippi River (see Range Map for United States). As with sugar maples and many other plants that grow in northern temperate zones, ginseng's southern range is limited because some extended exposure to cold is required over the winter months to stimulate its seeds and roots to break dormancy and to sprout in the spring. Although there have been reports of wild 'sang (as ginseng is often referred to throughout much of its range) growing as far west as the Texas Panhandle, its western spread is probably curtailed by the drier climate and the lack of hardwood shade trees.

The shaded area of the range map displays ginseng's present wild range in the United States as determined by the Department of the Interior's International Convention Advisory Commission and published by the World Wildlife Fund. Within its natural range, ginseng is being cultivated successfully on sites with good soil, shade, and drainage. Indeed, it has been grown commercially in eastern North America since the late 1880s.

Outside its native habitat, cultivation of *Panax quinquefolius* has been difficult until very recently because so little was known about its horticulture. Since the 1980s, however, two extensive plantings of enormous commercial significance have been established: one in the northeastern provinces of China and the other (less successful one) in the arid interior of British Columbia, Canada. In addition, a few small-scale growers are now farming American ginseng in temperate climates all over the world. There are, for example, successful farmers in Oregon, Washington, Idaho, and North Dakota. In Europe, I know of growers in Switzerland, Sweden, England, France, Italy, Belgium, Poland, a prospective grower in Hungary, and a hydroponic grower in Berlin. I have also supplied seed for an experimental operation in the treeless Golan Heights of Israel. Even in the Southern hemisphere—in Argentina, Chile, New Zealand, and Australia—enterprising individuals are attempting ginseng cultivation. (Chapter 2 covers more about the history of farming American ginseng.)

Related Species

American ginseng, *Panax quinquefolius*, is one of approximately 700 plant species in the ancient Araliaceae family, which also includes English ivy, schefflera, and sarsaparilla. The 700 modern species of Araliaceae are grouped into approximately 70 genera, one of which is *Panax*. (*Panax*, incidentally, translates as "panacea," or cure-all, which is what ginseng is believed to be.)

The *Panax* Genus

Depending on who is doing the taxonomy, there are anywhere from 5 to 13 species of the *Panax* genus—all forest plants. The five species about which there is little debate are the following:

1. *Panax ginseng* C. A. Meyer, found (now rarely in the wild) in northeast China, the Korean peninsula, Manchuria, and extreme eastern Russia near the Chinese border (where the only sizeable populations remain). It is usually referred to as Oriental or Asian ginseng, or sometimes as "true" ginseng.
2. *Panax quinquefolius* L., found in eastern North America, and commonly called American or Canadian ginseng, or colloquially, "'sang" in its southern range and "shang" in its northern range. The North American Indians used it in a similar manner to the ancient Chinese use of *Panax ginseng.*
3. *Panax trifolius* L., found in North America, and called dwarf ginseng.
4. *Panax notoginseng* Burkill, found in southwest China and Vietnam, and sometimes called Sanchi ginseng.
5. *Panax japonicum* Nees, found only in Japan, and called Japanese ginseng or bamboo ginseng.

Of these five ginseng species, *Panax quinquefolius* and *Panax ginseng* are thought to have exceptional curative properties, and they have the greatest commercial value. (As raw root, *P. quinquefolius* is the more valuable per pound.) They have similar, but distinctive, chemical compositions and are used differently in traditional Chinese medicine. Thus, they do not compete directly with each other in the Asian marketplace. Their foliage is strikingly similar in appearance, as are the roots. The best way to tell the two apart is to break a root in two and look at the cross section. The vascular bundles in *P. quinquefolius* are round, while those of *P. ginseng* appear jagged and irregular, which contributes to its more fibrous quality. Like trillium, mayapple, and other flora that have close counterparts in eastern Asia, American ginseng probably did not evolve into a separate species until the ancient land bridge between Alaska and Siberia disappeared.

Modern chemical analysis shows *Panax notoginseng* has pharmacological properties similar to the two more widely valued species, which it resembles, and its popularity and commercial value in the world of medicinal herbs is increasing. *Panax japonicum* is used in some regions of China and has modest economic value. *Panax trifolius* is distinctively different in appearance from other ginsengs and has virtually no medicinal use or worth.

Several other Asian species (or perhaps only subspecies or geographical variations of *Panax japonicum*) have been identified—some fairly recently. These include three species found in western China: *Panax pseudoginseng* Wall, or Tienchi ginseng; *Panax zingiberensis* Wu and Feng, or San qi ginseng; and *Panax stipuleanatus*, or Pingbiann ginseng. None of these is widely used medicinally, and none has significant commercial value at present.

Other "Ginsengs"

The plant commonly called "Siberian ginseng," which has been widely marketed as ginseng, is also a member of the Araliaceae family; however, it is not a true ginseng, as it is not a member of the *Panax* genus. Its proper botanical name is *Eleutherococcus senticosis*, and it is a shrub, not an herb. Traditional Chinese medicine uses *E. senticosis* as a sleeping aid and to treat acute bronchitis, but never as a substitute for ginseng. Both the bark and the root of *E. senticosis* do produce some medicinal effects similar to ginseng, and in the 1960s, Soviet scientists touted it as a useful, cheap substitute for Asian ginseng. An American importer, in the process of persuading a customs agent to allow his shipment of *E. senticosis* from Siberia into the United States, explained that it was similar to ginseng. The agent, who apparently could find no guidelines covering *Eleutherococcus*, solved his dilemma by labeling it "Siberian ginseng" and letting it through, thereby setting a precedent. Since then, when sold in Europe or the United States, much of *E. senticosis* was misleadingly labeled as "Siberian ginseng" or even just as "ginseng." Federal legislation, enacted in 2002, now prohibits such false labeling in the United States.

Another member of the Araliaceae family, *Echinopanax horridum*, or devil's club, is found in wet areas all over northwestern North America and is sometimes referred to as Rocky Mountain or Alaskan ginseng. Although a medicinal plant in Native American culture and related to ginseng, it does not have the same medicinal properties.

There are at least ten other plant species from all over the world that are sometimes marketed as ginseng, though they have no botanical relationship to the *Panax* genus or even the Araliaceae family.

Finally, anyone shopping for ginseng is likely to encounter "red ginseng" and "white ginseng." Red ginseng is made from high grade, usually six-year-old, Asian ginseng (*P. ginseng*) roots that are steamed (sometimes with other ingredients) and dried at high temperatures for at least eight hours. This process produces a translucent reddish brown root with the look and feel of hard candy. When sold (usually at high prices) as whole root, red ginseng is separated into three grades: heaven, earth, and good—with each grade having nine size categories. American ginseng and other species can be processed in this same way, but little market has been developed for such products. Asian ginseng roots are also the source of white ginseng, traditionally made from roots that are of lower grade than those processed into red ginseng. Scraping or removing the outermost layer of root tissue before drying lightens the appearance of the roots. However, sometimes the term "white ginseng" refers to fresh roots or to any ginseng roots—regardless of species—that are dried normally. For example, *P. quinquefolius* that is grown in China and dried normally is sometimes misleadingly labeled as "China White."

Government Regulation of the Ginseng Trade

Convention on International Trade in Endangered Species

The Convention on International Trade in Endangered Species of Wild Fauna and Flora (CITES) identifies *Panax quinquefolius* as one of the species that needs the protection of an international trade agreement. (Only ginseng *roots* are included under CITES; seeds and leaves are not.) The United States and Canada are two of more than 160 countries that are party to the Convention, having signed on in 1977. CITES monitors, controls, and restricts trading in the identified species to prevent adverse impacts on their populations and to insure the continued existence of those species in their natural habitat.

In the United States, obligations under the CITES agreement are the responsibility of the United States Fish and Wildlife Service (USFWS), more specifically the responsibility of two divisions of USFWS: the Division of Scientific Authority (DSA) and the Division of Management Authority (DMA). Under the authority of CITES (Article IV), the USFWS will only allow export of American ginseng—both cultivated roots and roots collected from the wild—if the DSA advises the DMA that such export will not be detrimental to the survival of the species. In addition, the DMA must be satisfied that the specimens intended for export were legally collected or cultivated. (Ninety percent of our ginseng is eventually exported—see next chapter.)

In accordance with CITES, the DSA has chosen to use a state-by-state basis in determining whether or not ginseng export will be detrimental to the survival of the species. As of this writing, the DSA has determined that the export of cultivated American ginseng roots would not be detrimental to the survival of the species if a state has a program in place to certify the roots for export. The following states have such a program: Alabama, Arkansas, Georgia, Idaho, Illinois, Indiana, Iowa, Kentucky, Maine, Maryland, Michigan, Minnesota, Missouri, New York, North Carolina, North Dakota, Ohio, Oregon, Pennsylvania, Tennessee, Vermont, Virginia, Washington, West Virginia, and Wisconsin. Of these states, Oregon, Washington, Idaho, Maine, and Michigan export only cultivated ginseng. The remaining states have all established laws and state programs, including a legal foraging season, that regulate the harvest of wild ginseng and require the certification of ginseng roots as either wild or cultivated prior to their export. At present, the DSA finds that in all these remaining states the export of both wild and cultivated ginseng would not be detrimental to the survival of the species.

Every year the DSA reconsiders its nondetriment findings based on information from each state, such as pounds of wild ginseng harvested, average number of roots per pound, average age of harvested plants, and trends in abundance of wild ginseng populations as measured in field surveys. In 1999, the DSA found that throughout all states the continued harvest of wild plants younger than five years would be detrimental to the survival of the species. Therefore, all states must now prohibit the harvest and sale of wild roots less than five years old (as evidenced by the number of scars on the neck, or rhizome). In anticipation of future ginseng harvests, the DSA continues to seek trade and biological information concerning the impact of ginseng harvest and international trade on wild populations of the species. The DSA seeks input from the public, the ginseng industry, and scientific authorities, as well as from conservation groups and other interested parties.

For its part, the DMA requires that each state monitor all commerce in American ginseng (wild or cultivated) within its borders. Beginning with the 1978 harvest season, all states seeking export approval for wild or cultivated ginseng roots were required to have legally mandated ginseng programs that included the following: (1) state registration of dealers who purchase ginseng in the state; (2) requirements that such dealers maintain records and submit annual reports to the state government concerning their purchases and sales of ginseng; and (3) inspection by state officials and the issuance of accompanying State Certificates of Origin for each lot of ginseng being shipped out of the state, documenting that the ginseng was legally foraged or grown within the state. In addition, the DMA issues its own CITES permits, which must be obtained in order to ship American ginseng out of the United States.

In all of Canada, the export of wild ginseng has been prohibited since 1989. In Quebec, the harvest of wild ginseng was prohibited since the species was listed on Appendix II of CITES in 1973. The harvest of wild ginseng (but not the export, since 1989) was allowed in Ontario until June 30, 2008, but both the harvest of, and the trade in, wild ginseng is now prohibited there. Moreover, to be exported, roots can now only be cultivated in open fields under artificial shade on land licensed (with a fee) by the Ontario Ginseng Grower's Association under the Farm Products Marketing Act.

All shipments of field-grown ginseng artificially propagated in Canada must be accompanied by valid CITES documentation. Exports of woods-grown ginseng are currently assessed on a case-by-case basis by the Canadian Scientific Authority. According to Adrianne Sinclair of Environment Canada's Canadian Wildlife Service, no Canadian export permits are being granted for woods-grown ginseng, due to concerns related to habitat disturbances associated with site preparation and maintenance, the introduction of seed-borne pathogens that are common in cultivated seed sources, and the potential for genetic contamination of wild ginseng populations. Also of concern is the difficulty in differentiating between the roots of wild and woods-grown ginseng. Not surprisingly, there is now very little commercial woodland ginseng farming in Canada. [Author's note: Despite, and perhaps in part because of, these regulations, wild ginseng in Canada is under increasing pressure.]

United States Department of Agriculture

The USFWS works closely with the United States Department of Agriculture (USDA) Animal and Plant Health Inspection Service (APHIS) to enforce and implement ginseng regulations. APHIS is responsible for inspecting all exported and imported ginseng to make sure that it is properly certified as to state of origin, is accompanied by the required CITES permit, and is at least five years old. Since the necks (which are needed for proof of age) of many dried ginseng roots easily break off during shipping and handling, it is fortunate that so far the inspectors are not being too rigorous and technical in their assessments (because every container of roots has individual roots with their necks broken off and therefore of unverifiable age and subject to rejection for export). In addition, a general export permit must be obtained from the USDA in order to export any agricultural product. [Author's note: Contact PPQ-APHIS-USDA, Permit Unit, listed in the Comprehensive Resource Directory under USDA.]

Impact of Government Regulation on the Individual

Because state laws vary slightly, the impact of CITES regulations on the individual will differ from state to state. To determine what the laws are in your state, ask your county agricultural extension or conservation agent about ginseng regulation and what department of state government is administering your state's program. If your agent does not know, then you can contact the United States Fish and Wildlife Service, Division of Management Authority. The DMA will advise you whom to contact in your state. I urge you to learn your state's law, whether you are a digger, a grower, or a buyer. Contact information for the DMA is listed under USFWS in the Comprehensive Resource Directory.

To hunt wild ginseng, you will need to know your state's legal season and any other state laws, such as a license for hunting 'sang, or requirements that you only take plants old enough to bear seeds or that you immediately plant some of the seeds on the site where you dig the plant. Selling wild roots out of state requires a State Certificate of Legal Take. This document will accompany the roots on any resale because ginseng roots (alive or dead) cannot be exported without state certification. An in-state buyer will have certification forms available himself.

If you are interested in just growing ginseng, then CITES will probably affect you only when you are ready to sell your roots. Furthermore, if you always sell to an in-state buyer or to an out-of-state dealer who has registered as a buyer in your state, then you will likely never have to deal with permits or certifications of any kind. (All you have to do is grow the roots.) In any case, contact your state regulatory office, as they may be able to put you in touch with fellow growers and other knowledgeable people in your area. In addition, if your state should ban the collection of wild ginseng sometime in the future, you may need support from a representative of the state to verify that your roots were grown from seeds that you planted. Along this line (and, of course, for tax purposes), keeping records of your purchases of planting stock is important. A few states require growers to acquire nursery licenses and meet other reporting regulations.

If a grower wishes to sell his roots out of state, he will have to comply with regulations. Like wild roots, cultivated roots must have proper documentation before they can be bought and sold. All ginseng sold across state lines is required to have a State Certificate of Origin accompanying it. State personnel must inspect the roots and determine whether they are wild or cultivated and then issue an appropriate certificate documenting the state of origin. My experience in North Carolina has been that this documentation is convenient to obtain. (Note that seeds or live roots intended for transplanting in the United States need no certification, even if sold out of state.)

In addition to the documentation needed to ship out of state, a grower who wishes to directly export his roots must also obtain a USDA General Export Permit and a CITES permit from the DMA. The grower must then ship or hand carry the roots, along with the necessary documentation, to a designated port of export for APHIS inspection. (Contact the USDA for a list of ports.)

If you live in a state where there is no regulation of ginseng commerce as mandated by CITES, then there can be no legal ginseng buyers in your state, and any ginseng you grow (or forage) cannot be legally exported directly from your state. You will have to ask your county agricultural extension agent or some other state official to write an informal certificate of origin on some official state letterhead, which will allow you to move your roots (accompanied by the informal certification) out of state. Then you will be able to sell your roots to a registered dealer in a state that complies with CITES, and, in turn, that dealer can legally export or resell them as long as he documents buying roots from your state in his annual report to his state's administering office.

While compliance with government regulations is no great burden for the 'sang digger or most growers, anyone who wishes to buy and resell ginseng is destined to fill out a lot of paperwork. Ginseng buyers must register with their state as dealers and are required to fill out and submit forms supplied by the state, recording all root purchases and sales. In addition to knowing state law and becoming a state-registered dealer, a buyer must thoroughly understand and comply with the CITES regulations (which can be obtained from the DMA—see the Comprehensive Resource Directory under USFWS). To export ginseng, a dealer must follow the same procedures as a grower. This includes acquiring CITES permits and a USDA General Export Permit, and shipping or hand carrying roots (along with the necessary documentation) to a designated port of export for APHIS inspection.

2

History of the Ginseng Trade: Ancient China to the New Millennium

Ginseng, which means "man root" or "essence" in Chinese, is so named because the root sometimes grows into a man-like shape and because its medicinal qualities are believed to benefit the whole man. Ginseng has been a central component of Chinese traditional medicine for thousands of years, used to balance the body and prolong the quality of life in old age as well as to treat a broad spectrum of more specific maladies. Commerce in ginseng probably began soon after its medicinal reputation was established. As it became greatly valued as a life-enhancing herb, its monetary value attracted diggers, traders, and even smugglers. Ginseng has often been worth its weight in silver or even gold, and enterprising people have long made money in its trade and, more recently, in its cultivation. This chapter traces the fascinating history of the ginseng market, from ancient times in the Orient up to our present-day international trade.

Ancient China

Many thousands of years ago, *Panax ginseng* was surely known to hunter-gatherers in mountainous Manchuria in China. It probably was first used as a food and then became revered for its strength-giving and rejuvenating powers. In ancient rural China, the occasional, truly man-shaped root was regarded as divine order implanted in the soil and as immensely valuable. Wealthy warlords wore the dried root around their necks as an amulet of long life and power.

The first unifier of China, Shih Huang Ti of the Ch'in (or Qin) dynasty (221–207 BC), built long stretches of the Great Wall of China and then sought to assure himself of immortality or at least of notable longevity. Legend has it that he sent 3,000 young men and 3,000 young women to the most remote mountains in search of the "divine herb"—ginseng. Unfortunately, none of these young herb seekers ever returned to the emperor, and he remained without a supply of the root and the means to greatly prolong his remarkable life.

Ginseng first appears listed as a medicinal herb in the *Classic Herbal* of Shen Nung, which appeared during the Han period (from 206 BC to AD 220). Among other wonderful properties, ginseng was said to expel evil effluvia and, if taken for some time, invigorate the body and prolong life. The *Classic Herbal* is thought to be a preservation of much older writings and recommendations, including those made by the legendary Shen Nung, founder of Chinese herbal medicine, agriculture, and animal husbandry, who is said to have lived around 2800 BC.

Insistent demand for ginseng nearly exterminated wild ginseng in China and stimulated international trade in the revered root beginning in the third century AD, when envoys from Korea brought the Chinese emperor ginseng and other valuable gifts in exchange for Chinese silk, medicine, and culture. It was about this time that the first test for true ginseng was devised. Ginseng has always been a rare and highly valued herb (still held in highest esteem by practitioners of traditional Chinese medicine), and apparently there have always been those who sought to pass off substitutes as the real thing. Today there is DNA analysis, but the *T'u-ching Pen-ts'ao*, the history of the Sung Dynasty (around AD 700), records a simpler analytic method (though undoubtedly less certain):

> In order to test for the true ginseng, two persons walk together, one with a piece of ginseng root in his mouth, and the other with his mouth empty. If at the end of three to five li [about a mile-and-a-quarter] the one with ginseng in his mouth does not feel himself tired, while the other is out of breath, the ginseng is genuine root.

It also was early in the history of the ginseng trade that smuggling began. This probably first occurred when the Manchurian province forbade export of the herb. The roots brought a great price, so travelers across the mountains began smuggling ginseng by twining the roots into the braids they wore at the back of their heads. This early contraband practice came to be called "pigtailing." Today, to avoid tariffs, American ginseng is being smuggled into Mainland China.

Korea

Korea claims the title of "The Ginseng Country," and surely deserves it. Since before recorded history, Symmani (which translates as "ginseng man") have hunted wild ginseng after making ritual offerings of food and colored ribbons to the mountain gods. Tales of ginseng as the medicine taken by an immortal, supernatural hermit were being told as early as 4,000 years ago. Records of finding "mountain ginseng" in Korea date back as far as 2137 BC. Geography and climate provide excellent habitat and growing conditions. Korea has been exporting wild mountain ginseng to China since the third century AD and has been cultivating ginseng for export since the beginning of the 20th century. Moreover, research undertaken at the Korean Tobacco

A traditional ginseng farm in Korea, where *Panax ginseng* is an important aspect of the economy and the culture. Thatched roof hutches facing east provide the plants full sun in the early morning but full shade the remainder of the day. Photo courtesy of General Nutrition Corporation.

and Ginseng Research Institute, originally founded in 1899, has established much of the world's understanding of ginseng's medicinal properties.

The thriving ginseng export business and the destruction of its forest habitat gradually reduced the availability of wild Asian ginseng on the Korean peninsula. Today there are almost no truly wild *Panax ginseng* plants left in Korea (or anywhere else in the world, other than limited populations in eastern Russia). To meet China's constant demand, the Koreans learned to cultivate ginseng, first under natural forest shade and more recently in open fields under artificial shade. (Of course, the Chinese began growing ginseng as well.)

Cultivated Korean ginseng plants (*Panax ginseng*) are botanically identical to wild Korean ginseng as well as to the Asian ginseng plants native to China and Russia. Cultivation on a small scale began in the provinces of South Korea in the 16th century, but it was not until the early 1900s that cultivation was undertaken on a scale large enough to produce a surplus for export. During the early 1900s, Korea produced as much as 700,000 pounds of dried ginseng

Harvesting ginseng berries for seed in Korea. Photo courtesy of General Nutrition Corporation.

root per year from wild and cultivated crops. Most of this was exported to China, Japan, and other Southeast Asian countries.

Over time, South Korea gradually increased its production until, in the late 1990s, it was utilizing over 24,000 acres for ginseng farming (almost all of it under artificial shade). Traffic, an arm of the World Wildlife Fund, reported an annual harvest of 20 million pounds of roots in 1998, with about 4.8 million pounds of that exported. Ginseng is a major agricultural product in South Korea. It is shipped to some 60 foreign countries, but primarily to Hong Kong, Mainland China, Japan, and Taiwan, with smaller quantities going into the United States, Spain, Canada, Vietnam, and other Southeast Asian nations.

The Korean Tobacco and Ginseng Corporation (KT&G, which now funds the Research Institute) oversees ginseng cultivation as well as marketing, especially of the red ginseng that the KT&G produces (a value-added product, see Other "Ginsengs" in chapter 1). The corporation attaches a registration trademark, called Cheong-Kwan-Jang in Korean, to its red ginseng products in order to guarantee quality to the consumer. Each year, ginseng and tobacco exported under the KT&G bring in as much as three trillion won ($2.4 billion) in revenue to the country's budget. Much of this is from sales of tobacco products, but ginseng, especially red ginseng products, makes an annual contribution of up to $200 million.

In recent years, there has been a decrease in export revenue from ginseng, but the Korean Ministry of Agriculture and Forestry has plans in place to reverse that trend. Almost one billion dollars will be injected into the Korean ginseng industry to foster "global strategic export." Moreover, approximately 200,000 acres bordering the demilitarized zone will be set aside for potential Asian ginseng production. The plan is to increase the average size and yield of the typical ginseng family farm (of which there were approximately 22,000 in 2001) from 1.3 acres with yearly ginseng sales of $4,000 to 2.4 acres with annual sales of over $14,000. Furthermore, the Koreans are increasingly interested in producing "mountain ginseng," which seems roughly equivalent to our term, "woods-grown," because, according to Dr. Hoon Park of Chung-Ang University, "mountain ginsengs are high in efficacy and very rare and price is very high."

Japan

Wild Japanese ginseng, *Panax japonicum*, has grown in the mountains of Japan since ancient times. It is different botanically from the wild ginseng of Manchuria, eastern Russia, and Korea (*Panax ginseng*). Its roots have a more bitter taste and are not nearly so valuable, though they have a traditional market in China. There is ongoing research in Japan to determine if *P. japonicum* has unique medicinal properties.

Nearly all the ginseng cultivated by the Japanese is the more highly prized *Panax ginseng*. Japan began importing seeds and seedling roots from Korea in 1607. After three hundred years of experimentation, ginseng farming had become a good business; in 1907, ginseng grew in 43 counties, and thousands of Japanese farmers were involved in its cultivation. In the early 1900s, Japan's ginseng suffered a disease epidemic (probably Alternaria), but with no understanding of the causes, nothing could be done. Most of the crops perished, and the farmers plowed their gardens under and started other crops. Today, *Panax ginseng* is cultivated primarily by Japanese rice farmers on ginseng farms, which are small, individual gardens. According to latest figures, the annual crop is only about 21,000 pounds, most of which is exported to Hong Kong.

Asian ginseng growing in Japan under traditional thatched roof shade. Photo by Al Oliver.

French Canada

Knowledge of ginseng spread to the Western world beginning with Marco Polo's reports of the root's use throughout China. He recorded in 1274 that ginseng was "...powdered, cooked, and used as a tea, syrup, or food condiment, or even burned as incense in the sickroom." Canadian historian Brian Evans has cited the Dutch as being the first to import ginseng roots into Europe in the 1600s. However, the French, with their Jesuit missionaries in northern China, were the more successful early exploiters of the herb's commercial potential in the West. The French were also more appreciative of ginseng's medicinal benefits, recommending its consumption with a little white wine to cure asthma and stomach problems and to promote fertility in women.

In 1702, a French Jesuit priest, Father Jartoux, went to China to help survey the area of Manchuria. There he observed the use of Chinese ginseng and its healing powers, and was able to furnish an account of the plant, including a detailed physical description (presumably with an accompanying sketch) and a description of the environment where the herb flourished. Father Jartoux's information fascinated another Jesuit missionary farther west, Father Joseph Francis Lafitau, who was living among the Iroquois Indians in North America. Father Lafitau reasoned that the environment of French Canada was much like that of Manchuria and that there was a good chance he would find this wonder herb growing there. Indeed, he did discover the closely related American ginseng, *Panax quinquefolius*, near Montreal in 1716.

Soon after this discovery, the French realized ginseng's value to the Chinese, and French Canadian fur traders expanded into ginseng, paying the Iroquois Indians to dig all they could find. Trade with China began in 1717 and prospered from the beginning. According to Professor Evans, the Cantonese merchants who purchased the first Canadian roots off the clipper ships lightly mixed those roots into lots of Asian ginseng, which look very similar, and were able to sell the mix as lots coming exclusively from either Manchuria or Korea. In time, as wild Asian ginseng became increasingly scarce, American roots took on great legitimate value of their own.

The trade with China went well, until the year 1752, when the price peaked, and the temptation for quick riches proved too much for shortsighted French Canadian traders. They began collecting huge quantities of roots without considering their size or age. This significantly depleted Canadian ginseng, which only grows naturally in southern Ontario and Quebec, and wild ginseng has never again been a major Canadian export. Also, in their rush for profits, the fur traders dried the roots too quickly at excessive temperatures in ovens, destroying their value. The Chinese knew what they wanted and quit buying the undersized, poorly processed roots. Trade with China fell from $100,000 in 1752 to $6,500 in 1754. It took many years to restore the reputation of ginseng from Canada.

The Role of Ginseng in the Settling of America

At the same time that the Chinese stopped buying Canadian ginseng, knowledge of the herb was spreading to the American colonies. Wild *P. quinquefolius* was discovered in western New England in 1750 and in central New York, Massachusetts, and Vermont in 1751. When settlers with knowledge of the root's value spread out from central New England, they discovered ginseng growing throughout the eastern deciduous forests. Though history books largely ignore it, ginseng was one of the two major exports of colonial America, the other being raw furs. Later, as citizens of the new nation continued moving west to settle the land from the Appalachian Mountains to the Mississippi River, wild ginseng often played a critical role by supplying pioneering homesteaders with the immediate cash they needed to buy necessary supplies until their first crops came in.

Early United States-to-China Export

John Jacob Astor of the American Fur Company financed one of the first American shipments of *Panax quinquefolius* to China in the late 1700s. The story goes that ginseng started the Astor fortune. Astor sank his entire fur-trading capital into ginseng export. His plan was to sell "all the ginseng root I can get" to China. This early gamble paid off. Shortly after his ship, which had carried only ginseng as cargo, returned to New York from China, he had several small heavy kegs brought ashore and delivered to his home. He opened these kegs in front of his young bride, who was delighted to find them filled with silver! Astor had made a $55,000 profit—a sizeable fortune in those days.

Daniel Boone

Just as had been the case with the French in Canada, the fur traders in colonial America were usually the ginseng traders as well. They brought many of the roots to Philadelphia for sale, and the port of Philadelphia became the principal site of ginseng export for the early years of the American ginseng trade. By 1802, ginseng was one of the few products that was profitable enough to absorb the cost of overland transportation to Philadelphia for export. (Over 90 percent of United States ginseng has traditionally been exported to the Orient.)

In Kentucky in 1788, Daniel Boone gathered a few roots for himself and his family and then purchased 12 large barrels for export. He had bought ginseng for export the previous year but had lost his entire cargo when his boatload of roots overturned in the Ohio River en route to Philadelphia. However, ginseng was such a valuable commodity and such a good potential money-maker that he repeated his effort the next year and made his own fortune.

Ginseng "Saves" Minnesota

Truly abundant ginseng stands that allowed constant digging for extended periods grew only in virgin deciduous forests. Thus, brief local ginseng "gold rushes" moved westward with the frontier, as hunters discovered easily accessible ginseng populations and rapidly extracted the bulk of the roots. Despite the need to find new virgin woodlands to harvest every few years, ginseng export from the United States was remarkably constant during the mid-1800s. In 1821, Department of Commerce statistics document that the total ginseng export was 352,992 dried pounds, at a price to the digger of 48 cents per pound. Records show that harvests totaling over 300,000 pounds per year continued for most years through 1888, when ginseng exports reached 308,365 pounds, valued at $2.13 per pound.

William E. Lass, a history professor at Mankato State College in Minnesota, has researched and written an interesting history of the ginseng gold rush in southeastern Minnesota (one of the last). He writes about the important role that ginseng played in the early settlement of that state:

> 'Sang, as it was called on the frontier, was first gathered by colonial pioneers, and it continued to provide a valuable supplement to the meager incomes of later frontiersmen.... In 1859, [there was a deep economic depression with no market for farm products, and] Minnesotans thronged into the woods in search of ginseng.... In time, the ginseng rush became legendary as pioneers reminisced about the weed that "saved"

> Minnesota.... Thousands marketed their first cash crop in two years.... Mankatoans sponsored a "ginseng ball" which was to make diggers "oblivious to the musquito [sic] bite or toil of delving for the bulbous root, whilst tripping the light fantastic" to the music of the Ginseng Polka.
>
> After four productive years, sanging declined sharply.... On February 13, [1865,] the [Minnesota] legislature approved "an Act to preserve and protect the growth of ginseng."

According to Professor Lass, Minnesota exported 245,434 pounds of ginseng in 1860, followed by 208,650 pounds in 1861. This was a significant portion of total United States exports (395,909 pounds in 1860 and 347,577 in 1861). However, by the time the state legislature acted in 1865, only 80,259 pounds were harvested in Minnesota. Unfortunately, the Ginseng Act was ineffective, and the annual harvest continued to decline, in part because the legal digging season was set early in the summer before the seeds matured, leaving no way for the herb to propagate itself during the year of its harvest. Today, only a few thousand pounds of "shang" (as it is now called in Minnesota) are dug each year in the state.

After the Civil War in Southern Appalachia

For a hundred years before ginseng "saved" Minnesota, knowledge of ginseng's value and where to find it was being passed on through generations of mountain families throughout southern Appalachia. (In the South, ginseng only grows in abundance in the cooler mountains.) Ginseng was an available source of supplemental income to folks in rural areas, and, when times were hard, entire families would wander the rugged hillsides, searching the woods for anything they could sell, but mainly for ginseng. Because it provided folks with the cash to buy what they desperately needed, even land on which to start a farm, ginseng became an important part of the mountain culture. During the decades preceding the Civil War, there were regional ginseng rendezvous to which diggers trekked through the mountain passes, leading oxen or horses loaded with roots for sale. In addition to selling their 'sang at these rendezvous, they engaged in games of chance and skill, shooting, wrestling, and sometimes a good fist fight.

The Civil War changed things. In those parts of Appalachia where the armies of both the North and the South along with irregular partisans ravaged farmland and confiscated crops and livestock (not to mention committed acts of violent brutality), a prolonged economic depression suddenly made a large segment of the population almost wholly dependent on diggin' 'sang. Luke Manget, a graduate student in history at Western Carolina University, has written a paper documenting this sudden shift to a ginseng-dependent economy in Cherokee County, North Carolina. As a result of the war, the county lost over half its livestock, over 40 percent of its improved farmland, and 10 percent of its population.

It started producing over 75,000 pounds of ginseng annually (nearly twice as much as the entire state produces now). Manget makes a compelling argument that, spurred by civil strife and global trade, a significant subculture of itinerant 'sang-digging families developed in every southern mountain state and that it was these independent folks whom the "color" writers from eastern newspapers found so fascinating and unfairly characterized as a "wretched class of ignorant poor whites who never seem to have a settled habitation." Thus was created the southern Appalachian stereotype, sometimes referred to as the hillbilly.

History of Cultivation in the United States

First Successful Attempts at Ginseng Farming

As the eastern half of the country was settled, the supply of wild ginseng eventually succumbed to overharvesting. By 1895, less than 200,000 pounds were being exported annually to China, most of it entering through the port of Hong Kong. Depletion of the wild populations was accelerated by the practice of digging the plants early in the summer; even more damaging in the long term was the intense cutting of hardwood trees and the trampling and grazing of the forest floor by fenced livestock, which destroyed the forest habitat and prevented the plant from reestablishing itself. The dwindling supply of wild 'sang encouraged farmers to attempt its cultivation.

Undoubtedly many farsighted early ginseng diggers benefited from immediately planting back the berries from the roots they dug; however, the first efforts at growing ginseng using standard farming methods failed so frequently that it was widely believed *P. quinquefolius* could not be cultivated. This belief still lingers today. One acquaintance recently told me that planting seeds from ginseng wouldn't do any good, because 'sang seeds wouldn't sprout unless "they had been run through a bird first"!

Eventually a few folks figured it out: ginseng requires a cool, shaded environment, a soil rich in humus, and good drainage but with a moisture-retaining mulch; it is adapted to moderately fertile soil, low in nitrogen; and the seeds from ripe red ginseng berries will not sprout for a year and a half. In the 1870s, Abraham Whisman of Virginia became the first American known to successfully cultivate ginseng. Soon afterward, George Stanton, a tinsmith from Apulia Station in Onondaga County, New York, began to grow green gold in some quantity, and he is generally recognized as the first successful commercial ginseng farmer in America. His technique was to plant in an open, tilled area and then build a shade structure over the planting using narrow strips of wood lath spaced a few inches apart, thereby simulating the light and shade pattern produced by a forest canopy of leaves. (Other early growers were successful creating plowed beds in the woods; they had to hack through tree roots, but avoided the time and expense of shade construction.) Stanton's original ginseng plot was only 129 square feet and eventually produced only five pounds of dried roots,

Early one-tenth acre ginseng garden farmed by Mr. Frink of Scott, New York. George Stanton's wood-lath roof design, simulating the forest leaf canopy, provided Frink's plants with partial shade throughout the day. Photo courtesy of Bob Beyfuss and Cornell Plant Pathology Herbarium.

which sold for $16.83. He learned to greatly increase his yields, and Stanton's records document one garden that produced 106 pounds of dried roots, which sold for $575.

Cycles of Boom and Bust

George Stanton loved his gardens and genuinely wanted to share the rewards of profitable ginseng farming. Several publishers laughed at him, but he managed to have a small circular on ginseng farming printed and a New York newspaper article written about his success. Interest boomed with remarkable speed. By 1895, enthusiasm for growing 'sang was so high that the United States Department of Agriculture published its first bulletin on ginseng cultivation: George V. Nash's *American Ginseng*, which was revised and reissued in 1898 and became widely distributed. Nash, and others in farm and garden journals, also wrote glowingly about the Chinese market and reported that large cultivated roots were often selling for a higher price than wild roots. At the turn of the century, nearly every outdoor magazine ran ads for ginseng seeds, many of them touting 'sang farming as a fabulous get-rich-quick business.

With effective open-field cultivation practices now known and the enterprise receiving great encouragement from both authoritative sources and unprincipled huck-

sters, ginseng farms sprang up throughout the eastern and midwestern states. Department of Commerce records document total exports of 149,069 dried pounds (at $5.38 per pound) in 1901, increasing to 160,101 pounds (at $5.20 per pound) in 1903. Most of this increase was probably due to the addition of cultivated ginseng, although it is impossible to be certain, since no distinction between wild and cultivated ginseng was made at that time. Indeed, no official export records distinguished wild from cultivated roots before 1980.

This first cultivation boom started to die about 1904 with a widespread outbreak of Alternaria blight, a foliar fungal disease for which there was then no treatment. Discouraged by the loss of most of their seed crop, many growers did not replant.

Early 20th-century wood-lath shade structures. Photos courtesy of Bob Beyfuss and Cornell Plant Pathology Herbarium.

By 1909, the thriving ginseng farming business had dwindled, and that year's census of United States agriculture reported only 23 acres under cultivation.

Just as the last remaining growers were struggling to hang on, Dr. I. C. Curtis of Fulton, New York, reported in *Special Crops*

Magazine his success in treating the blight with Bordeaux mix (half copper sulphate, half lime). As word of this preventative treatment got out, ginseng acreage dramatically increased again, and it became widely recognized as an important alternative cash crop. By 1929, 434 acres were being cultivated by small farmers throughout the Midwest and eastern United States. Demand was so high that exporters could not meet it, and prices exceeded $12 per pound.

Marvin Dickman, a modern Missouri 'sang grower, gives this first-hand account of the American ginseng trade at the beginning of the Depression:

> I saw ginseng being grown commercially for the first time back in summer of 1929, when I was five years old. An old man grew it on terraces on a steep hillside near Hunter, Carter County, Missouri, about 90 miles east of here. Near a very rugged area of south Missouri called the "Irish Wilderness." He had moved the plants in from the woods, and cultivated them with a mule-drawn cultivator....
>
> Apparently the old man was successful, because that fall of 1929 he dug his ginseng and shipped it in a flour barrel, and got over $1000 for the barrel of roots. It was around $12 per pound that year, at the start of the Depression. He hauled the barrel of dried roots to the railroad station in his mule-drawn buggy with iron-tired wheels that were warped so that they went "in and out" as he drove along.

Then, just as both production and price were peaking once again, a different kind of problem arose: war in the Far East. Between 1929 and 1949, first the Sino-Japanese War and then World War II reduced and eventually blockaded all American trade with China. Without the Chinese buyers and consumers, there was little market for American 'sang, and most folks got out of the business. Only in Wisconsin did a few growers hang on hopefully, storing their dried roots for eventual sale and maintaining planting stock. Today, Wisconsin remains the center and by far the largest producer of cultivated ginseng in the United States.

After the hiatus of the war years, as communications and general trade with the Far East improved, demand for American ginseng was renewed. Growers sought to fill an ever-growing demand, many of them starting with seed obtained from the four Fromm brothers of Marathon County, Wisconsin. Most ginseng, both wild and cultivated, was still first shipped to Hong Kong (under British governance), rather than directly to Mainland China, and this trade through Hong Kong has continued even though the island has reverted to Chinese control.

The scale of ginseng farming grew slowly but steadily during the 1950s and '60s. Sometime during the 1970s, annual cultivated production (almost exclusively from open fields under artificial shade) began to exceed the wild harvest in poundage. In 1972, President Richard Nixon's visit to China resulted in a boost in ginseng sales.

In 1975, the United States government, as part of re-establishing relations, presented the Chinese government a gift of American ginseng. By 1977, the year when our country instituted CITES regulations (see chapter 1), the export of American ginseng, both wild and cultivated, reached a value of $26.5 million. The majority of that value was in the wild roots because, as soon as cultivated production resumed in quantity after World War II, the buyers began paying much more for wild roots than for field-grown roots. In 1977, wild roots brought about twice as much as roots grown in open fields under artificial shade.

United States trade missions continued to promote American ginseng, and the demand increased further. According to the United States Fish and Wildlife Service's Division of Management Authority, 1,600,000 pounds of dried American ginseng roots, including 140,000 pounds of wild roots, were certified for export in 1992. Most of this ginseng, worth approximately $70,000,000, was resold through the Hong Kong auction market. Much of the cultivated root made its way into Mainland China, and most of the wild root either stayed in Hong Kong or was resold in upscale markets throughout the rapidly developing Pacific Rim countries (especially those with large populations of Chinese origin).

United States ginseng production peaked in 1996 at over 2.3 million pounds of dried root (2.2 million of that being field grown). Unfortunately, this coincided with dramatically increased production of American ginseng in China and especially in Canada (see below). Not surprisingly, the rapidly growing world supply of field-grown root led to a rapidly deteriorating per pound price for field-grown roots. Prices fell and stayed under $20, which is barely above the cost of production for a small-scale, artificial-shade family ginseng farm like those in Wisconsin, so most United States field growers have gotten out of the business. Only about 150 shade growers remain in business (down from 1,500 in the early 1990s), producing only about 500,000 pounds of root for export. There has been, however, a rather dramatic upward trend in cultivated prices over the last few years (see Recent Prices at the end of this chapter), and interest in artificial-shade farming has started to return. Presently, the United States is third to Canada and Korea in ginseng exports (all species), and is the fourth largest producer of ginseng, following the People's Republic of China, Korea, and Canada.

The Value of Wild vs. Cultivated

For a long time, buyers paid as much for American ginseng grown in gardens as they did for roots foraged from the forest. A very small percentage of roots from a ginseng garden may be hard to distinguish from wild roots, but tilling the soil preparatory to planting, weeding, and fertilizing inevitably yield faster-growing roots that are on average noticeably smoother, lighter colored, denser, bigger for their age, and less bitter tasting than wild roots. These differences are more marked when

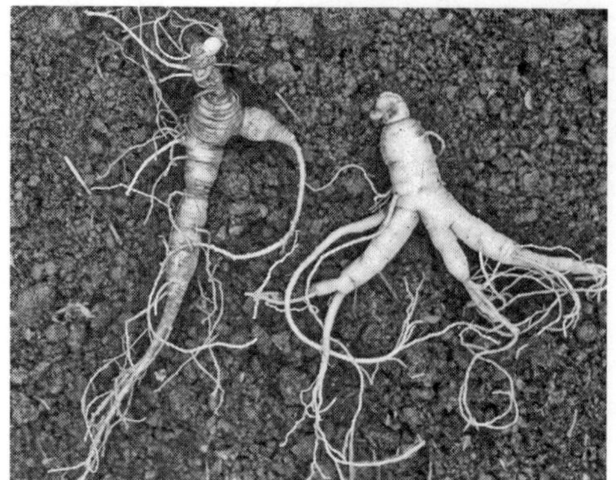

Freshly dug roots: wild (left) vs. field cultivated (right). Photo by Mark Haskett.

Dried roots: wild (left) vs. field cultivated (right). Photo courtesy of Dennis McClintic.

the roots are grown big and fast under the ideal light conditions provided by artificial shade than when the ginseng is nurtured more slowly in a woodland garden, where all conditions are less uniformly ideal and where the shade trees are competing for soil nutrients.

Probably, until the last half of the 20th century, buyers routinely mixed cultivated roots with the relatively more abundant but variable wild roots before export. Little mixing occurs today. It became impractical as ginseng farming moved to mechanized, intensive cultivation practices in open fields under artificial shade. Increasing quantities of large three- and four-year-old roots were produced, and separate pricing and marketing emerged.

Wild American ginseng is now greatly valued over cultivated American ginseng by Asian consumers, mainly because it much more closely resembles the revered wild Asian root (although scarcity is probably also a factor). In addition, wild roots are valued because they have grown without human manipulation, and it is believed that the slower-growing wild roots (Asian or American), which are harvested at an older age, absorb more curative power from the forest soil. There are some pharmacological studies that show more biologically active chemicals and greater concentrations of them in wild roots—especially older ones—than in cultivated roots, although other studies do not support this.

The relatively little ginseng grown in tilled woodland gardens has never been historically or commercially significant, and most of it was almost certainly mixed with wild roots. Today, such woods-cultivated (see chapter 5) roots are finding their own market niche, although some of the very highest grades are still mixed with wild ones, and some of the lowest grades are dumped into field-grown lots.

[Author's note: The chapters on ginseng that follow this history are based on the author's opinion that the future of small-scale American ginseng farming is probably in the forest. Here the grower can produce roots nearly indistinguishable from wild roots, which are certainly not in oversupply and whose price is currently 10 to 25 times the field-grown price.]

History of Cultivation in Canada

Ontario: A Parallel Story

In 1896, not long after George Stanton harvested his first plantings in New York, Clarence Hellyer sowed wild seed near the small town of Waterford in southern Ontario. Whether or not he was aware of, and benefited from, Stanton's experience is unknown, but Hellyer was also able to cultivate ginseng successfully in open fields under wood-lath shade. He soon formed a partnership to grow ginseng with his brother, Albert. In 1917, two of Albert's sons, Audrey and Russell, formed Hellyer Brothers, whose last crop was dug in 1970. They contracted to sell their production to a Colonel I. E. York, also of Waterford. Colonel York, a druggist, world traveler, and wild ginseng buyer, paid the Hellyers well for their roots and made nice profits for himself on the resale. To promote ginseng farming in his buying area, York published a pamphlet on the cultivation of ginseng, and by 1929, between 30 and 40 farmers were producing ginseng in southern Ontario.

During the next 25 years, which encompassed the Depression, the Sino-Japanese War, and World War II, most of these Ontario growers abandoned their gardens as did their counterparts in the United States. However, the Hellyer brothers (like the Fromm brothers in Wisconsin) persevered, dealing with the adversities of the times, which included $58,000 worth of unpaid-for ginseng left sitting on the dock the day Pearl Harbor was bombed. When the war ended, and trade with the Chinese community in the Far East resumed, mainly through Hong Kong, the Hellyers prospered.

Unlike Stanton, and later the Fromm brothers, they did not sell their seed widely or promote the growing of ginseng. Even though root prices were rising, there was only slow steady growth in Canadian production outside the Hellyer family. In 1962, no more than eight ginseng farms of any size existed in Ontario, but by 1983, between 50 and 60 Ontario growers were cultivating approximately 150 acres total. Then dramatic changes began to occur.

Just as ginseng was being more widely appreciated as an alternative crop, tobacco farmers in southern Ontario were having difficulties. They turned to green gold as an alternative to tobacco, and the Canadian government supported them. The Ginseng Growers Association of Canada (GGAC) organized in 1984, and in 1988 the Ginseng Research Gardens were planted at the Agriculture Canada Research Station in Delhi, Ontario, with start-up money from the government's Tobacco Diversification Program. To help farmers succeed in changing crops from tobacco to ginseng, a team of researchers from the Ontario Ministry of Agriculture, the University of Guelph, and Agriculture Canada began ongoing studies of disease control, optimal nutrition, root-drying methods, etc. With ever-improving, more efficient cultivation practices, Ontario ginseng production soared during the 1990s so that, entering the new millennium, there were approximately 4,500 acres of ginseng in the ground in Ontario, and 2.5 million pounds are now being dug and exported annually.

When field-grown prices began to drop precipitously in 1994, many of the relatively small-scale family ginseng farms in Wisconsin were unable to maintain profitable operations and, because these small growers stopped replanting, three-fourths of United States production has been lost. In Ontario, on the other hand, ginseng farms are typically larger and better able to mechanize operations and achieve economies of scale. Furthermore, research has helped to optimize cultivation practices and maximize production. Growers struggled to survive in a difficult market but are now in position to do very well indeed, if recent prices hold.

British Columbia: Ginseng Farming on the Stock Exchange

Out in the barren regions of south-central British Columbia (BC), where most of the land looks like hilly scrub desert and agriculture is limited to a few river valleys, another green gold rush began. Indeed, that is what Al Oliver, retired BC Ministry of Agriculture's Ginseng Specialist, entitled his history of ginseng in BC: *The "Green" Gold Rush.* No wild 'sang has ever been found in British Columbia. It certainly seems too dry and too hot for ginseng; but, when provided with artificial shade and occasional irrigation, *Panax quinquefolius* adapted here to an environment far different from its native cool, moist, acidic deciduous forest habitat. Utilizing dry, open plateaus of land bordering the rivers, enterprising folks were able to produce annual crops totaling as much as two million pounds of dried root.

In 1982, John Latta, a former real estate investor, planted five irrigated acres of American ginseng under artificial shade in Botanie Valley near Lytton, BC. Some years earlier, while visiting Hong Kong, Mr. Latta had seen a prized, wild Asian ginseng root sell for $64,000. He researched the plant and its cultivation and convinced himself and a number of investors that they could become wealthy by growing ginseng in British Columbia. They purchased seed and shade technology from Ontario and Wisconsin, and the dry climate of the province's interior helped control disease. Latta's operation flourished, expanded rapidly into the world's largest ginseng producer, and was listed on the Toronto Stock Exchange in 1989, and on the NASDAQ in 1999. He called it Chai-Na-Ta, which is phonetic Cantonese for Canada. At the height of its early success, in 1992, the Chai-Na-Ta Corporation grossed $7.9 million, with an estimated value at maturity of its existing plantings in excess of $100 million. Of course, that was the time to sell stock in Chai-Na-Ta, not to buy it.

Latta's success and his promotional efforts quickly attracted other growers. In 1986, The Associated Ginseng Growers (TAGG) of British Columbia formed and

Growing ginseng under polypropylene shade cloth with irrigation piping, which is needed in the arid interior of British Columbia. Photo by Al Oliver.

the provincial Ministry of Agriculture began to strongly support ginseng farming. Planting peaked around 1994, when 129 listed growers were farming almost 3,000 acres of ginseng in BC. The model John Latta provided, and which others copied, was one of a large-scale corporate farming operation, very different from the family ginseng farm that is the model in the United States.

The dramatic drop in prices in the mid-1990s forced most of the small-scale growers out of ginseng, and even the big corporate farms began to struggle to remain profitable. By 2002, TAGG listed only 40 growers. By 2012, there were only two ginseng farms still operating in the province. The innovative risk takers who figured out how to grow ginseng in an alien environment got rich quickly but just as quickly had their fortunes reversed. It is a story of boom and bust, in some ways similar to the Klondike Gold Rush that occurred just to the north 100 years earlier.

American Ginseng Farming in China

Chinese farmers have been growing Asian ginseng for a long time, but it was not until 1975 that they first imported large quantities of American ginseng seeds (from Canada). With their accumulated knowledge of Asian ginseng production techniques, the state-controlled farms in the northeastern provinces—Liaonong, Jilin, and Heilongjiang—were successful with *P. quinquefolius* from the beginning and have been

Planting on a large scale in British Columbia. Photo courtesy of Chai-Na-Ta.

An aerial view of Chai-Na-Ta's Schellers Ranch, a 92-acre ginseng farm under polypropylene shade in Lillooet, British Columbia. Photo courtesy of Chai-Na-Ta.

improving the quality of their roots ever since (though there seems to be some concern today about the vigor of their planting stock). In 1996, a visiting group of American agricultural extension specialists found 16 Ph.D.-level scientists at one university in Jilin Province studying the physiology and ecology of American ginseng. Officials from the People's Republic of China Ministry of Agriculture stated their intention to become self-sufficient in field-grown American ginseng by the beginning of the new millennium.

American ginseng growing in China under shade made with bamboo or dried grasses bound together with wire and laid in strips below and above a layer of rain-shedding translucent polyvinylchloride. Photo by Tom Konsler.

Labor-intensive bed making in China, where donkey carts are taken into the mountains to carry back down loads of woods dirt. Photo by Tom Konsler.

Good production figures are difficult to come by for China, but the best information is that over 1,000 acres of American ginseng are planted and harvested annually in open fields under artificial shade. Nonetheless, a great deal of American ginseng that has been field grown in Canada and Wisconsin is still being exported to China, where it is preferred for its higher quality and commands a premium price relative to the local product.

The Chinese are already exporting some of their American roots. They have the capacity to produce them in increasing quantity and very cheaply due to low labor costs. It is not clear whether North American shade growers will ever experience an extended booming market again (though, as of this writing in 2013, many are optimistic).

Fortunately for those who grow American ginseng in the woods, most of the trees in China were cut down over the centuries to provide heat and cooking fuel. More recently, some were cut down to clear rich mountain land so the soil could be more easily removed and transported to open fields to grow ginseng in. Despite the fact that ten years ago the Chinese government banned the cutting of forest, few fully mature hardwood forests with their accumulated humus-rich soils and cool moist environments remain, except in the Changbai Mountains of Jilin Province in northeast China near the North Korean border. This is the prime area where Asian ginseng was traditionally gathered before it was essentially extirpated in China. Though no production figures are available, numerous wild-simulated (called "mountain ginseng" in China) and woods-cultivated ("forest garden" ginseng) growing operations can be found in the Changbai Mountains, but they are all producing Asian ginseng (*Panax ginseng*), not American, probably because Asian ginseng produced in that area will bring more money. According to Dr. Eric Burkhart of Pennsylvania State University who visited the area in 2013, these woodland growers are experiencing the same problems with poaching that their American counterparts face and are using cameras, motion detectors, dogs, and even 24/7 guards to protect their green gold.

The Present Market for American Ginseng

Ontario is now by far the largest producer of American ginseng—over 2.5 million pounds annually, all of it field grown. Wisconsin currently produces over 90 percent of the field-grown roots exported from the United States annually (between 300,000 and 600,000 dried pounds, which at $30 per pound would be worth $13.5 million). Between 45,000 and 65,000 dried pounds of wild ginseng are exported annually from the United States, which at $675 per pound are worth roughly $37 million. Kentucky, Tennessee, and North Carolina are the largest suppliers of wild 'sang. In Kentucky, the single largest producer of wild 'sang, the ginseng industry is valued at $5 million annually, with an estimated potential in the current market to reach $25 million each year if the wild-simulated and woods-cultivated growing methods presented in this book (see chapters 4 and 5, respectively) are more widely utilized.

China (especially Hong Kong) and the Pacific Rim countries (particularly people of Chinese extraction living in Taiwan, Malaysia, Singapore, Indonesia, and the Philippines) consume almost 95 percent of exported American ginseng, while Australia, New Zealand, South America, North America, and Europe provide extremely small (but growing) markets.

The United States Domestic Market

The United States Food and Drug Administration classifies ginseng as an herb, not as a medicine. Until very recently, Korean products have almost exclusively supplied the growing health food market for ginseng. (The *Smithsonian* magazine reported that Americans spent about $170 million on ginseng and ginseng products in 2001.) However, American and Canadian entrepreneurs are beginning to compete with the Koreans for the potentially rich United States market. They are introducing product advertising along with new products developed for American tastes, such as American ginseng capsules marketed as "energy enhancers" or "high performance formulas" for athletes and folks with active lifestyles or simply in need of a little extra pep. Most of these products come from low-cost, field-grown roots, but the American public is beginning to appreciate high-grade American ginseng as well.

The Hong Kong Auction Market

Over 90 percent of American ginseng production goes to the Orient, where much of it is sold in bulk and goes through Hong Kong for resale at auction. After it is graded and processed, distributors supply retailers in Hong Kong, Mainland China, and in countries throughout the Pacific Rim. Grading and processing of American ginseng in the Orient is a labor-intensive activity, usually done in Hong Kong or southern China, and a "lot" may be sorted into 30 or even 40 grades. This grading is extremely subtle, though anyone can quickly learn to distinguish field-grown from wild roots. The highest grades are old, stubby, corky, wild roots that retain their long necks (or rhizomes) showing more than 20 growth scars and having "the wrinkle"—extremely thin, concentric, black rings around the circumference of the root, which are sometimes referred to as growth or stress rings. The lowest grades (other than broken, damaged, moldy, or diseased roots) are three-year-old long, smooth, carrot-shaped, dense, field-grown specimens. Field-grown "octopus" roots (see photo), in particular, are extremely undesirable. Whenever a dealer purchases a "lot" of American ginseng from a grower or another dealer, he inspects the roots and offers a per pound price for the entire lot based on his estimate of what percentage of the roots will fall into which grades.

"Octopus" or "spider" roots are considered very low grade and bring a much lower price than well-formed roots with a large trunk and no more than two branches. Photo by Al Oliver.

The Hong Kong auction market, however, involves much more than simply distinguishing between grades of cultivated and wild roots. It is a high risk, high reward commodity exchange. There are several different auction companies in several different locations. Agents who represent these companies oversee the ginseng through the importation and selling process. After the agent clears the roots through customs, he puts them in cold storage to wait for the best time to sell. Knowing the right time to sell depends on being able to predict the future market, so an importer's profit on his ginseng can vary greatly.

Dennis Chan, an agent representing companies in Hong Kong, Singapore, and Taiwan, describes the selling process:

> For... inspection, the roots of the specified sample drum will be dumped out on the rattan mattresses. [The buyers] will inspect right on the mattresses or will gather some roots on rattan trays for a closer examination. Then they will

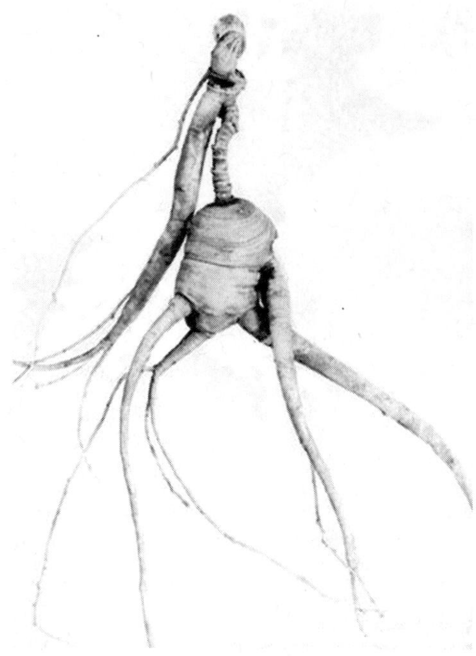

The highest grade roots have multiple "stress rings" around a large "bulby" trunk and a long neck proving their age. Photo courtesy of Bob Beyfuss and Cornell Plant Pathology Herbarium.

> figure out the price and write the bid on a piece of paper provided by the auctioneer. After gathering all the bids, we will open the bids in a separate room. If the price is up to our expectation, we will announce the [buyer] with the highest bid.

If the agent decides that the price is not good enough, the barrels are immediately resealed and returned to storage, and the agent waits for a better day to sell his green gold.

Over one billion dollars' worth of ginseng from all over the world comes into Hong Kong each year. Some is consumed there, but much is redistributed throughout the Indo-Pacific markets. The government of Hong Kong (now under Chinese control) has decided to become a world center for the development of pharmaceuticals based on Chinese medicine. To that end, the government is supporting efforts to achieve a better understanding of these traditional medicines by analyzing them using modern biotechnology. Ginseng is at the center of these development plans.

Recent Prices

The wholesale price per dried pound of both wild and field-grown American roots is constantly fluctuating, sometimes dramatically. Prices vary in response to many factors: root quality and availability, the relative value of the US dollar, the state of Pacific Rim economies, proximity to the Chinese New Year, and occasionally factors that seem to defy economic analysis. Table 1 presents the range of ginseng prices paid for typical lots—neither damaged nor select—during each buying season from 1982–83 to 2012–13.

These are the prices received by the grower or digger before export; that is, these are prices paid for 'sang by buyers or agents in the United States, not the Hong Kong auction prices. (Direct export sales are somewhat difficult to arrange.) The Hong Kong auction price usually reflects a rather modest markup over the stateside price, but the retail price in, for instance, a Singapore store may have a profit margin of several hundred percent. This is somewhat misleading, however, since only the central trunk of the root is being offered, while the fiber and branches (as well as any grossly undersized, misshapen, or damaged roots) are often ground up to be sold in tea or soup at an actual loss to the retailer. Ginseng is most commonly sold to the retail consumer in taels (a traditional unit of measurement roughly equivalent to 1.3 ounces). If you visit a Chinese traditional medicine store, determine the price per tael in US dollars, then multiply by 12 to get a close approximation of the price per dried pound.

There are no prices given in Table 1 for woods-cultivated ginseng—that is, for roots produced in plowed or tilled beds under natural forest shade (see chapter 5). Historically, there has not been a great deal of ginseng planted in woodland gardens, and there have been no separate records kept of the number of acres under woods cultivation or of the annual amount of woods-cultivated ginseng produced and exported. Roots grown in this manner have been, and to some extent continue to be, simply added in with either field-grown or wild ginseng.

However, since about 1994, interest in woods cultivation has increased substantially, and there is now good information available from buyers on the prices they pay for woods-cultivated lots, though not on the total annual quantity bought. Roots grown big and fast in intensively cultivated, woodland beds usually are categorized with field-grown roots and sell at prices near the upper range of that market (see Table 1). On the other hand, roots nurtured slowly (for at least six years) in tilled woodland gardens with little or no fertilization

Table 1: Recent Prices for Wild and for Field-cultivated Ginseng

Year	Wild	Field-cultivated
1982–83	$133–152	$36–51
1983–84	$137–220	$32–49
1984–85	$143–171	$24–48
1985–86	$125–152	$20–36
1986–87	$118–160	$18–30
1987–88	$180–225	$28–52
1988–89	$225–315	$30–67
1989–90	$200–270	$28–45
1990–91	$220–270	$38–62
1991–92	$220–340	$40–65
1992–93	$210–350	$28–56
1993–94	$210–310	$28–43
1994–95	$225–375	$18–30
1995–96	$400–500	$18–30
1996–97	$260–425	$10–30
1997–98	$190–425	$ 6–18
1998–99	$240–360	$ 7–20
1999–00	$305–575	$12–18
2000–01	$320–500	$13–20
2001–02*	$220–400	$12–17
2002–03	$250–500	$14–32
2003–04	$300–400	$18–32
2004–05	$250–500	$18–30
2005–06	$250–550	$10–22
2006–07	$300–600	$12–22
2007–08	$400–1,150	$10–18
2008–09	$250–1,000	$8–12
2009–10	$350–600	$12–24
2010–11	$350–1,100	$12–40
2011–12	$320–750	$15–20
2012–13	$400–1,250	$12–42

* For several weeks after 9/11, no one was willing to buy roots for export, and it was virtually impossible to sell any ginseng at any price.

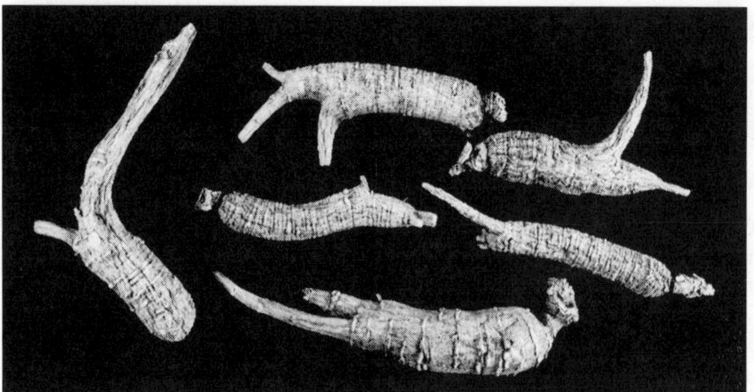

Sample of dry five-year-old woods-cultivated roots of a lower grade, showing little wild characteristic. Photo by Mark Haskett.

usually sell for between 30 percent and 70 percent of the going wild price (see again Table 1). Of course, roots that are grown from seeds simply dropped onto, or poked into, untilled forest soil in a manner simulating wild seeding (called wild-simulated planting—see chapter 4) and then left to grow naturally, develop as wild roots and have sold at the wild price.

PART TWO

A Ginseng Grower's Manual

The grading and pricing of American ginseng roots sometimes seems as much art as science, but the root's resemblance to the revered wild Asian root largely determines its value. Generally, the more slowly the root grows, the greater its resemblance to the ideal Asian root and the higher its price. Tilling of the soil and fertilization greatly speed development, while competition with weeds and trees for water and soil nutrients slows growth. Hence, cultivation techniques markedly influence the price paid per pound.

If you decide to grow green gold, there are three basic ways to go about it: (1) construct artificial shade over beds planted in the well-tilled soil of an open field and, within only three or four years, harvest large but low-value roots; (2) sow seeds in the woods with minimal disturbance to the forest floor, simulating wild growing conditions and, in 7 to 11 years, harvest rather small roots usually indistinguishable from truly wild ginseng; or (3) plant in tilled beds under natural forest shade, combining aspects of both (1) and (2) and, in six to eight years, harvest mid-grade roots. (Although the wild-simulated/woods-cultivated dichotomy mirrors common practice, there is actually a continuum of approaches in the woods, depending on the density of seeding and the extent of soil cultivation, fertilization, weeding, etc.) There are folks who have been successful with each of the three basic methods. You will have to decide what best suits you, your objectives, and your particular circumstances.

The next eight chapters are intended to serve as a detailed practical ginseng grower's manual. Chapters 3, 4, and 5 are each devoted to one of the three growing methods outlined in the preceding paragraph. Chapter 6 covers harvesting and processing seeds and roots. Chapter 7 discusses the future of the ginseng market and provides guidance on important business decisions that growers must make, from how to get started to where to sell their crop. Chapter 8 presents the story of a woodland 'sang grower told in his own words. Finally, chapters 9 and 10 provide useful resources and references for the ginseng grower.

3

Under Artificial Shade

Most of the world's ginseng is now grown in cleared fields under artificial shade, and most North Americans who have made their fortunes growing ginseng have utilized that approach. However, the profitability of a standard field-grown operation has become problematic since about 1996, especially for the small-scale grower. Nonetheless, this method does still have utility. It is, of course, the only method possible for those who have no wooded area available. Growing under artificial shade is the quickest way to produce roots suitable for personal consumption. It is arguably the best route to the dependable production of large quantities of quality seeds for sale or for extensive woodland planting (and the large, impressive seed-producing plants will be located in the open, where poaching is less likely). It can indeed be profitable on a small scale if the grower has direct access to upscale niche retail markets. Finally, the market for field-grown roots has recently returned to previous high levels and could stay there.

This chapter provides the basic information necessary for growing under artificial shade. Once the shade (either wood lath or polypropylene cloth) is in place, there is not a great deal of difference between small-scale ginseng farming under artificial shade and under natural forest shade in cultivated beds. Chapter 5 details woods cultivation, so I will emphasize here the unique aspects of artificial-shade growing, particularly as they relate to the small-scale grower. The highly mechanized, capital-intensive, technically sophisticated techniques employed by large-scale artificial-shade growers are fully described in the British Columbia Ministry of Agriculture's publication, "Ginseng Production Guide for Commercial Growers," as well as in other resources listed in chapter 10. [Author's note: The BC production guide is now only available on the web. Look under

Websites of Interest in chapter 10, for Internet address.]

Readers who have absolutely no interest in growing in open fields under artificial shade can certainly skip the sections of this chapter detailing the construction of shade structures. I would encourage you, however, to read the rest of the chapter for general information and comparison purposes before going on to the wild-simulated and woods-cultivated approaches.

Selecting a Planting Site

An artificially shaded ginseng garden will do best on gently sloping land (6–10 percent slope) that will accommodate well-supported shade construction but also promote drainage. The soil, itself, must drain well and be suitable for growing flowers or vegetables. Cleared woodlots, pastures, and lands that have previously been planted in corn, small grains, grasses, and legumes for hay are often used. If the site has lain fallow for several years, the accumulation of grass and root residues will increase the organic matter just as the annual leaf fall does in a hardwood forest. For a typical small-scale commercial venture, you will need to plant a minimum of half an acre each year. [Author's note: You can easily grow enough roots for your own consumption on a much smaller plot. Detailed instructions on creating and maintaining a very small-scale non-commercial artificial-shade garden are presented in *The Practical Guide to Growing Ginseng*, by Bob Beyfuss (see References).]

Bed Preparation

Planting is usually done in the fall, but pastures and fields are tilled periodically in advance of planting to reduce future weed problems. Immediately before planting, the soil is broken up again to make the beds; the shade-structure dimensions are staked out with as much precision as possible; and the posts, which will support the artificial shade for at least the next three years, are set. Ideally, the beds, which are usually four to six feet in width and slightly raised, are laid out so that the ditches run directly down the face of the slope (about 6 degrees is ideal) for quicker run-off during heavy or extended rains. I say this is the ideal orientation because the shade strips must run roughly north-south to properly shade the plants as the sun passes over from east to west.

Fertilization

Ginseng benefits from soil high in organic matter with a calcium level of at least 2,000 pounds per acre, but it does not require high soil fertility. In particular, ginseng usually does not respond well to soil amendments high in nitrogen. Forest soil, to which it is adapted, is typically low in nitrogen, because the decomposition of plant residues ties up most of the available nitrogen. A few shade growers mix in some humus-rich soil from the forest when they prepare their beds in an attempt to add whatever it is in the woods that makes ginseng grow so well there. We do not fully understand what it is in woodland soils that benefits ginseng, but there is a correlation between high organic matter content and healthy, vigorous wild ginseng populations. To add organic matter, large-scale growers in Ontario till in as much as 50 tons of animal manure per acre, but it is safer to till in two or three inches of shredded hardwood leaves and bark instead (preferably, already partially decayed). Leaves and bark are safer because they are not likely to alter soil chemistry, introduce weed seeds, or markedly accelerate growth. Some growers produce a cover crop of alfalfa, clover, peas, or soybeans and turn that crop under in the fall before planting.

Planting

In large artificial-shade gardens, sowing about 20 seeds per square foot is typical (roughly 90 to 100 pounds of seed on an acre). However, if the small-scale farmer wants to grow his roots longer than four years before harvesting, then a more appropriate amount is eight to ten seeds per square foot. Seeding by hand, with a small garden planter, or with a mechanized planter pulled behind a tractor are all options. I prefer a planting depth of about ¾", although depths varying from one inch to just below the surface are successful. If you hand cast your seeds, you would do well to rake or shovel at least a thin layer of soil over them to insure good contact with the soil and to protect them from drying out. Mulching immediately is also highly advisable.

Transplanting seedlings from a nursery area to permanent beds was once common practice, but most commercial 'sang farmers have abandoned transplanting roots because it is labor intensive, and labor has become increasingly expensive and hard to find. However, transplanting remains an option for the small grower primarily interested in the quality of the roots produced. For root harvest at four to eight years of age (for personal consumption or perhaps for sale to a niche market), transplanting at a 6" × 6" spacing is typical. If the purpose is long-term seed production, roots (preferably select ones) should be transplanted no closer than nine inches apart both ways.

Mulching

Since there is no leaf fall to naturally mulch your plants growing under artificial shade, you will have to do the job yourself. Straw is the choice of most commercial shade farmers, because it is cheap, readily available in quantity, and easily transported in bales and distributed onto the beds. Large growers use straw-spreaders that chop the straw and blow it out the rear as they are pulled behind a tractor over the beds. Oat straw, which has been rethreshed to get rid of remaining grain and weed seeds, is probably best. Seed-free mulch reduces both weed and rodent problems. However, straw decomposes quickly and is a siren song to slugs. Small-scale growers are advised to consider alternatives such as an inch of hardwood bark/sawdust mixture, compost, or shredded leaves.

Wood Lath vs. Cloth Shade

Small differences in the percentage of light penetration and in air and leaf temperature have been recorded beneath the two types of shade structures. Polypropylene shade cloth does seem much hotter, because your head is near the black cloth when walking under it. However, the heat absorbed and generated by the cloth rises away from the plants, which are well below at soil level. Experienced growers tell me that while growth is marginally better under lath shade, cloth is much easier to put up and take down. Once posts are set and cables strung (no quick job), three men can put up an acre of cloth shade in a day. Both work well if installed properly; after all, 100 years of trial and error have gone into their development. Nonetheless, wood lath shade is rapidly becoming obsolete.

Constructing Wood Lath Shade

Wood lath shade construction is a major undertaking. A height of over 6' is needed for working space and ventilation, so 8½' posts (cedar or locust 6" in diameter) are typically set 2' into the ground (before planting). There are sophisticated variations, but a 12' × 12' grid pattern works out nicely. Except on the field's perimeter, the posts are set 12' apart in the exact center of every other 4½' wide bed. Given 1½' wide drainage trenches, this works out perfectly and doesn't block the trenches (2¼' + 1½' + 4½' + 1½' + 2¼' = 12').

The tops of the posts are connected with 12' long 2 × 4s, which brace the posts and support the sections of lath shade. Each section is 4' × 12' and is constructed using three 12' runners (or slightly longer for extra support surface) with hemlock or cedar lath strips nailed or stapled across them. The lath strips are spaced apart about one-third the width of the lath, resulting in ideal 75–80 percent shade. A thousand of such 4' × 12' sections will shade one acre. It is of critical importance that, in position, the laths run roughly north/south, perpendicular to the sun's east-to-west movement. This provides alternating periods of

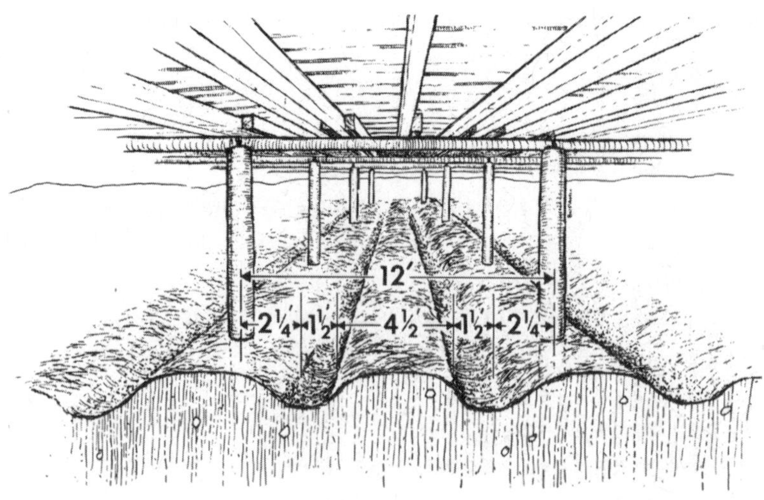

Spacing of beds, drainage trenches, and posts.

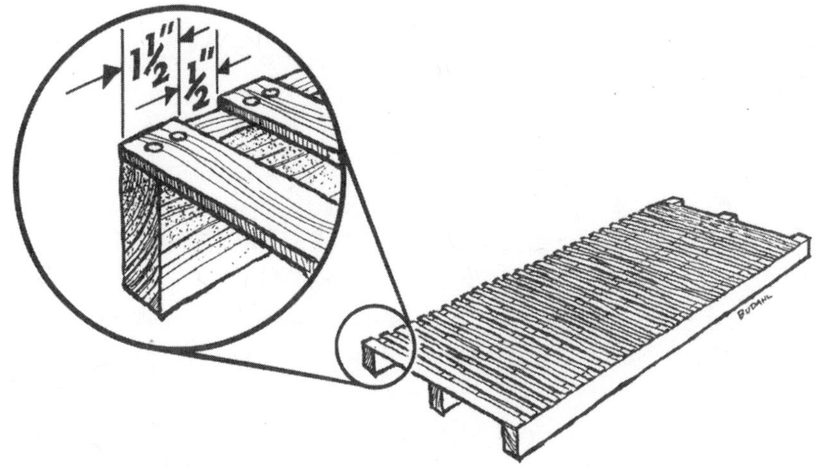

Wood lath section.

Stacking shade structure in winter to avoid breakage from snow load. Photo courtesy of Tom Condon.

light and cooling shadow on any one ginseng leaf.

The lath sections are placed in position in the spring before the plants emerge. (Be sure to walk in the trenches, not on the beds.) In northern climates, the lath sections must either be taken down or stacked double during the months when heavy snow loads can break them. Large growers often have tractor-mounted forklifts to help move the shade sections. A snow fence, which shades plants on the outside edges and provides a critter barrier, often surrounds the shaded area. Access is provided along two opposite sides so that a tractor and other equipment can easily enter and cross back and forth via the trenches. Some growers leave the sides open to improve air circulation and extend the overhead shade several feet beyond the planted area, especially on the south and west sides that are exposed to the hot afternoon sun. (The sun's heat, not its light, is what stresses the plants.)

If you decide to cover a large area with wood lath shade, consider visiting an ongoing operation for a first-hand look. The trip would be most helpful early in the spring when the shade is going up. You may be able to find a shade-grower in your region, but very likely you'll have to travel to Wisconsin or Ontario.

About Polypropylene Shade Cloth

Polypropylene cloth has been stretched over everything from ferns to Christmas trees. Its use with ginseng is recent and increasingly popular. The color is usually black because the polypropylene strands of fabric are impregnated with nickel and carbon black to block out the sun's ultraviolet rays, which would otherwise soon make the fabric brittle.

For ginseng, instead of a solid mesh, the cloth is woven in a thick and thin pattern, simulating the shading effect of lath strips and providing better air circulation than would a solid weave. When installed, the striped pattern is run north to south, just as with lath. Enough 78 percent shade fabric needed to shade an acre of ginseng weighs about 1,700 pounds and can by hauled in a pickup truck (when it comes folded by the manufacturer). Several brands of shade cloth are appropriate for ginseng, and improvements are occurring regularly. Suppliers are listed in Resources.

Installing Shade Fabric

In addition to the shade fabric, a supplier should provide you with an explicit set of installation instructions. Use them! I speak from experience. Also, use the various clamps, bolts, hooks, grommets, and possibly steel posts and anchors that he will sell you to go with his particular instructions. Time and expense spent in proper initial installation will be repaid in low damage and maintenance problems over the multiple years you will use the structure. Shade cloth that is not properly installed will tear or, much more likely, sag. Where it sags, water will run and drip on the beds below, greatly increasing the risk of root rot starting where the puddles form.

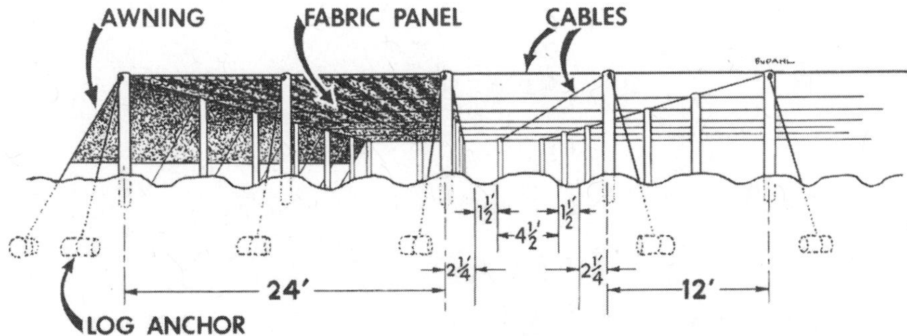

Fabric shade support system.

The cloth comes in long strips up to 24' wide, and that maximum width is used most often. Posts are set in a 24' × 24' spacing (or a few inches wider to insure stretching the fabric taut). All posts (except exterior ones) are centered in every fourth bed, keeping the trenches clear for access and drainage.

Either 3⁄16" or ¼" cable is stretched tightly between the tops of the posts in a grid pattern and firmly fastened to ground anchors around the entire perimeter of the garden. The 24' wide strips of shade are then unrolled on top of the grid and attached to the cable, usually with S-hooks.

In the late fall, each cloth strip is unhooked on the two short (24') sides and on one long side, rolled up to the remaining long side, bound every few feet by cord to the cable on that side, and allowed to hang furled in the field for the winter. Spring deployment simply reverses this process.

Shading the edges of the beds (especially along the south and west sides) can be accomplished by placing snow fence around the perimeter, or by stretching a five-foot-wide shade-cloth awning down along the ground support cables at a 45-degree angle from the top of the outside cable. An awning is easy to fold back temporarily for access. Alternatively, extending the overhead cloth well beyond the planted area so that the margins of the planting remain shaded, but the sides are left open will improve air circulation.

Advantages of Farming under Artificial Shade

There are aspects of artificial-shade growing that are significantly easier than woods cultivation. Bed preparation without tree roots to contend with is a comparative snap. Rodents and other critters are much less likely to munch your roots or gobble up your berries. A woods-grower must pick his berries as soon as they ripen, or his seeds for next year's planting will provide this fall's menu for the mice and squirrels. Because the berries ripen sequentially, he may need to pick several times. Under artificial shade the harvest will be much greater and can be completed in one step, since the

One advantage of shade growing is the ability, at harvest time, to turn up the roots with a modified potato digger and then simply pick them up. Photo by Al Oliver.

plants are more robust and productive, and the critters usually don't get there first (although wild turkeys have become a problem in Wisconsin).

Digging the roots in an open field setting with straight, level beds and no tree roots is also much easier. Digging is really a cakewalk for large-scale growers who attach a modified potato digger to their tractor and pull it down each row, bringing the roots to the surface to be simply picked up by hand. (Before digging, the shade structure is furled or taken down and stored for re-use over another planting.) Probably the biggest advantage to growing ginseng under artificial shade is that the roots grow much more quickly and are ready for harvest in only three or four years.

Problems with Field Cultivation under Artificial Shade

Up-front Cost of Shade

Either kind of shade structure will set you back about $1,000 for each tenth of an acre covered, not to mention the time and effort required to put it up and maintain it. (I put up cloth rather than lath, mainly because I have poor carpentry skills, and one of my good neighbors works for the power company and knew how to tighten cables.) Many growers have designed and built their own unique shade structures out of cheaper (or already in hand) materials. For plantings smaller than a twentieth of an acre, this is an option worth considering; a very small bed can be tried under some already existing structure like a grape ar-

Scott Persons and son in artificial-shade garden for long-term seed production under polypropylene cloth. Photo by Mark Haskett.

bor. On a larger scale, however, I have not seen any one-of-a-kind shade houses that I thought were clearly worth the savings.

Vigorous Weed Growth

Weeds, especially grasses, are usually much more serious in field cultivation. First- and second-year plants need weeding early in the summer. Weeds left unchecked will not only compete for nutrients, but they will also further reduce air circulation, creating a more humid, disease-favorable environment. Unchecked grasses can eventually choke out young ginseng. Most large-scale commercial growers apply chemicals the summer before planting to sterilize the soil and greatly reduce the number of future weeds, as well as to destroy potentially harmful nematodes and disease organisms. With careful usage, selective herbicides like Poast (Sethoxydim) and Fusilade (Fluazifop-butyl), which kill grasses but not broad-leafed weeds, can be sprayed over ginseng without damaging the crop. Growers also effectively spray herbicides on weeds just after the ginseng dies off in the fall. Weeding done by hand is a tedious job, even on a small scale. [Author's note: The regulation, approved use of, and labeling of pesticides are subject to change at any time (see Pesticide Use below). Their mentions in this book are provided only as information. It is always the pesticide applicator's responsibility, by law, to read and follow all current label directions for the specific pesticide being used.]

Disease-friendly Environment

Cultivation of wild plants outside their native environment and in a monoculture (only with members of their own kind) in order to accelerate growth rate often increases susceptibility to, and spread of, disease. Wild 'sang diggers rarely find diseased ginseng growing in humus-rich woodland soils amidst a mix of herbaceous plants, and wild-simulated plantings rarely suffer from blights and rots. But woods-cultivated gardens are usually monocultures of fast-growing plants, and they are often attacked by fungal pathogens. However, disease problems are much greater in field cultivation. Even small, thinly spaced, shade gardens, grown for either personal consumption or seed production, are at great risk of disease under artificial shade.

Ginseng is not adapted to a field environment. There are no trees to wick away excess moisture, and woodland soils are typically rich in humus, which not only supplies ginseng with appropriate nutrients but also contains soil fungi that are thought to be symbiotic (mutually beneficial) with ginseng. Microscopic fungal symbionts, called mycorrhizae, actually live within the cells of the ginseng feeder roots and utilize sugars from the root. In return, these mycorrhizae enhance the absorption of nutrients by the root hairs and produce antibiotics to some of the soil-borne pathogens that attack ginseng. It seems likely that the field soils of shade gardens rarely contain high concentrations of these fungal symbionts. In addition, the shade structure itself significantly reduces air circulation, and the open field setting gives fungal spores easy wind-borne access.

The typical shade grower who attempts to maximize the return on his considerable up-front capital investment in lath or shade cloth by planting as thickly as possible, creates a dense, humid, stagnant-air environment, which is ideal for the introduction of fungal diseases. Closer spacing not only reduces ventilation, creating higher humidity in which fungus diseases prosper, but close proximity also facilitates the spread of disease. Disease can spread over the surface of the soil from stem to stem, through the soil by root contact, between foliage when infected leaves touch or fall on healthy leaves, and by raindrops splashing spores from diseased to healthy plants. Crowding also may place the plants in competition with each other for sunlight, water, and soil nutrients, thereby stressing them and making them more vulnerable both to diseases and to unusually harsh weather conditions. Large densely planted commercial gardens are usually dug after only three years, because disease threatens plant survival; moreover, second plantings in the same ground usually result in disease and failure (see chapter 5, for more on this). I started a 2,000-square-foot polypropylene shade-cloth garden with three-year-old roots spaced nine inches apart both ways and maintained it for exceptional seed production for 12 years before having to dig the then 15-year-old roots to make room for an addition to my house; however, I had to strap on my backpack sprayer and spray fungicides every week for 12 summers.

The Latest and Best Information on Diseases

With the support of the Ginseng Board of Wisconsin (an organization of large-scale growers), Dr. Mary K. Hausbeck of the Department of Plant, Soil and Microbial Science at Michigan State University has developed what is by far the most up-to-date and complete information on managing ginseng diseases and other pests in the United States: "Pest Management in the Future: A Strategic Plan for the Michigan and Wisconsin Ginseng Industry." The information is intended primarily for growers in Wisconsin and Michigan, and it is so detailed in some respects that it loses user friendliness; nevertheless, it is a useful resource, although only a starting point if you live in another state. The December 2010 version is available at: ipmcenters.org/pmsp/pdf/MI_WI_ginseng_PMSP_2011.pdf. A more recent, less encompassing 2013 update, "Control of Diseases, Pests, and Weeds in Cultivated Ginseng," is referenced under "Ginseng Reports and Conference Proceedings" in References.

The Ontario Ministry of Agriculture, Food and Rural Affairs (OMAFRA) has produced an excellent, 88-page, well-illustrated booklet titled "Production Recommendations for Ginseng," which is based on the latest field research. It includes complete information on the diseases of ginseng and detailed recommendations for their prevention and control in artificial-shade gardens. The booklet concisely covers other aspects of shade production as well (including the mechanized equipment used) and is easily worth the price of $20 Canadian. It is available by calling 1-800-668-9938 or going to ServiceOntario.ca/publications and asking for Publication 610 (see also listing in chapter 10). In chapter 5, Woods Cultivation, I have tried to provide the basic knowledge a small-scale grower needs to cope with the diseases of ginseng; however, any grower with plans to manage more than half an acre of cultivated ginseng gardens (in the field or in the woods) would be well advised to purchase this booklet.

Pesticide Use

The more intensively ginseng is grown, the more likely it is that pesticides will be required. All field-cultivated ginseng grown under shade cloth is routinely sprayed with a regimen of fungicides. Other pesticides, including insecticides, herbicides, molluscicides (snail and slug poisons), soil fumigants, and rodenticides are utilized as well. Woods-cultivated ginseng generally requires some pesticide use, whereas wild-simulated ginseng can often be grown with no pesticides at all.

In order for any pesticide to be legally used on ginseng, that particular pesticide must be specifically registered for use on ginseng by both the EPA and the local state authority that regulates pesticide use. Only if both registrations are obtained can the product specifically mention ginseng on its label under the "for use on in a given state" section. Organic pesticides are still pesticides, and they must also be registered and labeled for use on ginseng.

Some pesticides approved by the EPA and used on ginseng in the United States are covered by a national registration for that product, but most aren't. Since ginseng is a specialty crop usually grown on small acreages, unlike corn or beans or cotton, few states have many specific registrations for ginseng. Wisconsin, the major field-grown ginseng exporting state, has a number of "special use" pesticide registrations for ginseng, but that makes them legal to use only in Wisconsin. (Canada also has a large field-grown ginseng industry and consequently has numerous pesticides registered.)

The reality is that many small-scale US growers, especially those outside Wisconsin, ignore the regulations (sometimes out of ignorance, sometimes out of desperation, sometimes in knowing disregard for regulations perceived as inappropriate) and use whatever products are readily available and known to be most effective on the same or related diseases that may also occur on vegetables, such as tomatoes. Or, they may use a product that they know is registered for ginseng in Wisconsin or Canada. This practice could have serious repercussions for the grower if any residue of the pesticide is detected on the roots or leaves. All pesticides leave residues that can be detected by a competent laboratory test. Some buyers of ginseng now have pesticide residue tests performed if their re-sale market is sensitive to the issue. Growers illegally using these products risk serious fines or, at best, loss of sales.

Defending against pests (weeds, fungi, insects, rodents, etc.) that threaten ginseng is discussed throughout the ginseng section of the book. Many of the fungicides and other pesticides commonly used by 'sang growers (and therefore mentioned herein) are only approved for ginseng on a state-by-state, year-to-year basis. Needless to say, the illegal application of unapproved chemicals can't be recommended here. Check with your local agricultural extension agent to find out which are legal in your state, and then confirm their advice by reading the label. It is always the pesticide applicator's responsibility, by law, to read and follow all current label directions for the specific pesticide being used. There are likely to be new fungicides and biological controls approved for use on ginseng in the future. [Author's note: The author is indebted to former Cornell extension agent Bob Beyfuss for help in explaining the complexities of pesticide registration, labeling, and use.]

The Bottom Line under Artificial Shade

Growing in open fields under artificial shade is obviously not a casual enterprise, being both labor and capital intensive. In Table 2, an experienced commercial grower estimated his profits and expenses over a four-year period to grow just one-tenth of an acre.

The figures in Table 2 are typical, although they do not include the cost of the land and the seeds (which in both cases were the grower's own), as well as the energy bill for drying the roots (roughly 50 cents per pound, or $100). The price of $40 per dried pound is current but certainly toward the high end of the recent market. A yield of 200 pounds per one-tenth acre assumes no serious disease problems and accurately reflects the past production performance of American growers. The large-scale, highly mechanized Ontario shade growers who remain successful in today's market are harvesting more than 300 pounds (3,000 per acre).

The Wisconsin grower did not include the value of the seeds harvested in the third and fourth years. On a tenth of an acre under artificial shade, a harvest of 20 pounds of seed each year (enough to plant almost an entire acre wild-simulated) would be a conservative expectation. To expand the projections to an acre, you would have to add the costs of a drying shed and probably some labor-saving equipment, such as a digger and a tractor set up high enough to roll over the beds with 18"–24" ginseng growing in them.

Table 2. Profit and Loss Statement for One-tenth Acre of Ginseng Grown under Artificial Shade

Income	
200 pounds of dried root**	**$8,000**
Expenses	
Shade and shade construction	$1,230
Straw mulch	$40
Fungicides and pesticides	$45
Fungicide sprayer	$100
Labor*	$6,000
Total Expenses	**$7,415**
Profit	**$585**

* (mostly weeding, but also seeding, picking berries, digging, washing and drying roots, etc.) 400 hours at $15/hour
** sold at $40 per lb.

4

Wild-Simulated Planting

The wild-simulated approach takes the longest time from planting to harvest, but it is by far the simplest method, the least expensive and time consuming, the most compatible with the natural ecology of your land, and it can be implemented on steep hillsides where nothing else is practical. Seeds are planted thinly with minimal disturbance to the forest floor and left to grow naturally so that extremely desirable roots, indistinguishable from truly wild roots, are harvested. In effect, the grower develops a naturalized population of wild ginseng within his woodlot. Wild-simulated growing requires only a modest investment in seed and labor, and the bulk of the labor is in the digging, which is not done until profit is nearly assured. It can be profitable on either a small or large scale. One person can plant and maintain many acres without mechanized equipment. In my opinion, it is the method of choice, if you have the right spot to do it, especially if that spot encompasses multiple acres. (If your woodlot is not ideally suited for ginseng, there is an alternative, more labor-intensive variation—see The Hankins Method at the end of this chapter.)

The Planting Site Is Critical

Because you are going to scatter your seeds and then just let the ginseng grow at its own rate with limited intervention until harvest, it is obviously necessary to own, lease, or share a planting site where ginseng will naturally flourish. The ideal location is woodland where wild 'sang is growing vigorously or is known to have flourished in the recent past; but, it's certainly worth considering on any site where there is a high shade canopy provided by mature mixed hardwoods, at least some slope for drainage, and rich woodland soil with high organic matter content.

Since ginseng has been hunted to near extinction in many locations, determining whether it once grew on your property could be difficult. You may want to find an old-timer who knows the local history of ginseng. If you live in 'sang country, such venerable resources are often available, and they will appreciate your interest. Asking for local advice, however, always incurs a risk. The major drawback of using the wild-simulated method is the usual 7 to 11 years your green gold must stand in the forest subject to predation, especially human, before it reaches maturity. The fewer unscrupulous folks who know about its presence, the better. If 'sang diggin' (as the old timers in the southern Appalachians call foraging wild ginseng) is still part of your area's culture, then very likely the boundaries of your property will not be respected. As one Kentucky grower told me: "You're growing green gold and everybody is interested in your claim!" You must consider whether poaching will be a problem on your site and, if so, whether you can prevent or at least minimize it. (See Security below.)

If you live where wild 'sang is virtually unknown (and poaching is of minimal concern), you still may be fortunate enough to find a woodlot that has the conditions ginseng needs. These microenvironments are not so rare. A trapper friend of mine found 11 of them inside the city limits of Greensboro, North Carolina, a large city in the central part of the state where 'sang digging is most certainly not presently part of the culture.

In addition to mixed hardwoods, slope, and rich soil, look for plants that are known to grow right alongside *P. quinquefolius*. If they are growing in your woods, then there is a good chance ginseng will like it there too. Many of these companion plants are found throughout much of the eastern United States. Consider at least an experimental wild-simulated planting if you have a few of the following species growing on your forest floor: maidenhair fern (*Adiantum pendatum*); rattlesnake fern (*Botrychium virginianum*), sometimes referred to as "ginseng pointer"; Christmas fern (*Polystichum acrostichoides*); baneberry (*Actaea pachypoda*), often called "doll's eye"; yellow lady's slipper (*Cypripedium parviflorum* var. pubescens); bloodroot (*Sanguinaria canadensis*); blue cohosh (*Caulophyllum thalictroides*); black cohosh (*Actaea racemosa*); wild ginger (*Asarum canadense*); Solomon's seal (*Polygonatum biflorum*); false Solomon's seal (*Smilacina* spp); jack-in-the-pulpit (*Arisaema triphyllum*); trilliums (*Trillium* spp); foam flower (*Tiarella cordifolia*); jewelweed (*Impatiens* spp), also known as touch-me-not; mayapple (*Podophyllum peltatum*); goldenseal (*Hydrastis canadensis*); sweet cicely (*Osmorhiza* spp); yellow mandarin (*Prosartes lanuginosa*); Dutchman's pipe (*Aristolochia macrophylla*); bedstraw (*Galium triflorum*); liverwort (*Hepatica* spp); or Canadian violet (*Viola canadensis*). Some of these companion plants are calcicoles (normally growing in calcareous soils) and need conditions nearly identical to ginseng. Maidenhair fern, yellow lady's slipper, baneberry, and blue cohosh are calcicoles. Many others inhabit a broader range of conditions than does ginseng—mayapple, jewelweed, and foam flower will tolerate wetter conditions; while jack-in-the-pulpit, Christmas fern, Solomon's seal, and false Solomon's seal will also do well in drier locations; and bloodroot often favors flood plains. However, these species often grow where the wild 'sang grows, so their presence is certainly propitious. An abundance of especially large and vigorous specimens of ginseng's companion species is particularly encouraging.

In the northern states, the ideal site would be mixed hardwoods, dominated by sugar maples (*Acer saccharum*); and, in the South, it would be a northeast-facing slope filled primarily with tulip poplar trees (*Liriodendron tulipifera*), known in some areas as yellow poplar. A few Carolina silver bells (*Halesia carolina*) and black walnuts (*Juglans nigra*) would be spaced here and there. Bigger, taller examples of any species are preferable because they usually have a broader higher canopy, allowing better air circulation over the beds. Trees with deep root systems also are more desirable, because they compete less for surface soil moisture and may even bring moisture up to the surface during periods of drought. Ginseng is not normally associated with woody shrubs; however, there are two exceptions: spicebush (*Lindera benzoin*), which is often a companion in the southern Appalachians, but not at high elevations and much less so as you move into the northern states; and red-berried elderberry (*Sambucus pubens*), which is a common associate in northern states.

Because heat from the afternoon sun can stress ginseng (especially in July and August), it is preferable for your slope to face north or east, so that the sun will set early on your plants. A northeast-facing slope is particularly desirable in the southern states. Southwest-facing land that receives direct sunlight long and late on hot summer afternoons is usually unsuitable anywhere south of Maine. However, ginseng sometimes grows well on southwest slopes, if the slope faces directly into an adjacent mountain that effectively blocks the afternoon sun. Where wild 'sang or its companion plants are thriving, the aspect of the slope is almost certainly satisfactory. (Incidentally, ginseng likes the cool sunlight of early morning. A potted plant placed by a northeast-facing window will grow nicely.) Chapter 5 includes a visual site assessment checklist.

Calcium Appears Crucial

If wild 'sang is not flourishing on your prospective planting site, but otherwise it looks promising, then a soil test is the next task before you jump to sowing seeds. A soil test may also be appropriate on sites where a visual assessment may not be terribly helpful: severely disturbed areas, sites that have been overgrazed by livestock or deer, sites reverting from fields to forests, sites that have been overgrown by invasive plants, and sites located outside ginseng's native range are difficult to evaluate. Your state's agricultural extension or soil conservation service will probably help you get a soil analysis done at minimal expense. However, they probably will not know what soil amendments to recommend for improved wild-simulated ginseng production. No one really knows for sure at this point, and amending the soil runs the risk of promoting growth so much that your roots will neither look like nor sell like wild roots. Nonetheless, when you get the analysis, check the calcium (Ca) level; hopefully, it will report at least 2,000 pounds available per acre. (If calcium is reported in parts per million, or PPM, multiplying by two will give a rough estimate of pounds available per acre.)

A great deal of sophisticated research has been completed on the nutritional requirements of field-grown ginseng, but even the most basic analysis of what wild ginseng needs in its woodland soil was not begun until 1997. In that year, our knowledge of the ideal soil conditions for the growth of wild American ginseng made an important and long overdue advance. In two completely independent studies, investigators in New York and North Carolina both found high-calcium, low-pH soils consistently associated with the most vigorous and healthy wild ginseng stands. They also found that the levels of other standard soil nutrients, such as phosphorus, which is associated with good root growth in other crops, and potassium seem to make little difference. However, preliminary findings from ongoing research do suggest that levels of phosphorus below 10 pounds per acre and levels of potassium below 250 pounds per acre are weakly associated with poor survival of one- and two-year-old plants.

It is not news that calcium is important to American ginseng. For example, in a pot study conducted by L. P. Stoltz at the University of Kentucky back in 1982, calcium was the only major nutrient whose complete withdrawal led to the death of the ginseng (by root rot). Nonetheless, it is surprising that calcium appears to play such a unique and crucial role in the vigorous growth and survival of wild populations.

The Beyfuss Study in New York

Robert Beyfuss, a recently retired Cornell Cooperative Extension agent for Greene County, New York, and a small-scale ginseng grower, realized that there was no data on the soil characteristics where wild ginseng grows, especially where it grows best. When an opportunity arose to get minimal funding, Bob jumped at it. Then all he had to do was obtain the cooperation of New York's wild shang diggers (in the north it's "shang," not "'sang"). He designed a baseball cap with "Shang Hunter" emblazoned on the front and offered it to reputable diggers who would, in exchange, provide a properly taken soil sample from their most

Cap distributed to New York ginseng diggers who provided a soil sample for analysis from their most productive foraging site. Photo by Bob Beyfuss.

productive digging site. (They were not asked to divulge the location of the site.) Bob distributed 144 caps and obtained an excellent sampling of soils, which he sent to the Cornell lab for analysis.

He found that the most vigorous wild ginseng populations in New York are growing where the pH is low (averaging 5.0) and the calcium levels are very high (averaging about 4,000 pounds per acre). No other soil nutrients consistently occurred in either above or below average concentrations in the samples. In a presentation made at the national conference, American Ginseng Production in the 21st Century, sponsored by Cornell Cooperative Extension of Greene County, Beyfuss concluded:

> The most interesting and puzzling result of the analysis was the positive correlation of very low pH and very high levels of calcium. This is the exact opposite of what one would expect in mineral soils. During the past 20 years as a Cornell Cooperative Extension agent, I have looked at soil test analysis from hundreds of soil samples, and I have rarely seen this correlation except in certain forest soils. The average pH for these samples was 5.0, ±0.7. Soils that are strongly acid such as this usually have calcium levels in the range of 400 to 500 pounds per acre or less. The average level of calcium in these samples was 4,014, ±1,679. It is my suspicion that this abnormality may, in fact, be the key to the limited range of healthy populations of wild ginseng. Duplicating [or being initially blessed with] this soil condition may be critically important to successfully cultivating American ginseng in a forested environment.

Because nutrient levels of both the major elements (nitrogen, potassium, and phosphorus) and the minor nutrients (boron, iron, sulfur, etc.) varied so widely in his samples, Beyfuss further concluded that these other elements are of little significance to the prospective woods grower.

Having found the calcium connection, Beyfuss puzzled over the source of all the calcium. In a survey, which the diggers filled out and returned along with their soil samples, he had asked them to indicate the dominant tree species at their site. Every single survey listed sugar maples. Being an old New York shang digger, Bob was not surprised that great ginseng was dug in the sugar bush; but now he started wondering why, and he researched *Acer saccharum*. He reported the following results:

> Sugar maple trees accumulate calcium in their leaves. They use their leaves… as biological "sinks." All trees use their leaves as biological sinks during the growing season, but most trees suck those nutrients out before the leaves drop, particularly oak trees. Sugar maples, however, do not do that. Sugar maples concentrate calcium in the leaf tissue, and when they drop those leaves there's a tremendous storage of calcium right in that leaf litter.… As a matter of fact, if you take sugar maple

leaves and analyze them, they are about 1.81 percent calcium. That is a high level of calcium.... It turns out that the trees you find ginseng growing under in the South, namely tulip poplar and black walnut, also use their leaves as calcium sinks. So, the tulip tree is basically the sugar maple of the South in terms of concentrating calcium. The calcium is coming from the leaf litter of these trees.

Table 3. Comparison of Average Calcium Levels Found in Soils with Wild Ginseng Stands of Varying Quality

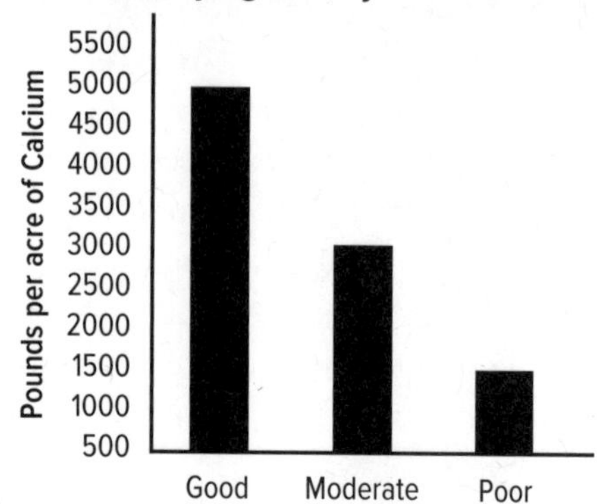

The Corbin Study in North Carolina

Almost simultaneously with Beyfuss' work in New York, Jim Corbin, Plant Protection Specialist with the North Carolina Department of Agriculture, was sampling soil and tissue from ginseng stands in the Great Smoky Mountains National Park, which runs along the North Carolina–Tennessee border. Corbin sampled from all ginseng stands, not just the most vigorous. Before sending the samples to the North Carolina State University lab for analysis, he rated each stand as good, moderate, or poor, based on plant size, age diversity of the stand (an indicator of reproductive success and survival rate), and plant health (lack of disease and chlorosis of the leaves).

From the bar graph in Table 3, it is obvious that there is a dramatic correlation between ginseng stand quality and calcium level in the soil. On the good sites, the average level of calcium per acre was 4,999 pounds; on the moderate sites, the calcium level averaged 2,598 pounds per acre; and, on the poor sites, the average calcium level was 1,315 pounds per acre. Moreover, calcium was high in all the tissue samples from good sites and was only slightly lower in samples from moderate sites, with a noticeable drop in calcium levels occurring in the samples from poor sites. The level of no other soil nutrients correlated with the quality rating of the ginseng stands (pH was uniformly low, as there were no limestone outcrops in the sampled area).

Corbin noted: "In ginseng, calcium deficiencies can be seen in stunted plants that lack general vigor. Growth buds are smaller and more fragile. In good ginseng stands, calcium on a per acre basis is consistently higher than in the other stand categories, and within the good stands there was better plant diversity, less disease, and a greater stem height in mature plants."

Add Gypsum

The work of Beyfuss and Corbin argues for selecting high-calcium, low-pH planting sites for wild-simulated production. The more interesting question is whether the addition of calcium to the soil can improve ginseng's survival and growth rates if the soil test shows a level of calcium below 2,000 pounds per acre on an otherwise promising site (a common occurrence). Very preliminary work with gypsum suggests that it can.

Gypsum (calcium sulfate) is the soil amendment of choice because it is a natural chemical that is widely available, inexpensive, and safe. Gypsum will increase calcium without significantly altering the soil chemistry, and it does not increase soil pH as calcium carbonate (lime) would. Circumstantial evidence suggests fungal diseases that attack ginseng, such as *Phytophthora cactorum* root rot, are suppressed by acid soil conditions, possibly because low pH is strongly associated with high concentrations of aluminum in the soil, and aluminum in high concentrations is toxic to fungi. Unfortunately, no one has undertaken the necessary research to test that hypothesis.

There is no completed research yet to establish how much gypsum to apply, when to apply it, and how often in order to raise the calcium level. Nor have we established how much calcium is optimal or for how long it should be added—and the answers will likely be very site-specific anyway. Until the needed studies are carried out and replicated, I think it is safe and conservative to add gypsum as follows based on soil test results: If your initial soil analysis shows calcium levels below 1,000 pounds per acre, then add 100 pounds per 1,000 square feet onto the bare soil, immediately after your initial fall seeding and just before mulching. Calcium leaches out of the soil, and the other woodland plants as well as your ginseng will probably utilize any added calcium. Take an annual soil sample in late winter. Whenever calcium is less than 2,000 pounds per acre, top dress your wild-simulated site with 50 pounds of gypsum per 1,000 square feet of planted area just before the plants begin their season's growth in the spring. If Ca is 2,000 pounds or higher, do not add gypsum unless either your plants are not growing vigorously, or you see evidence of disease in your planting. Calcium is a crucial element in fighting disease at the cellular level, and adding a little extra during the growing season at the 50-pound rate is not likely to hurt and might help significantly.

Site Preparation

No two wild-simulated growers do things exactly alike, and that diversity begins with site preparation. If you only intend to plant a pound or two of seed and you have acres of suitable planting space available, then your major task will be picking out the best spots where little site preparation will be required. On the other hand, if you have limited space available or even if you have a great deal of promising land available but want to eventually utilize most of it, then you'll probably want to spend considerable time in preparing your site for planting. The following advice assumes that you have a good site under mature mixed hardwoods and want to eventually plant most of it, avoiding small stands of thick shrubs or evergreens, rocky outcrops, old roadways, watercourses, slopes too steep to stand on, and, of course, low-lying wet areas, as well as soggy, leaf-filled depressions. If you intend to plant a large area in the fall, you may want to begin site preparation well in advance, during mild winter or early spring days, and avoid the heat of summer.

Cleaning and Clearing

Before you can plant, you need to rake aside the leaf litter, and it is difficult to rake efficiently where there are dead limbs and rocks lying around or lots of small saplings, vines, and shrubs growing. Even if you cast seeds on top of the leaf litter and then till them into the top inch or two of ground—an approach that some growers swear by—you'll need to clean and clear to make space for the tiller to operate.

I pick up and relocate most of the dead limbs thicker than my thumb and most of the loose rocks bigger than a baseball but smaller than a basketball. I remove this debris to nearby spots that are not suitable for planting, such as amidst three or four trees growing closely together or in a wet place. An abundance of shrubs, vines, and saplings should not be growing under a canopy of mature mixed hardwoods. Even if there are only a few, I usually take the time to remove them. They are not needed for shade, they will compete with the ginseng (especially young dogwoods and some species of maple with their shallow matted root systems), and they will get in the way of raking and planting. The preferred tool for the job is a pulaski, essentially an axe with the blade of a grubbing hoe on the back side of the axe blade. A pulaski allows you to remove dirt at the base of the undesirable plant and then spin the tool 180 degrees in your hands and slash it off just below ground level. This will keep you from raking, and tripping, over the stump. As with tree limbs, I look for nearby places to pile the debris.

Pulaski—the ideal tool for clearing saplings from a future planting site. Photo by Jeanine Davis.

Some things are better left alone and worked around. Experience has taught me that sawing up and removing large downed trees is often not worth the considerable time and effort required. If the tree is right smack in the middle of a prime planting slope, I may cut and move all of it or all but the main trunk. Most of the time I leave it alone and use it as one of my prime brush pile locations. Large rocks that cannot be easily rolled downslope and out of the way and large stumps are left and worked around. Similarly, there are always young understory trees too big for my pulaski that are not needed for shade but may not be worth the trouble to remove. If several are

Area prepared for wild-simulated planting. Photo by Mark Haskett.

grouped closely together, then they will become another brush pile site. I either work around loners or cut them down with a chain saw, depending on convenience. If I do bring a chain saw to the site, I often saw off low hanging branches in order to make working easier and improve the air circulation. Sometimes mountain laurels or mountain blueberries grow adjacent to wild 'sang, but I never find ginseng in amongst them, so I never clear them for planting. Also, any thick stand of woody shrubs indicates that the section of the site where they are growing is probably too sunny (and therefore too hot and dry), and I leave such sections alone.

If the slope that I am going to plant is steep, then I'll be concerned with holding mulch on the soil after planting, as heavy rainstorms may wash leaf litter and even seed downslope. Young saplings and dead limbs with lots of branches work well to hold the mulch if they are laid across the face of the slope after seeding and mulching. On steep slopes, I make small separate temporary brush piles that only consist of light, well-branched saplings and dead limbs.

A friend once described an area I had just finished prepping for planting as looking like a woodland park with brush piles. I do tend to be a little obsessive-compulsive about many aspects of 'sang farming. You will learn from experience on your own particular woodlot just how clean you want the forest floor before you start planting.

Applying Weed Killers

Ginseng has evolved to compete successfully with other woodland plants, and there is usually no reason to clear off any of the forest floor greenery before planting. Indeed, having other plants intermixed with the ginseng is highly desirable, as they inhibit the spread of disease, and their competition adds to the wild look of the root. However, I have seen a few extremely fertile sites where the herbaceous undergrowth, even under deep hardwood shade, formed a complete ground cover that was so dense during the growing season that it was questionable whether a young ginseng seedling could handle the competition for water and nutrients and would seem certain to be deprived of sufficient sunlight. Such rich woodlands are rare and are excellent candidates for woods-cultivated planting; but, if you have such a woodlot and want to grow using the wild-simulated method, then you will have to consider using an herbicide, because hand weeding is absolutely impractical. In addition, herbicides may be needed to rid a site of "exotic" weeds, such as garlic mustard. Usually, herbicide applications are made during the summer before seeding and/or just before the ginseng emerges in the spring, but some herbicides have been used successfully on top of the ginseng during the growing season (see Weeding in chapter 5).

Growers sometimes use Roundup (glyphosate) or Touchdown (sulfosate)—potent synthetic chemicals—to kill weeds. Roundup is currently labeled nationally for ginseng, although some states require additional approval, but Touchdown and many others (including natural chemicals like vinegar—acetic acid—see Applying Weed Killers in chapter 4) are commonly used on the farm and garden but are not nationally approved by the Environmental Protection Agency (EPA) for application on ginseng (see Pesticide Usage in chapter 3). Check with your local agricultural extension agent to find out which are legal in your state, and then confirm their advice by reading the label. [Author's note: The application of pesticides should always be done with caution.]

Synthetic chemical herbicides are anathema to many people interested in growing ginseng. I am certainly not enamored of them myself, although I will admit to occasionally utilizing Roundup to eradicate poison ivy. However, they do work and could be your only good alternative if your woodland weeds are terribly thick. Alternatively, there are natural herbicides available. Vinegar (acetic acid, but in higher concentrations than you buy at the supermarket) is popular, available in most states from organic suppliers, and often reported as effective, especially when applied in sunny weather. However, concentrated vinegar can burn your hands or eyes, and it will corrode the non-plastic parts of a sprayer, so you have to rinse equipment thoroughly after every application.

If you intend to grow organically certified ginseng roots, then I suggest using only natural pesticides that are on the Organic Materials Research Institute's approved list (OMRI, see Comprehensive Resource

Directory), and then follow up by clearing it with your particular certifying agency. Having a wild-simulated planting certified as "organically grown" is usually an available option, though rarely utilized. (A bit more information on natural pesticides is provided in the following chapter, which details woods cultivation.) Since wild-simulated ginseng will be sold as wild on the wholesale market, there is usually no financial incentive to grow organically, unless you are planning to sell retail to a niche market that appreciates organic certification.

Measuring and Marking Off the Area

When you hand cast seed, as I do in the wild-simulated seeding method recommended below, it is difficult to control the spacing of your seeds and therefore the density of your planting. The best solution I have found to the problem is to weigh out only small quantities of seed and hand cast them over known small areas. I usually spread no more than two ounces (between 800 and 1,000 seeds) at a time over 200 square feet, resulting in an average of four or five seeds per square foot. Weighing only an ounce or two of seed accurately requires a digital or small postal scale, but you may be able to borrow one for this once-a-year task or use one at your local produce stand. In order to determine the square footage of the area you are planting, you will have to measure it off.

I like to plant 40' wide sections (or only 20' wide, if 40' isn't practical), beginning at the lowest part of the planting site and moving directly up the face of the slope to the top of the site. Then I go back to the bottom and plant an adjacent and parallel section, and so on, until the entire area is sown. To prepare the site for this procedure, I usually cut thumb-sized branches or saplings in 2' lengths with one sharp end. I stake parallel lines of them 40' apart running directly up and down the slope. I place a stake every 30', but that is not critical. Getting the lines of stakes parallel on a rolling hillside is a little tricky (a 50', preferably cloth, tape measure and a partner certainly help), but perfection is not the goal, just a good approximation.

Stakes, of course, are just one means of marking off your planting area. They'll last for a few years; but, if you're going to be spreading gypsum in the spring for the next nine years, you'll probably want something more permanent such as surveyor's flags or stakes cut from treated 2 × 2's to guide you in distributing it at the proper density. On the other hand, if you are only concerned with the initial planting, you can just spray-paint parallel lines on top of the leaf litter or drop thin trails of gypsum 40' apart. One grower I know runs string up and down his hillsides, and some folks just walk it off adjusting and estimating as they cast seed. There is nothing sacred about 40' or 20' sections either; they just result in good workable areas and make the math simpler. This also simplifies the process for any hired labor or family members you get to help you. Regardless of exactly how you measure and mark it off, the desired planting density will be a great deal easier to achieve when you have a fairly accurate estimate of the area you are seeding.

Planting

I've tried a number of different wild-simulated planting methods and had at least some success with most of them, but the one described in this section is efficient, applicable on most any kind of slope, and requires no mechanized equipment. Most importantly, I have been able to consistently get good initial stands using it. (Understand that I have not been spending my time and effort on poor planting sites.)

In my book *American Ginseng: Green Gold*, I recommended a wild-simulated planting procedure that involved first raking aside the leaf litter to the bare ground, then hand casting the seeds onto the soil, and finally putting the leaf litter back on top of the seeds as mulch. One of my readers was a fellow grower whom I will call "Jeff." Jeff saw the inefficiency of first moving the leaves off the area and then putting them back on again. He found a better way that reduces the total labor of seeding by about 40 percent. It is certainly not rocket science, and other growers may have developed a similar approach independently, but the basic procedure described below did not occur to me until I saw Jeff utilizing it. I wish to give him credit…even, at his request, anonymously.

Raking off leaf litter prior to seeding. Photo by Mark Haskett.

Rake Off the Leaf Litter

Take a large rugged leaf rake, measure five feet from the tip of the tines to a spot on the handle, and make a highly visible permanent mark there. Then, start at the bottom of one of your measured sections, face directly uphill, and, using the mark on your rake handle as a guide, rake the first five feet of leaf litter down and off the soil across the entire width of the section (or across the width of several adjacent sections, as in the above photo). Once the ground is bare, quickly scratch it with the rake again to give the soil a little texture. The rough surface will help the seeds make good contact with the soil as they are cast and will keep them from rolling downhill on a steep slope. You now have a 5' × 40' strip of bare ground at the bottom of your planting site, running perpendicular to the slope, that is ready to be seeded and then, if needed, dusted with gypsum.

When you have exposed the bare ground, you may have the bad luck to discover that the surface is pockmarked with holes about the size of your thumb, which connect tunnels running just below the soil surface. This is evidence of woods voles, *Microtus pinetorum*, a mouse-like creature that relishes the tubers of woodland herbs, including ginseng. They have never been documented eating wild ginseng in any of the numerous wild stands being closely monitored in the Great Smoky Mountains National Park; but, on a successful wild-simulated site, ginseng may be the most abundant food available, and a few wild-simulated growers have had problems with them. One successful grower I know, who has dealt with voles for years, finds that it is worth his time to stick poison bait down their holes just before he casts his seed. Because they love tunneling through loose loamy soils, woods voles are more often a problem in the well-tilled soil of woods-cultivated beds, and dealing with them is discussed in the next chapter. However, if you rake aside the leaf litter during initial seeding and discover what appear to be spaghetti junctions of their highway system, then refer to Dealing with Rodents, Insects, and other Critters in chapter 5. You can be confident that your soil is loose and loamy and rich enough to promote excellent growth of your ginseng; but, if you want to be the one to harvest the roots, then consider taking preemptive action against the voles.

Seeding Rate

From my own experience and that of other growers, I am convinced that at least one, but no more than two, mature plants per square foot is best. Anything denser courts disaster by facilitating the spread of disease. Anything thinner does not optimize the financial potential of the operation, though I'd certainly recommend erring on the side of caution. To achieve a final plant density of between one and two per square foot, the optimal seeding rate is between four and five seeds per square foot, assuming good seed on a good site. Roughly 50 percent of seeds do not germinate for many reasons: consumption by birds, rodents, slugs, and insects; insufficient contact with the soil; desiccation when wind or rain removes the mulch; and the inevitability of some nonviable seeds. During the first few growing seasons, there is often additional, less predictable, attrition. Once the plants have sufficient root mass to support two prongs, the rate of natural attrition usually slows down dramatically.

There are usually between 6,400 and 8,000 ginseng seeds to the pound, or between 400 and 500 to an ounce. Thus, if you sow one ounce of seed on 100 square feet of land or, as I prefer, two ounces (between 800 and 1,000 seeds) on 200 square feet, you can achieve an average planting density of between four and five seeds per square foot, or about 25 pounds of seed to the acre. (Buying and/or producing and stratifying your own seed is covered in chapter 6.)

A cautionary tale: My friend Jeff got a bit greedy a few years ago and seeded at a high

Dropping a measured amount of seed onto appropriate area of raked-bare ground. Photo by Mark Haskett.

rate. It was an excellent site, the weather the following spring was favorable, and he estimates that he got a truly exceptional (for a wild-simulated planting) 85 percent germination rate. Recently, when he showed me the area after four years of growth; it was difficult to walk through without damaging a ginseng plant. Even after normal attrition, there were three or four standing on nearly every square foot, just getting to the size where their foliage was overlapping. We both realized that it was a disaster ready to happen. Indeed, both foliar and root diseases are already present and could spread rapidly. Jeff is sweating that site. It has at least three more seasons to go before what could be a remunerative harvest; now, he plants a maximum of five seeds per square foot.

Casting Seed

Just before going into the woods to plant, weigh out your seeds into multiple plastic bags and place them in a small cooler, or at least a covered bucket. Assuming most of your sections are 40' wide, weigh your seed out into two-ounce units. For smaller sections (hopefully, close to 20' long), weigh out a few bags with only one ounce of seed in them. My experience has been that germination is not quite as good on steep slopes, so I now weigh out a heavy two ounces when I am seeding such sections.

After you have raked down the leaf litter from the bottom 5' × 40' strip, take a two-ounce bag of seed and walk down the length of the strip, dropping the seeds left and right on the bare ground as you go. If you are new at it, hoard your seeds a bit, and turn and go back over the strip, dropping the seeds remaining in the bag onto spots you missed initially. If you are spreading gypsum, do it after sowing the seed, because the gypsum will mottle the ground and you will have great difficulty seeing where your seeds are lying.

When a portion of a 40' strip is not suitable for planting because you've run into a ravine, rock outcrop, brush pile, etc., then you'll have to estimate what percentage of the strip you can plant. Fifty percent (20 linear feet of strip) is simple: just cast all of one of your one-ounce bags of seed. Otherwise, use your judgment and try to leave the same percentage of seed in the bag as you leave length of the strip unplanted. Put the bag containing the leftover seed back in the cooler or bucket. You can consolidate with

Raking one strip bare for seeding while mulching the just-seeded strip immediately below.
Photo by Mark Haskett.

other leftovers and weigh it out the next day before planting, or you may be able to add it to the contents of a one-ounce bag in order to get roughly the right amount of seed to plant, say, 30 linear feet of strip. Because the planting areas are so small, if your estimates and judgments are even close to accurate, you'll get close to the right density of seed on the ground, and close is good enough.

Rake Down and Mulch

After casting two ounces of seed and spreading perhaps 10–20 pounds of gypsum, stand on the strip that you've just sown, reach straight uphill with your rake, and, again using the five-foot mark on the rake handle as a guide, rake down the next five feet of leaf litter directly onto the strip you've just seeded. It is not necessary to get uniform depth of mulch coverage. Just make sure that there is no bare ground showing and no piling of leaf litter thicker than about three inches. Although you may have a natural instinct not to, it is perfectly okay, even desirable, to walk all over the seed you have just sown while you are raking down the leaves to mulch them.

If you find that you have barely enough leaf litter for good coverage, then in the future you may want to delay your planting until late fall or even early winter (as in the photos in this section). The cooler weather causes fewer exposed seeds to dry out and snowfall will soon help to protect your seed and stick the leaf litter you do have firmly to the soil. Some growers insist that, if they plant before the autumn leaf fall, they end up with too much mulch on their site and

Placing branches perpendicular to slope on top of mulch after planting on steep hillside. Photo by Mark Haskett.

reduced emergence in the spring due to the sprout's struggles to work its way through the thick mulch of leaves. I pass this concern on because I have heard it from several successful growers. It seems to me, however, that you are going to get the same total amount of leaves on your site regardless of when you plant. I have certainly had success planting in September or early October, well before the leaf fall. On the other hand, I have had poor results planting just before my seed were about to germinate in late March. I think the seeds may not have had time to work their way into the soil before germinating, and I know that on one occasion, without any snow to stick the leaf litter back to the soil, heavy spring rains washed away both mulch and seeds. (Late winter planting in general is a second choice. I only plant late what I am unable to finish planting in the fall.)

After raking down five feet of leaf litter as mulch on the first strip you planted, you now have another strip of bare ground ready to seed. Walk along and drop seed from a plastic bag over this ground just as you did on the bottom strip, add gypsum if needed, then stand on the newly seeded strip and rake down another five feet of leaf litter from above as mulch. Repeat the process all the way to the top of the 40' section, where a bare strip will remain unplanted. Then move down to the bottom of the next section, and so on, until you have planted all of your prepared land. Note that you have not moved any leaves more than once, and with the exceptions of the bottom and top sections, every rake stroke has accomplished two tasks. Finally, on steep sections, lay limbs and saplings from one of your dedicated brush piles across the face of the slope to hold the mulch to the soil when the wind blows and the rain pours.

A hardworking fellow can expect to seed an acre in this manner without using up more than 30 hours of his time. However, when you do it on rough uneven ground, raking is deceptively hard work. If you are not used to constant physical labor, I'd recommend you operate on half-day shifts and give yourself a day off if your body complains loudly. A pair of gloves will come in handy if you normally wear a white collar. For many reasons, I do not recommend beginning growers planting more than one-half acre their first year, and one-tenth acre is sufficient to get started.

Alternative Wild-simulated Planting Methods

For many years, I have tried to find or develop a practical reliable method for seeding that mechanically injects the seed directly into the soil through the leaf litter. I have had no success, but I have successfully used a technique employed by a number of growers in which seeds are cast on top of the leaf litter and then tilled or disced into the soil. Since much of the leaf litter is tilled into the ground, seeding is usually done just before the leaf fall in order to have some mulch on the site. I was skeptical of this approach, until I tried a small planting directly below a section that I had seeded with the method described above. I cast seed at the usual density and then made one pass over them with my tiller at its shallowest setting. Germination rate, survival, root quality, planting time, and effort were all similar. There was a little more vigorous weed growth on the tilled plot, but weeds were not thick to begin with, so that was not a problem. On gently sloping, relatively rock-free ground, this is certainly an available alternative, though not a superior one. I have not had much opportunity to use it, but I encourage you to at least experiment with it on suitable sites.

On large areas of woodland that are relatively open and gently rolling, a few growers are using small tractors to disc seeds into the ground, which saves considerable time and physical labor. However, operating a tractor on a steep, uneven hillside in the sometimes tight quarters created by randomly spaced trees and rocks can be dangerous. Other potential drawbacks to this method also need to be considered: tilling produces early nitrogen availability, which encourages the growth of competing shallow tree roots and weeds; dormant weed seeds will be helped to germinate as well as the ginseng; by more drastically altering the natural state of the forest floor, the risk of disease is probably increased (this is speculation based on a few observations—there is no research); and if the discing is too deep and the roots grow too big too fast, then they may not grade as wild, or you may have to risk extra years of growth in order to sell your roots as wild.

If you have plenty of cheap labor, then you can plant the seeds one at a time. Photo by Mark Haskett.

Of course, if the time and cost of labor are not a primary consideration, each seed can be individually hand poked into the ground, which should result in excellent germination. I know of parents and kids who do it this way as a family project. At the opposite extreme, the least labor intensive method is to hand cast the seed on top of the leaf litter, ideally just before a rainfall, and walk away. Randy Brunn, a former ginseng crop consultant in Marathon, Wisconsin, reported encouraging results with this approach. At my request, he put his thoughts in a letter:

> For years, we have monitored the conventional [rake aside, hand cast, mulch] method of wild-simulated seeding in woodlots. The method is almost always successful, but, with today's labor and time challenges, we have implemented a new labor-saving approach to establishing ginseng plants in wooded areas.
>
> Basically, we go to the woods in late October in western Wisconsin and hand "scatter" ginseng seed before the majority of the leaves have fallen and preferably just before a rain event. Seeding rates are about 80 pounds to the acre, and we can easily plant one acre in a day. You can walk along a ridge and scatter seed 50' down a hillside. (We have our own supply of seed, which reduces capital expenses.)
>
> We started plot trials in the fall of 2001 to compare the conventional method with the "scatter" method. Here are our observations from the first growing season: (1) the conventional method has a more uniform stand but only a slightly higher rate of germination and emergence; (2) the "scatter" method has satisfactory germination (as high as 65 percent in some sections), so that with our dense seeding rate we are typically finding five to eight plants per square foot; (3) there is a fairly even distribution of plants with the "scatter" method, but clustering is definitely more common, and we anticipate the need for eventual thinning and possibly transplanting; and (4) the loss of seedlings during the first growing season is approximately 33 percent with both methods.
>
> On other "scatter" sites, we have let the plants establish themselves until the fourth year, when we go in and thin any plants growing underneath the foliage of dominant three-pronged plants. We think the time and labor saved with our seeding process more than compensates for the time needed to thin. We intend to look at the cost effectiveness of transplanting the thinned plants into any gaps in the stand.

When Randy told me of his initial success with this scatter-on-top-of-the-leaves approach, I was extremely skeptical. I tried the "scatter" method many years ago with almost no plants emerging, and I have heard of similar negative results from other sources. Though I have no doubt that this method worked for Randy, I still urge caution on the reader. It may be that, in the colder climate of Wisconsin and with the right timing before a rain, seed do not dry out as quickly, and the snow soon comes to protect them and help work them down into the soil. Also, the density of Randy's planting (he began as a shade grower) concerns me. I certainly would recommend that, if you do try his method, you scatter no more than 40 pounds of seed to the acre. Given the amount of labor saved, this alternative is certainly worth trying.

Maintenance

A low-density wild-simulated planting does not require a great deal of inspection and maintenance. Though it is slow going, you can carefully step your way through the thinly spaced plants, and often you can utilize old logging roads and other unplantable features to get around more quickly. If you do plan to patrol regularly, then it is best to map out walkways before you start seeding. Especially in the early years, before the plants are large enough to be at risk from poachers, it is usually unnecessary to enter the site often and thereby facilitate the spread of disease. Never go (walk, ride, etc.) from any bedded planting, where spores of fungal pathogens are likely to be present in quantity, to your wild-simulated site. It is very easy to unintentionally transport spores from diseased plants onto healthy ones. During the growing season, inspections or any passage through the site should be limited to once every two weeks, unless there is a problem that needs attention. You may need to replace some mulch or defend against slugs and other critters, but walking through your site is a hazard, so I often refrain for many weeks, particularly during the middle years of the growing cycle. Some attrition in the early years is to be expected. Let nature take its course and minimize the labor you expend.

If you are unlucky enough to find diseased plants (or even just what you think are diseased plants), clip the stems at ground level and remove the tops from your property. It is wise to send a tissue sample to your state lab to identify the fungal pathogen (see Joe-Ann McCoy's Disease List in Appendix 5); but, it is generally not cost effective to try to control disease with a regular spraying program in a thin wild-simulated planting, especially if the site is not readily accessible. The area of infection is likely to be small and isolated, so it may be worth your time to spray the foliage of the surrounding plants with either a synthetic or natural chemical fungicide on a weekly basis until the disease stops showing up and perhaps one more application the next spring, just as the ginseng is emerging. If you suspect root rot, remove the tops, but do not dig up the roots (except to send a sample to the lab for identification). Just move aside the soil and inspect a few. Again, if the area of infection is small and isolated, and you think you have a root rot, then you might consider drenching the soil with bleach or hydrogen peroxide (or with Ridomil, if you know it's *Phytophthora cactorum*). If your site is well drained, disease spread will likely be minimal in the natural setting.

Fortunately, diseased wild ginseng plants are rare (or at least rarely found), and thin wild-simulated plantings seem to possess the same resistance. I know of

A 'sang poacher hangs in effigy before being taken to the fields to warn would-be offenders. Photo courtesy of Bob Beyfuss.

none that have been devastated by disease. There seem to be fewer and less virulent pathogens in woodland soils, and the actions of the trees, themselves, probably also reduce disease. Trees constantly respire, which keeps the air cool, and their roots constantly wick up water, which, during wet periods, reduces the excess water around the ginseng roots. (This is also true in woods-cultivated beds, where the feeder roots of shade trees quickly grow back and permeate the tilled soil within two years after bed preparation is completed.)

A few wild-simulated growers do weed, irrigate, and/or fertilize, either by top dressing or by foliar feeding. In so doing, they may enhance root weight, but they increase the risk of disease and may produce roots dissimilar in appearance to the wild and therefore less valuable. I use the category "wild-simulated growers" loosely to distinguish them from woods-cultivated growers, who till the soil deeply before planting and therefore rarely produce roots that sell as wild. (Obviously, there is a continuum of approaches to woodland 'sang farming.)

Security

Poachers pose a potential threat to all ginseng farmers, but especially to those simulating the wild conditions. Their individual plants are more valuable, their sites are often more remote and difficult to monitor, and they must protect their green gold for a longer period before harvest. Poaching 'sang is a felony in some states (grand larceny if enough value is stolen), but an entire hillside of mature plants is terribly enticing to a digger who may have never before seen so much ginseng and knows that he can sell it immediately for cash. There are instances of thievery everywhere, but in the southern Appalachian region, poaching is the limiting factor for 'sang growing. Large-scale ginseng farming under artificial shade in the Midwest might never have amounted to much if the folks who live in the rich hollers of the Appalachians were able to grow 'sang without fear of theft. I'm one of them, and, in retrospect, I'm lucky I wasn't fully aware of the risks when I got started or I might not have planted my first seed.

Initially, I did know enough to keep my efforts as quiet as possible. Being a 'sang digger myself, I knew that any forager traversing my woods would spot the ginseng if he got in sight of it, so I placed my plantings in areas where I knew traffic would be minimal. Nonetheless, in time, the news got out. Ginseng farming is an interesting topic of conversation, and friends, relatives, neighbors, and especially anyone you hire to help you could spread the word. When it became obvious that my 'sang was no longer secret, I took the opposite approach. I told my good neighbors all about it and they voluntarily helped me keep a lookout. I cheaply fenced off my older more valuable plants with chicken wire to alert essentially honest folks that my 'sang was planted and not just standing there in the woods going to waste, and I nailed up "No Trespassing" signs and registered my intent to exclude at the county court house. Finally, I put up not-so-well-concealed fake cameras, and got a story in the local newspaper making mention of the ferocity of my dogs and my intention to catch and prosecute trespassers.

The poachers stayed away for the first 12 years, but, as the price of wild ginseng rose dramatically, they did begin to dig some of my plants. By that time, fortunately, I was operating on a large-enough scale that one poacher in one night could not do me a great deal of financial harm. (In poor light conditions, it can easily take a man more than three hours to dig enough wild-simulated roots to make one dried pound.) Their unintentional vandalism of what they did not dig was actually more costly. Moreover, my multiple-acre operation could afford some real security by then, and eventually I caught a couple of would-be poachers. Though I did not have the evidence to prosecute, I think they were made sufficiently uncomfortable that they chose not to try again, and I suspect that word got out. Now that North Carolina has passed new, more forceful anti-poaching laws, and it has become known that technology for marking and tracing roots is available, fewer poachers seem to be active.

These changes in the poaching climate in my part of the ginseng world are in no small part due to the determined efforts of Jim Corbin who, in addition to researching ginseng's propensity for calcium, developed a marking technique for the protected plants in the Great Smoky Mountains National Park. (He tried his ideas out first on some of my ginseng, before he risked killing any in the park's population.) Jim has probably been responsible for catching, prosecuting, and thereby deterring more ginseng thieves than anyone else alive. Here is a synopsis of the advice Jim Corbin gives to private independent 'sang growers:

> You are unlikely to catch many poachers. Your emphasis should be on deterring them and minimizing your losses. If you want to mark your plants, contact me at 828-421-5445, and I'll send you some silicon microchips, color coded for your farm. (The chips are nontoxic and the codes have held up in court without exception.) Select a small percentage of your most impressive and accessible plants that you can refrain from

digging, remove most of the soil around them, rub the microchips (which have the consistency of flour) onto the roots, replace the soil, and some of the chips will adhere to the root, eventually being permanently pinched within the root's concentric growth or stress rings. These roots are now uniquely traceable to your farm, if an enforcement officer can catch the thief with them.

[Author's note: Jim will tell you how law enforcement can recognize your microchips, but he does not want to widely disseminate that information here.] Also, when the dealer who buys your roots exports them out of state, they must be inspected by the state agency (it varies from state to state) that monitors ginseng exports in accordance with the CITES treaty. If you alert that agency that your marked roots have been stolen, they may be able to find them during inspection; then, the dealer, who is required by law to keep records of all his purchases, should be able to identify the thief.

Unless your roots cross state lines, federal laws can't help you much, so you will have to rely on the relevant laws of your state (and they differ from state to state). The law enforcement people most knowledgeable about plant laws are usually within the state conservation or wildlife departments. Find out which officer in your local area has the knowledge and interest needed to catch plant poachers, get to know that person, and discuss with him (or her) the value of your ginseng and what steps you are taking to protect your plants. Also, if you make it to court, you will need a paper trail documenting your ownership of the roots. In North Carolina, this means obtaining a nursery license for growing ginseng and having your plants inspected annually. In your state, your local agricultural extension agent should be able to advise you on acquiring the appropriate documentation. [Author's note: He may also be able to help you lay down a paper trail in case you need to document a loss for insurance or tax purposes or in case ginseng digging is made illegal in your state and you need proof that your wild-simulated roots were grown from seed and not foraged from the wild. Contact the nearest USDA Farm Service Agency for prior documentation in case disaster strikes your planting and you qualify for money from a government disaster relief program.]

Marking your roots is not as important as putting up signs that say that you have. Signs stating "WE ARE PARTICIPATING IN THE PLANT MARKING PROGRAM UTILIZED BY THE GREAT SMOKY MOUNTAINS NATIONAL PARK" have proven to be effective deterrents. Establish a perimeter around your ginseng and nail up such signs, along with NO TRESPASSING signs, close enough together that no one is likely to get to your ginseng without seeing them. Then there are a number of devices on the market that

> you can set up both along your perimeter and in the midst of your ginseng that are designed to make the poacher lose his sense of comfortable privacy, which is the key to deterrence. In the park, we use strategically placed smoke alarms rigged with refrigerator switches connected to monofilament trip wires, because they're cheap. You can learn to rig these, but for about $50 you can purchase motion-sensitive alarms like the "Critter-Gitter," which will cover a 40' × 50' area and can be heard for half a mile. Cheap radios protected from the elements and tuned to a 24-hour talk radio station, fake video cameras with a battery-operated flashing red light (about $20), barking dogs, and any suggestion of a human presence are effective as well. Finally, to stop a truly hardcore poacher (who is the exception) you'll need to spend time patrolling your property, but you must do this on an unpredictable schedule and slip into the woods unseen and unheard.

There will be little interest in your ginseng until some of your plants are three-pronged, so you don't need to set up deterrents before you have real value to protect. At least until then, keep a low profile. Growers often find that they need to hire help when it comes time to dig. Be careful whom you bring into your woods if you're going to keep growing ginseng. If you are in a part of the country where 'sang diggin' is not part of the culture, you may never have to worry about poaching. On the other hand, once you get poached, the odds are good that it will happen again. Unless you want to harvest your crop immediately and get out of the ginseng business, you should definitely consider the measures Jim Corbin recommends. If you have enough value growing to warrant the expense, there are reasonably priced, higher tech security devices to consider as well: motion-sensitive cameras that can even take pictures at night; real video cameras designed for outdoor use that look just like the fake ones; camouflaged motion-activated systems that will set off a remote alarm at your house or ring your cell phone; even wireless systems that will send a picture up to two-and-a-half miles away. Also, be aware that the poacher may be watching you. A great deal of green gold is stolen when the grower and his family are out of town or just gone to church or to a funeral.

Please understand that, unless he is breaking into your house, you cannot legally hurt someone who is trespassing on your property. Whether you agree with it or not, the courts have ruled that if you shoot a thief or wound him with a booby trap or even if your dog wounds him, you are the one likely to go to jail, and you can be sued. You can shoot a gun into the air and verbally threaten him all you want in order to scare him to death and contain him until a law enforcement officer arrives, but bluffing is basically the only legal way you, yourself, can catch a poacher.

The Bottom Line

The eventual worth of the first wild-simulated planting in a woodlot is highly unpredictable. The grower has to learn

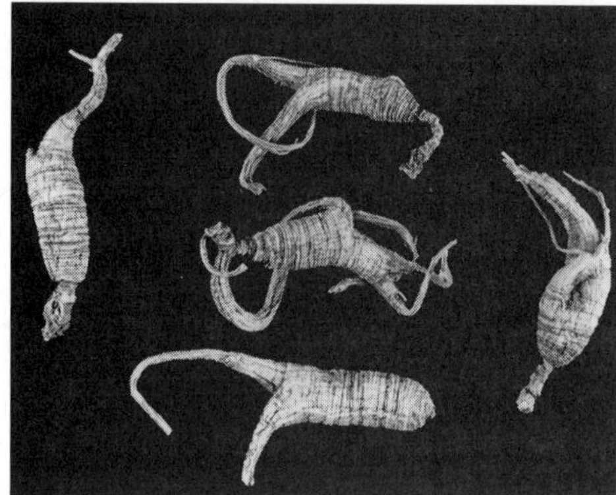

Sample of dried seven-year-old wild-simulated roots, selected to show desirable horizontal rings on trunk. Photo by Mark Haskett.

through experience and by trial and error. The natural fertility of your particular planting site, which is left essentially unchanged, and the exigencies of Mother Nature will determine both the quantity and the quality of the ginseng that remains to be harvested there 7 to 11 years after seeding. An associate once harvested 413 dried pounds from one acre after only seven years, but $0 is a far more likely outcome than that remarkable success. If you can't accept the real possibility of a total loss, then you probably ought to pursue a more conservative enterprise.

Table 4 is based on my own experience and the reports of other growers. It provides a basic budget for a typical, reasonably successful wild-simulated planting. There is no hint in this table of the great variability associated with some of its numbers. (And I have chosen not to even include costs for gypsum or for security measures, because those expenses are sometimes not needed and vary so widely.) I am operating on the

Table 4. Projected Nine-year Budget for One-half Acre of Wild-simulated Ginseng

Seed*	
12.5 pounds at $135/lb.	**$1,690**
Labor	
Site preparation and planting: 25 hours × $15/hr	$375
Inspection and troubleshooting: 200 hours × $15/hr	$3,000
Digging roots: 350 hours × $15/hr	$5,250
Total Labor	**$8,625**
Materials and Equipment	
Rake, pulaski, and digging tool (assume some equipment already on hand)	$50
Backpack sprayer, disease, and pest controls on hand for troubleshooting	$300
Total Materials and Equipment	**$350**
Drying	
Addition of insulation and drying racks to existing room or shed	$400
Energy cost to heat (50¢/lb. of dried root)	$40
Total Drying	**$440**
Total Cost	**$11,105**
Total Expected Yield : 80 lbs. of dried root	
Gross profit: 80 lbs. × $675/lb.	**$54,000**
Net profit at end of 9 years	**$42,895**

* The per pound price of seed varies with quality and quantity, and from year to year with supply and demand. The best seed comes from disease-free gardens of fifth-year and older plants. There are roughly 7,000 seeds in a pound. A successful grower may eventually produce his own seed.

premise that a little uncertain guidance is better than none, as long as the reader is aware of the uncertainty.

Table 4 assumes nine years to harvest, because that is probably the single most likely outcome, but other times to harvest are almost equally likely. Growth is so slow and uneven on some sites that the best approach is to harvest plants as they reach maturity, beginning after about eight growing seasons and digging the largest 25 percent every two years. On one such site, I have begun reseeding near each root as I dig it, with the hopeful intention of creating a perpetual stand of ginseng and harvesting bi-annually indefinitely. On any successful wild-simulated site, by harvest time you will have some young plants growing that plants you harvest have seeded. A grower can supplement this natural reseeding by sowing additional seeds after he has dug the mature plants. I have not done, or seen, enough of this to judge how well it works. If more good land is available, I prefer to go to a new planting site rather than replant an old one, because there are often serious disease problems with a second planting in bedded situations, and I'm sure the likelihood of problems is increased at least marginally when a wild-simulated site is replanted.

The hours of labor required are highly variable depending upon the layout of the site, the problems that arise, and upon the grower's experience and work rate. The $675 per dried pound projected as payment for your roots is based on the price paid for wild ginseng over the last few years (see Table 1, Recent Prices). Price is highly variable, depending on supply and demand as well as on the kind of soil and growing conditions your site possesses. All wild roots do not grade the same. Small young roots dug out of clay soils in the Ozark Mountains of Arkansas might bring $550 on the same day that older larger roots from the Catskill Mountains in New York are bringing $850. (Digging, drying, and selling your roots are covered in chapter 6.)

Also, the figures are based on a one-half-acre planting. More precisely, they are based on the seeding of 20,000 square feet. Since there are roughly 21,750 square feet in half an acre, this assumes that about 90 percent of the surface area is plantable. There are plenty of woodlots where the grower will need an entire acre to get 20,000 square feet that he can seed. It may be helpful to interpret the figures in Table 4 as simply a projected budget for planting 12½ pounds of seed on 20,000 square feet using my wild-simulated planting method.

If you extrapolate this budget by, for instance, a factor of ten to a five-acre planting, unfortunately you do not achieve great economies of scale, whether you seed it all in one year or plant half an acre for ten successive years. The bulk of the cost is in the labor to plant and especially to dig, and getting bigger does not make those jobs go any quicker. Still, there are some savings inherent in a larger operation: you will probably be able to reduce the per pound seed cost by about 15 percent if you buy in bulk; you'll probably spend proportionately less time inspecting, since some of that time is spent in the going and coming; materials and equipment costs might only double or triple; and you won't have to construct any additional drying area if you're just harvesting half an acre each year. The greatest advantage of increasing your scale is that you will be better able to afford security devices and patrol time if you need them.

Similarly, if you extrapolate this half-acre budget downward to only one-tenth of an acre, you do not lose a great deal of profitability. Bed preparation, planting, and digging will cost about one-fifth as much. The other expenses will be proportionately higher, but they are moderate expenses to begin with and will certainly not become prohibitive. The per pound price of seed will rise about 15 percent with a smaller quantity purchased; a larger proportion of your inspection time will be spent in traveling to and from your site; your materials and equipment requirements will not be reduced much and will thus be a more significant expense; and if the anticipated 4,000 to 5,000 roots are harvested, you'll probably still want to construct a drying area, although on a one-time basis you could make do by utilizing a warm dry upstairs bedroom or attic for much of the fall.

Finally, although many things can happen over such an extended growing period that could lead to failure, in a wild-simulated planting your potential losses are limited. As noted in the opening paragraph of this chapter, most of the expense is in the labor, and most of the labor is in the digging, which does not occur until a profit is nearly assured. Even if your crop is poor (not a rare occurrence), the labor needed to harvest a reduced number of plants will be less, and your venture may still be marginally profitable. If you perform all the labor yourself, then your potential reward is great compared to your risk of capital.

The Hankins Method

Andy Hankins, the Extension Specialist for Alternative Agriculture at Virginia State University for many years until his recent untimely death, encouraged landowners in his state to grow ginseng in a wild-simulated manner. He was frustrated when reports came back to him from the ginseng-rich poplar coves of Virginia's western mountains that hard-working folks, whom he'd helped to get nice stands of ginseng started, had been poached. Andy became increasingly interested in growing ginseng where there was no culture of poaching, and he helped many small woodlot owners in the central and even in the tidewater areas of Virginia to establish wild-simulated plantings.

As you head east from the mountains, ideal planting sites where wild 'sang would naturally flourish are rare, so Andy developed a simple method that compensates for the deficiencies of a marginal site but still produces roots that sell as wild. His approach takes about five times as long as mine to plant the same amount of seed, but it has produced ginseng where it wouldn't grow naturally. More importantly, it can produce ginseng at a profit. If your prospective woodlot is only marginal and you are interested in obtaining supplementary income, then the Hankins Method is probably your only viable option.

Even if your site is not marginal, you might want to try small experimental plantings using all the methods described in this chapter and see which one works best for you on your particular piece of woodland. If you only intend a one-time wild-simulated planting on a limited area, then time and labor are less important considerations, and Andy's more careful planting technique, which increases the probability of success, might be the logical choice.

Andy assisted woodlot owners all over Virginia to establish wild-simulated plantings. Ten of the sites were part of a pilot study, and therefore are well documented. The Virginia Cooperative Extension (VCE) has published a complete presentation of the Hankins approach in a handsome, eight-page pamphlet with four-color photographs titled "Producing and Marketing Wild Simulated Ginseng in Forest and Agroforestry Systems, Publication 354-312," which you can obtain at no charge by calling or going to the VCE website given in chapter 10. I have paraphrased extensively from that pamphlet and relied on personal communications with Andy to provide the condensed and updated description of his method that follows.

The Planting Site

Even using the Hankins Method, your planting site must approximate the shady cool moist, but never wet, conditions ginseng absolutely requires. That means at least some slope to the land with well-drained soil (definitely not heavy clay) and a dense shade canopy provided by deep-rooted mature deciduous trees. Of course, the ideal site would have wild 'sang and all its companion plants listed above, but the minimum requirement is just some herbaceous woodland plants growing on your forest floor. If species like Virginia creeper (*Parthenocissus quinquefolia*), rattlesnake plantain (*Goodyera pubescens*), pipsissewa (*Chimaphila umbellata*), various ferns, and even poison ivy (*Toxicodendron radicans*), which do not need quite as cool and moist a setting as wild ginseng does, are growing on your site, then you've got a shot at growing 'sang there with Andy's approach.

His method is appropriate on proper sites in the central and tidewater areas of Virginia, in the Piedmont of North Carolina, in northwest South Carolina, northeast Georgia, northern Alabama, middle Tennessee, central Kentucky, central Ohio, eastern Maryland, etc. In the low, flat country that is found in some of these regions, suitable woodlots might be limited to the northeast-facing banks of rivers. The grower may also have to be more selective within his woodlot, looking for the spots with the most herbaceous undergrowth and at least a few tulip poplar or sugar maple trees providing some of the canopy. The approach is also suitable to areas in the heart of ginseng's native range that are a little hotter and drier than ideal (maybe just on the other side of a hill from a stand of wild 'sang).

Seed

There are about 20 experienced commercial growers who offer ginseng seed for sale. I'm one of them, and I'm certainly pleased to let Andy make the case for quality for me: "Most experienced growers have bought seed that failed to germinate in the spring after fall planting. Growers are not advised to buy cheap seed. It is rarely a good investment. A great deal of meticulous care is required to produce viable, stratified ginseng seed. Seed producers who do the job the right way are not likely to sell their seed at cheap prices."

Planting

Following the Hankins method, the only tools you will need to plant are a rake and a garden hoe or mattock. Lay out defined sections 5' wide and 50' long, running up and down the hillside, rather than across the face of the slope, for better air and water drainage around the plants. These 5' × 50' sections should be separated by undisturbed, three-foot-wide walkways. It is uncommon on a marginal site with dense shade to have areas that are thick with weeds; but if that is the case, it's probably simplest to avoid those weedy sections, because it will be tough for the young ginseng plants to compete. Also, while ginseng will grow well near ferns, it's advisable not to plant in amongst them, because the roots of many ferns secrete allelopathic chemicals, which inhibit the growth of neighboring plants. Because it is desirable to disturb the site as little as possible in order to minimize the spread of fungal diseases, remove shrubs, saplings, tree stumps, rocks, dead wood, etc., only as needed. Rake aside all the leaves down to the topsoil and then, using the corner of either your hoe or your mattock, dig three narrow furrows 18" apart the full length of each 50' section. Ideally, the depth of the furrows should be no more than 1" and no less than ½".

Place ginseng seeds individually by hand at the bottom of each furrow spaced four or five inches apart. (Roots, rocks, etc. get in your way and your spacing will not be precise.) This means planting approximately 135 seeds per 50' furrow, just over 400 seeds (about one ounce) in the three furrows of a 250-square-foot section, and thus 16 ounces, or one pound of seed in 16 sections, which constitute 4,000 square feet (roughly one-tenth of an acre) of actual planting space. In extrapolating these figures to larger areas, do not forget the three-foot-wide walkways. [Note that, because he plants in rows, or "furrows," 18 inches apart, Andy's planting density is less than half of mine. Despite the excellent germination he gets as a result of hand seeding, he ends up with only about one plant per two square feet within his sections. This seems very conservative, but it may be appropriate on marginal sites, particularly when plenty of land is available for planting.]

Next, cover the seeds with soil so that they are about ¾" deep, and then carefully step down each furrow to firm the soil over the seeds. Finally, rake leaves back over the sections as mulch, walk away, and trust to the vagaries of nature. Because excessive mulch can prevent the young sprouts from emerging in the spring, do not rake on more than a loose and fluffy three inches of leaves (no more than an inch, if the leaves are predominantly oak). Rain will flatten the three inches to one inch by mid-winter, which is all that is needed. Woodland weeds will compete with the ginseng for water, nutrients, and sunlight. Insects and rodents will attack some of the plants. Fungal diseases could defoliate a few, and severe weather may reduce growth. All of these stressful conditions produce roots indistinguishable from truly wild ginseng.

On a marginal site, you may need supplemental mulch. One of the advantages of seeding in wide rows is that the grower can come in after the plants have emerged or whenever conditions merit it and put on a couple of inches of extra mulch between the rows without damaging the growing plants. If the grower seeds in late fall after the leaves have come down, he can control the amount of mulch surrounding his plants from day one. Mulch helps keep the soil and thus the plants cool and moist, which is important since heat and drought stress are common problems when plantings are attempted outside the heart of 'sang country, particularly in the southern states. Wheat straw and leaves are commonly used as mulch, but David Cooke and John Scott, of the West Virginia University Extension Service, found that shredded tobacco stalks also work well on ginseng.

Gypsum

As soon as you have identified a planting site, get a soil test done to determine the calcium level. On a marginal site, calcium is likely to be around 500 pounds per acre. If indeed it is low (less than 2,000 pounds per acre), hand cast gypsum immediately after seeding and just before mulching, at the rate of five pounds per 100 square feet (or 12½ pounds per 5' × 50' planting section). Then, to monitor calcium, get an annual soil test done in the late winter. Whenever calcium tests below 2,000 pounds per acre (including in the late winter after a fall planting), apply gypsum in the spring about two weeks before the plants emerge, hand casting at the same rate on top of the mulch. It would be difficult to overemphasize the importance of these applications.

Conclusion

Although both time and labor intensive, this planting technique puts every seed in cool, moist conditions, which maximizes both initial germination and long-term survival rates. The little bit of added tilth that the furrow creates probably helps the root grow bigger the first critical year, but it does not alter the mature root's wild characteristics, which happens in a tilled bed.

Andy summarizes some of the advantages of his method:

> Nearly every seed produces a plant, so I start with a good strong population that naturally thins itself to a healthy stand of wild ginseng. On average, only 125 plants survive to maturity per 5' × 50' section. I can create the furrows and plant a pound of seed (an ounce in each of 16 sections) in 16 hours by myself.

That's two days of work. Later, I can harvest the surviving plants in a section in less than an hour and a half.

[Author's note: Digging is a little quicker for Andy, since the plants are neatly aligned in rows.] For 16 sections, it takes me about 24 hours, or three days, to dig, wash, and prepare about 2000 roots for drying. I use a large heavy flat-headed screwdriver for digging, and I'm careful not to tear up the ginseng. That's a total of five days or 40 hours of labor. If I let the ginseng grow for nine years before digging, I will average about 7 pounds of dried root per pound of seed planted. [Author's note: 2000 ÷ 7 = 285 roots per dried pound, which is typical of wild ginseng.] At $675 per pound, 7 pounds of root will bring $4,725. That works out to almost $120 an hour wages.

There are some important additional costs, such as the pound of seed and probably considerable gypsum, but clearly the Hankins Method is potentially profitable, especially if you're doing all the labor yourself. The time required to plant is the only major drawback. Even if your work rate is comparable to Andy's, it will take you 160 hours to plant 10 pounds of seed. How many pounds would you be able to plant each fall? But, on a small scale, if you only want to plant a few pounds of seed every year, it makes a lot of sense.

5

Woods Cultivation

Cultivating ginseng in beds under natural forest shade is the way I got started growing green gold. During the late 1980s and early 1990s, when the price of field-grown roots was around $40 per dried pound, it was possible to intensively cultivate woodland gardens and produce big roots that would sell for between $60 and $80 per pound. In only four or five years, I could harvest 500 pounds on half an acre. Unfortunately, the overproduction of low-grade, field-grown roots brought the price of all low-grade roots way down. The grade of root I used to sell for $70 a pound has, until very recently, been bringing about $25, which is below the cost of production in the woods.

Since 1994, my woods-cultivated beds have been set up to grow smaller, higher quality roots over a six- to eight-year period. These kinds of roots, which have wild characteristics but will not pass for wild, have risen in value over the last decade as the wild price has risen. They generally sell for between 30 and 70 percent of the going price for wild roots. Opting for quality over quantity seems the most prudent approach now (and is the only one discussed in this chapter).

Producing ginseng this way requires hard work and the acceptance of risk over at least a six-year growing period, plus there is a learning curve to master. A densely planted ginseng garden in the woods requires conscientious attention and regular maintenance. The necessary care and protection are only slightly less demanding than for a vegetable garden of the same size, and a grower's neglect can lead to substantial losses. I understand why the majority of new woods growers are choosing the wild-simulated method. The outcome is more problematic (since you basically just seed and walk away), and it has lower potential annual gross return per acre, but a wild-simulated operation requires much less time, labor, and experience.

Nonetheless, woods cultivation still has its place. Woodland gardens have the

Intensively cultivated woodland ginseng growing on Scott Persons' farm at Tuckasegee, North Carolina, in the late 1980s. Photo by Mark Haskett.

greatest gross profit potential per square foot per year, and therefore may be the method of choice for the small-scale grower or the grower who has only an acre or two of woodland suitable for ginseng. There are special circumstances where it is appropriate as well: A woods-cultivated garden can be established for long-term seed production (see chapter 6). If the purpose is to raise roots for personal consumption, then small-scale thinly spaced well-cared-for woods-cultivated beds can produce roots for the dedicated grower even in woodlots not ideal for wild ginseng. Also, unlike growing under artificial shade, it is possible to grow roots naturally in cultivated woods gardens without using synthetic chemicals, and a few growers of woods-cultivated ginseng are making considerable additional profit by having their roots certified as organically grown.

Since you do not have to risk a large upfront investment in shade construction, your major cost in woods-cultivated ginseng will be the value of your own labor. It will be six to eight years before you harvest your first crop; but, on a good site, you can produce over $12,000 worth of roots on one-tenth of an acre, or better than $60,000 on half an acre. You have to be optimistic, patient, and willing to take the extra time to do things correctly and thoroughly, even when the pay-off is years away. Obsessive-compulsive personalities make good woods-cultivated ginseng farmers.

I have written a book and several magazine articles on woods cultivation, and occasionally I receive a letter from a reader beginning: "You make it sound so easy; why isn't everybody doing this?" Well, it isn't easy, or quick, or certain, and I don't want to give that impression. Woods cultivation really is labor intensive. One acre is about the maximum one man can take care of and still hold down a full-time job. The labor won't break your back, but it will bend it... often. Moreover, it is complex agriculture. This chapter and others in the book will help, but there is a great deal to learn, some of it inevitably by hands-on trial and error. There is no substitute for experience gained on your own spot in the woods.

Selecting Your Planting Site

If personal consumption or just a horticultural challenge is your goal, then with enough effort, imagination, and perseverance, a well-tended woods garden can be made to succeed on most any modestly fertile woodland with adequate shade, soil moisture, and soil drainage. The grower can create the rest of what ginseng needs (see chapter 32, Ginseng—A Horticultural Challenge). Understand, it may not be cost effective to do this unless you are able to sell your roots at retail prices.

Minimal Planting Site Requirements

Good drainage is critical for successful ginseng production, and it is extremely difficult to provide unless your planting site has at least two degrees of slope—more than five degrees is highly desirable. Grades almost too steep to till are often used.

Research has shown that ginseng requires a minimum of 65 percent shade, with 78 percent being optimal over most soil types. More than 90 percent shade will often produce spindly plants and extremely slow root growth. Too much sun can burn the leaves and directly kill the plants, or the excessive heat may put the plants in stress and make them vulnerable to disease. Without technical equipment, you cannot walk out into your woodlot and measure the exact percentage of shade provided by your trees. Fortunately, natural growth succession usually spaces mature hardwood trees at distances that produce shade within the limits acceptable to ginseng. A dense understory of young trees or undergrowth that hinders walking indicates too little shade. The lack of any undergrowth indicates too much shade. The leaf canopy should produce a mottled pattern of light and shadow on the forest floor that continuously changes during the day so that no areas are in continuous light or shadow. There also should be a noticeable cooling sensation on a hot summer afternoon when you walk from the open into your woodlot. A maximum temperature of no more than 80°F is desirable. If drainage is adequate, your best planting site will likely be where the snow stays longest on the ground, which is usually on a north- or east-facing slope.

Undoubtedly the best soils are humus-rich sandy loams, but ginseng grows in a wide variety of soil types, from black sand, through clay, to heavy black loam—as long as excess surface water leaves the root area immediately after heavy rains. Good results are obtained in heavier soils only when there is a decidedly sloping terrain or porous granular subsoil. Ginseng hates wetness, yet loves moisture. Thus, it does poorly and is often attacked by root rot in depressions that hold water after rains or in heavy, soggy soils—especially bottomlands. On the other hand, ginseng requires at least 20 inches of annual rainfall and a good mulch to prosper. Lack of available soil moisture greatly reduces germination, root growth, and seed production. On naturally drier south- and west-facing slopes look for some clay content in the soil. On other slopes, loamy soils usually retain sufficient moisture even during dry spells. Check your prospective site during a dry period to make sure the soil under the leaf litter retains some moisture.

Even if your prospective site lacks some of these basics, it's still not mission impossible if you're truly determined. David Charlet of Reno, Nevada (far removed from ginseng's native habitat), learned to grow green gold at a 4,500 foot altitude, where the rainfall averages less than nine inches annually (mostly during the winter), the summer temperatures commonly exceed 100°F, the porcupines nibble the tops, and the quail dig up the roots. He constructed lath-covered planting boxes beneath the natural deciduous tree shade, and he watered every five or six days during the hottest periods. David told me that the San Francisco herb dealers, to whom he sold his first crop, said the roots were as good as any they had ever seen.

Optimal Site Conditions—A Check List

My 30-plus years of growing experience have taught me that, if a densely planted woodland ginseng garden is established on a marginal site, it will be extremely difficult to keep it healthy long enough to produce high yields of quality roots. If you are hoping to have a commercially successful woods-cultivated operation of any size, then you need to have the same kind of promising site that the wild-simulated approach requires. Almost every successful moderate-to-large-scale grower of woods-cultivated ginseng that I know has a site that wild 'sang would have to love. Thus, most of the characteristics of an optimal site for woods cultivation are the same as those described for an ideal wild-simulated planting site in the last chapter.

By far the single most important condition is having the appropriate companion plants (see The Planting Site Is Critical in chapter 4) growing on your site. Table 5 provides a form for making a visual assessment of your potential site that includes an evaluation of the native plants that are present. Robert Beyfuss, Cornell Cooperative Extension agent (now retired), developed this form with a great deal of input from ginseng diggers. Although it is intended for use by potential 'sang farmers in only one part of the country, it has wider applicability. I have been in ginseng woodlots from one end of the Appalachian mountain range to the other, on up to Maine, and across the upper Midwest from Ohio to Wisconsin, and I believe this form would

Table 5. Visual Site Assessment and Grading Criteria for Potential Woodland Ginseng Growing Operation for a Northern Forest*

Circle only one choice for each category

	Points
CATEGORY A: Dominant tree species (50% or more of mature trees)	
1. Sugar maple (add additional 5 points more if average circumference is greater than 60", add 2 additional points if there is a presence of butternut)* In areas south of New York, tulip poplar is equivalent in value to sugar maple as an indicator tree species.	10
2. White ash or black walnut (add additional 4 points more if average circumference is greater than 60", add 2 additional points if there is a presence of butternut)	8
3. Mixed hardwoods consisting of beech, black cherry, red maple, white ash, red oak, basswood	5
4. Mixed hardwoods as above plus some hemlock and/or white pine	5
5. Red and/or white oak	3
6. Ironwood, birch, aspen	1
7. All softwoods, pine, hemlock, spruce, fir	0
CATEGORY B: Exposure (orientation)	
1. North-, east-, or northeast-facing	5
2. South-, southeast-, northwest-facing	2
3. West-, southwest-facing	0
CATEGORY C: Slope	
1. 10% to 25% slope	5
2. 2% to 9% slope	3
3. <10% or >25% slope	0
CATEGORY D: Soil and site surface characteristics	
1. Site dominated by mostly very large trees, more than 20" in diameter, few surface rocks, 75% of site plantable	10
2. Site dominated by medium-sized trees, 10–20" in diameter, some surface rocks, 50% plantable	8
3. Small trees less than 10" in diameter, very stony, 25%–50% plantable	5
4. No large trees, saplings and shrubs dominate or large rock outcropping, many boulders, less than 25% tillable	3
5. Soil too rocky to plant anywhere, poorly drained, standing water present	0

	Points
CATEGORY E: Understory plants (select highest score entry only)	
1. Reproducing population of wild ginseng	15
2. Sparse wild ginseng	10
3. Maidenhair fern or rattlesnake fern or red or white baneberry	8
4. Christmas fern or blue cohosh or red-berried elderberry or foamflower or stinging nettles *(spicebush is often found with wild ginseng in southern or midwestern sites and is considered a good indicator plant there)	6
5. Jack-in-the-pulpit or other ferns or trillium or bloodroot* (bloodroot is a much higher scoring indicator plant south of New York) or jewelweed or mayapple or herb Robert (a type of wild geranium) or true/false Solomon's seal	5
6. Wild sarsaparilla or Virginia creeper or ground nut or yellow lady's slipper or hepatica	3
7. Club moss or princess pine or bunchberry or garlic mustard or pink lady's slipper	0
8. Woody shrubs such as honeysuckle, mountain laurel, witch hazel, barberry, maple leaf viburnum, arrowwood, shrubby dogwoods, alder, lowbush and highbush blueberry	0
CATEGORY F: Security	
1. Very close to full-time residence of potential grower, with planting site within view of residence (noisy, outside dogs housed nearby add 5 points)	10
2. Forested land less than 440 yards (one-quarter mile) from grower's residence, patrolled regularly	8
3. Regularly patrolled woodlot within one mile of residence	3
4. Nonresident grower or remote woodlot	0
Total Score:	

Results:
50 points or above: Excellent site, great potential
40 to 50 points: Good site, do complete soil analysis
30 to 40 points: Fair site, test soil
Less than 30 points: Poor site, look elsewhere

* Prepared by Bob Beyfuss, Cornell Cooperative Extension Agent, Greene County, New York (Revised 02/04). Comments following asterisks were added by author for relevance in other parts of the country.

give good, if not perfect, guidance in all of them. Some of the less common companion species have not been included, and there are a couple that I think are slightly overrated; but, even so, this assessment form is a useful tool. Note that value is also assigned for what are primarily practical considerations that affect the grower more than the ginseng, like steep and rocky sites and the closeness of the grower's residence to his woods. This is as it should be, especially for a woods-cultivated operation.

On a site that scores anything less than excellent, a soil analysis is called for. Ginseng will thrive over a wide range of most soil nutrients, but you should try to find soil with high organic matter content and, most importantly, high levels of calcium. If not, you will need to add some (see Fertilization below).

Several sources in the literature advise the prospective grower not to plant under oak trees. Anyone who has hunted ginseng knows that it will grow under oaks, and I have seen beautiful gardens in oak groves. However, oaks sometimes grow in clay soils that have only a thin layer of topsoil, hold moisture poorly, and are very acidic. Naturally, ginseng will not thrive there.

Preparing Your Site for Planting

The hardest work a woods grower does is preparing his beds for seeding. Like writing a book, the first time you try, it takes a great deal longer than you anticipate. The extra effort spent in proper and thorough bed preparation, however, will later prove worthy of your fatigue and soreness.

Clearing

You need to clear the land to allow your tiller (or possibly a small tractor) room to operate. Remove everything that you can—small trees (roots and all, if practical), saplings, vines, briars, dead limbs, even old stumps (which can really test your back)—leaving only the large trees needed for shade. To maximize air circulation, clear areas around the perimeter of your beds as well, especially at the bottom of the slope. That is the likely direction of airflow, and humid air will back up the hill if there are no breaks in the undergrowth at the bottom. If you can reach low limbs on the shade trees, cut those off during the winter. The cold temperatures will minimize the chance of diseases infecting the cuts you make on the trees.

When you finish clearing the site, you will likely have the makings of many brush piles, some of which you may want to clean up by using a chipper/shredder to create mulch. One of the unexpected bonuses of a woods garden is the look of it, and piles of tangled brush diminish the aesthetic appeal of your woodlot. Rather than destroy the forest as most agriculture does, woods cultivation of ginseng creates a cool, green carpet beneath the trees, dappled in patterns of sunlight and shadow. It is pleasing to the eye, particularly when punctuated by crimson berries in the late summer.

The last step, which is not necessary on most sites, is to reduce your future weed problem. You cannot fumigate the soil, because some of your shade trees might die if their roots took up too much of the fu-

Nicely laid-out woods-cultivated ginseng beds at Harding's Ginseng Farm in Maryland. Photo by Eric Burhart, courtesy of Larry Harding.

migant. A few woods growers who have a thick stand of invasive weeds use herbicides such as Roundup (glyphosate). If weeds are not too abundant, spray only once in July according to the directions on the bottle. If weeds are truly thick, especially poison ivy, spray once in late June and again in mid-August. Alternatively, if you get your land cleared during the winter, way in advance of fall planting, you can till repeatedly throughout the summer whenever weeds emerge and before they go to seed, thereby killing many of them and reducing their abundance the following season when your young ginseng plants are trying to establish themselves.

Creating Beds

Breaking up woodland soil for the first time is a far different undertaking than tilling a vegetable garden in your backyard. There are trees that must be avoided, and tree roots, which you cannot avoid. The biggest roots often stop a tiller dead in its tracks. Always take an axe or pulaski into the woods with your tiller as you may have to chop the root in half to free the tiller. The smaller roots will wrap themselves around the tiller's blades (or tines) and gradually reduce its efficiency until you cut them out with a knife or pruning shears. Before I begin tilling, I try to find the largest shallow roots, chop them free at least one foot from the trunk to protect the tree, dig them out, and remove them. I have seen one grower kill his trees by tilling very deeply, right up to their bases with heavy equipment. At the same time, I also remove all rocks I find on the surface.

A note on tillers: the most common, least expensive garden tiller has its cutting tines located at the front end, where they not only break up the soil but also provide propulsion. I have prepared many a bed with five horsepower front-tined tillers, but they rattle my bones and wear me out. They have to be horsed around, and my hands and forearms are sore for weeks from muscling the tiller through the soil and absorbing the vibrations. The more expensive rear-tined tillers are better designed for most woods cultivation because the tines do not provide propulsion and can rotate fast, thereby cutting up the soil and roots better while the tiller rolls along slowly and under control. Keeping the tines sharp and in good condition also makes the job less difficult. You need at least five horsepower,

The woods-cultivated beds of a successful Kentucky grower laid out in parallel rows among the trees. Note a rodent barrier (arrow back of bed) of aluminum flashing that has been placed around the perimeter of the beds. The flashing is 30" wide, with about 12" sunk into the soil to prevent tunneling into the beds and 18" above to prevent ground access. Photo by Tom Konsler.

but heavy tillers (over eight horsepower) are difficult to manage on steep, uneven slopes. Since you only need a tiller for a limited period of time, consider renting a rear-tined model when you need it rather than buying one.

Rake aside the topmost layer of leaf litter just before you till, and pile it where it will be convenient to use later for mulching. Do not rake all the way down to the bare soil. Leave some of the already partially decomposed litter on top of the soil to be turned under and provide organic matter. Starting at the bottom of the slope, till directly up the hill, breaking up the soil six to eight inches deep. If the grade is at all steep, instead of turning around and tilling down the slope, I'd strongly advise you to back the tiller down the hill and then cultivate again as you return to the top. I go over the entire site twice—first, primarily to cut the roots and, second, to finish breaking up the soil. Do not worry about removing all the small roots you have cut up. They will eventually break down and enrich the soil. Incidentally, do not till in the rain or when your soil is very wet. It is a muddy mess, and, if there is any clay in the soil, it will cause clumping and reduce water drainage.

I create my beds between the trees, leaving uncultivated islands of land where the trees grow. Many other successful woods growers lay out their beds exactly parallel to each other, letting trees stand right in the middle of their beds and adjusting bed width so that trees don't block the trenches between beds. Whichever design you choose (and you can combine the two), run the beds and trenches up and down, not across, the face of the slope to promote the immediate drainage of surface water during rain storms.

If your slope is greater than 15 degrees and your soil is extremely loose and light, drainage trenches between beds can cause excessive soil erosion. In these rather exceptional conditions, do not dig trenches, but leave level walkways between beds to provide access on both sides. (Normally your trenches will provide walking access as well as drainage.) Beds should be between four and seven feet wide; anything wider is difficult to weed or access without walking on them and compacting the soil. Intervening trenches or walkways should be about one-and-a-half feet wide. Make beds as long as you want, or can.

At the top of each bed, join the bordering trenches to form a point, or inverted V. Dig the trenches to the same depth that you have tilled, and throw the dirt onto the center of your beds. Then rake the beds so that the center is at least two or three inches higher than the edges. This arch, or crown, on your beds will further facilitate drainage and is critically important if your land has little slope. At the bottom of the site, run the trenches a few feet below where the beds begin.

If you are farming where the annual rainfall is more than 35 inches, consider digging a water diversion ditch. This should also form an inverted V along the entire top of your bedded area and direct rainwater around and down, past your beds. The ditch needs to be at least a foot deep. (This practice also works well for artificial-shade beds on cleared slopes, and I know of one wild-simulated grower who has found a need for it.)

Cultivating with a Tractor

If you are preparing a large area, a small tractor can save time and your back. The danger of turning over a tractor in the woods on a steep slope, however, has to be weighed against the potential savings in time and effort. A tractor is a less sensitive tool than a tiller. You can tear out deep roots or otherwise damage or kill the trees you must have for shade. As you plow downhill dragging roots behind you, you can easily pull large amounts of topsoil down the slope and off the upper sections of your beds. Skill in operation is requisite.

I am aware of two types of plows that have been used successfully to prepare ginseng beds in the woods: a moldboard plow and a small brush and bog disc. In maneuvering either of these, you will have difficulty getting closer to the trees than one foot. That is just as well, since if you kill or severely damage a tree, the loss of its shade will require you to either construct a small patch of artificial shade or else lose the ginseng in that area. If you exercise reasonable caution, your trees will soon vigorously re-establish their shallow roots throughout your bed, which helps wick away excess moisture from the root area. Moreover, their competition with the ginseng for water and soil nutrients slows root growth, which helps woods-cultivated 'sang to more closely resemble wild, and hence to bring a higher price.

Fertilization

Wild ginseng grows where trees, saplings, shrubs, and other herbaceous plants compete for the available soil nutrients and where the soil is low in nitrogen because it is "locked up" in the decomposition of leaves and woody tissue. It is not surprising, then, that fertilizing a moderately fertile soil does not greatly help a plant that is accustomed to making do with limited nutrition.

If you have good, rich woodland soil, it is probably not wise to alter it dramatically. The exception, of course, is calcium. Just as with a wild-simulated planting, if wild 'sang is not flourishing on your prospective planting site, a soil test is appropriate. Should the calcium level test below 2,000 pounds per acre, then incorporate 50 pounds of gypsum (the non-pelletized form is certifiable organic) per 1,000 square feet of bed space when you till your beds for fall planting. Test again in late winter and, if calcium is still below 2,000, then top dress with 50 pounds of gypsum per 1,000 square feet of bed space in the spring. Repeat in subsequent springs whenever late-winter soil tests show the calcium level below 2,000. If your first soil test shows less than 1,000 pounds of calcium per acre, consider selecting another site. If that is not practical, then initially add 100 pounds of gypsum per 1,000 square feet as you till your beds in the fall before planting. Continue to monitor calcium, and top dress in the spring as needed.

If you have dense soil or very sandy soil, or a soil that is low in organic matter (testing below 5 percent), then it should benefit from tilling in two or three inches of shredded or partially decayed leaves and bark—preferably poplar or sugar maple, but leaves and bark from most any deciduous trees other than oak are acceptable. (Oak leaves contain tannic acid, which will lower pH and alter soil chemistry.) Peat moss is a good source of organic matter as well, but it is relatively expensive. Add two or three loose fluffy inches as mulch, after you have seeded. This is the simplest, safest, most natural way to increase the organic matter content of your soil. Do not get carried away. Adding manure or compost risks enriching the soil too much and accelerating root growth, which can both reduce the quality (and value) of your roots and make them more susceptible to diseases. Adding excessive organic matter of any kind runs the risk of creating a compost pile instead of a planting bed, and the heat generated may kill your ginseng. The heat will also help the soil microbes to rapidly decompose the very organic matter that you have added, which is self-defeating.

Obviously, on a site with humus-rich soil, it is not a great hardship to avoid synthetic chemical fertilizers, which leaves you the option of having your green gold certified as organically grown. (An interview with Oscar Wood, a grower who used no synthetic chemicals, is presented in chapter 8.)

If you know your soil is not rich and fertile, adding fertilizer can help you to grow ginseng anyway. As you prepare your beds, in addition to adding organic matter and calcium, you may want to incorporate as little as five pounds of a balanced fertilizer, such as 5-20-10 (ginseng doesn't like much nitrogen) per 1,000 square feet. If you are not well versed in plant nutrition and soil chemistry, consult your county extension agent before doing any more than this. You can easily overdo it. Excessive or inappropriate fertilizer will cause problems. For example, fertilizers—especially nitrates—applied in dry weather (without irrigation) will result in heavy concentrations of sodium (Na, salt) at the root level. The salt will attract water out of the growing plant, thereby stressing it and making it susceptible to secondary blight infection. An excess of commercial fertilizers or lime can raise the soil pH too high, and a high pH is associated with ginseng root rot. Where the pH is in the range of 5.0–6.7, the nutrients that ginseng needs are available and bacterial diseases are frustrated, leaving only fungi to threaten the plants.

Impressed with the yields artificial-shade growers were obtaining, many years ago I followed the fertilizer recommendations for shade gardens on some of my woodland beds. My ginseng did indeed grow much faster, but on balance, I was not pleased with the results. There were disease problems that I had never had before, and the roots were of a significantly lower grade. Over-fertilization can force growth and upset natural balances, thereby lowering the plant's disease resistance. While manure, natural soil amendments and foliar sprays, or synthetic chemical fertilizers may indeed increase growth rate, fast-growing large roots bear less resemblance to wild roots and will therefore be worth less per pound. An increasing amount of quality research has been done on the nutritional needs of ginseng, but nearly all of that research has been done on artificial-shade gardens, little on woods cultivation. References to those studies are listed in chapter 10.

Planting

When to Plant

Traditionally, ginseng is planted in September, October, and early November. Some growers prefer to plant as late as possible in the fall, so that the first snowfall quickly protects their seed from the critters. Actually, as long as the ground is not frozen, seeds can be planted at any time. Spring planting in late February and early March is an option in the South, where the ground thaws early; but, once the weather starts to warm up, seeds quickly begin to sprout. (If anything, refrigeration only accelerates germination.) Thus, spring planting is possible within only a small window of time. For that reason, growers generally do most of their seeding in the autumn.

Sowing Seeds

I've found that the optimal planting depth is between ½" and ¾". Seeds sown on the surface are subject to rodent and slug predation, while seeds planted deeper than one inch may have difficulty reaching the surface after sprouting. Two seeding methods are available to a woods grower.

You can delay digging the trench/walkways, hand-cast the seeds over your freshly tilled beds, and then dig the trenches, shoveling the dirt to uniformly cover all the seeds. Rake smooth if necessary. The depth you get this way will be close to one-half inch, but, because you distribute the soil from the trenches evenly, you lose the ability to crown your beds along the center line (which improves drainage). When you hand-cast seeds, getting the right planting density is a bit tricky. It helps to divide your planting area into sections of, say, 200 square feet and plant each section separately. Determine how many seeds you wish to sow per square foot (discussed below), multiply by two hundred, weigh out the appropriate amount of seeds (there are between four and five hundred seeds per ounce), and then cast them as uniformly as possible. Be careful not to use up all your seeds before you complete the section. If you have some left over, you can always go back and fill in thin places. Use the walkways where the trenches will be, and do not step on the beds—in soft ground your footprints will create lasting indentations where water will later accumulate.

The other method, which I prefer because it yields a much more even spacing, is to go ahead and dig the trenches, crown your bed, rake the surface smooth, then sow with a hand-operated garden seeder that has adjustable depth and spacing. I get the best results pushing the seeder up the slope. Garden seeders are available at

Disc for garden seeder with selected holes enlarged. This particular disc was designed for an Earthway garden seeder to plant radish seeds but will now drop a ginseng seed approximately every 4½ inches. Photo by Kim Fadiman.

most farm supply stores or garden shops, but they don't come equipped with seeding discs designed for ginseng. Select a disc intended for radishes or some other small seed. It should have cups barely large enough to pick up a single ginseng seed and small holes through which the seed drops to the ground. You will have to enlarge the holes—an electric drill and quarter-inch drill bit works well. File your new holes smooth. If you enlarge all the holes, you can alter spacing by taping some of them closed.

Results of a Spacing Study

The distance between ginseng plants markedly affects their rate of growth, and as they reach maturity, their seed production. Closely spaced plants grow more slowly and, other things being equal, are more vulnerable to disease. However, while dense planting reduces individual root size, it increases the total potential yield in root weight obtained on a unit area of land.

Dr. Tom Konsler, Dr. Jeanine Davis' predecessor at North Carolina State University's Mountain Horticultural Crops Research Station, conducted a six-year study on ginseng plant spacing, and his results support the growers' belief that the law of diminishing returns sets in when plant populations get too dense. Table 6 presents Dr. Konsler's findings regarding the effect of spacing on individual root weight. He planted all his test plots in rows six inches apart with different spacing within the rows. As early as the second year, roots only an inch apart were smaller than those more widely spaced. By the fourth year, plants six inches apart, or closer, showed effects of the competition and were smaller than those at nine inches. Clearly, the growth rate for individual roots is much faster when they are spaced widely apart. (Do not expect to grow roots averaging two ounces, as Dr. Konsler did, at any spacing. He is a first-rate professional horticulturist, and his test plots were small, grown under artificial shade in near-perfect conditions, and tended meticulously.)

Dr. Tom Konsler inspects his test plots at North Carolina's Mountain Horticultural Crops Research Station. Photo courtesy of *The Mother Earth News.*

Table 6. Ginseng Root Weight Response to Four Plant Spacings at Ages 2 through 6 Years

Spacing in Inches	Ounces* per Root				
	2 yrs	3 yrs	4 yrs	5 yrs	6 yrs
1" × 6"	.09	.09	.28	.45	.46
3" × 6"	.14	.44	.63	.87	1.13
6" × 6"	.15	.64	.89	1.08	1.26
9" × 6"	.15	.67	1.29	1.77	2.10

*Roots were weighed when freshly dug.

Optimal Seeding Rate

Even with Dr. Konsler's results in hand, the optimal seeding rate is not obvious for several reasons. Your germination rate may vary from 50 percent to as high as 90 percent, and your attrition rate (though usually significantly less than on a wild-simulated site) is initially unpredictable as well. Thus, for any given seeding rate, it is difficult to project your final plant density until you have had several years of growing experience on your particular site. Moreover, there is a difficult choice to be made between close spacing, which has the advantage of greater yield potential and wide spacing, which has the advantages of lower seed cost and a less dense, healthier garden that can be grown an extra year or two to produce a more desirable root. After tilling several thousand square feet in the woods, you will probably be even more interested in maximizing the return on a given area of prepared bed space. However, I strongly urge you to avoid anything like a 1" × 6" spacing. This has worked well for artificial-shade growers, which is why Dr. Konsler was interested in it, but I know of no successful woods-cultivated grower who is planting that densely. Indeed, shade growers may have been wasting seed, because Ontario Ministry of Agriculture, Food and Rural Affairs (OMAFRA) research indicates that ginseng tends to thin itself to no more than eight to ten plants per square foot. A 3" × 6" spacing means sowing eight seeds per square foot, and that's as dense as you should consider planting if you wish to grow for at least six years to get a wild look to your roots.

Many experienced farmers of woods-cultivated ginseng now sow at the rate of six to eight seeds per square foot, since the necessary care will be similar to a thinner seeding and the potential yield greater. (There are factors other than spacing that influence yield, of course.) If you sow at the rate of more than two seeds per square foot, you will probably need to undertake a regular disease prevention program (see Disease Prevention below). With the accelerated growth that tilled soil encourages, and in the monoculture of a bedded planting, diseases may show up before your plants reach maturity, even with only two plants growing per square foot.

If you want to try to grow organically without synthetic chemical fungicides, a low-density garden of only two or three plants per square foot is definitely advisable (though perhaps not absolutely mandatory). This means sowing only four or five seeds on each square foot of bed space and then conscientiously using organic methods (some are mentioned below) for disease prevention.

Replanting

When ginseng is replanted on the same site after a first crop has been raised to maturity and harvested, sometimes the results are disastrous. Losses of anywhere from 30–100 percent are reported. The exact cause of the difficulties is still unknown, but second plantings seem more susceptible to any kind of stress, especially diseases. Sometimes problems do not arise until after the plants are three or four years old and the grower has a great deal of time and effort invested in them. Because the labor of tilling new beds in the woods is so great, I have often tried replanting in the same beds, and I have found that it is only

Such a densley planted monoculture of ginseng usually means the grower must conscientiously undertake a disease-prevention program or else risk heavy loss sometime during the many years before harvest. Photo by Eric Burkhart, courtesy of Larry Harding

likely to be successful where I have had vigorous growth and absolutely no disease problems the first time around. I know of one grower who succeeded with a second crop, then had serious losses from his third planting in the same ground. If you have any disease problems at all with your first crop, I would certainly recommend moving to new ground. Try a small test plot on the used site. If it grows to maturity, then it is probably a reasonable risk to replant the entire area.

Mulching

As soon as seeds are planted, mulch them! Mulch is essential to helping the plants stay cool and retain moisture during hot, dry weather. With adequate shade and good mulching, your plants will survive all but the most extreme drought conditions without irrigation (assuming, of course, that you have selected a cool, moist planting site). This blanketing of the soil surface also greatly reduces weeds, curtails erosion, and protects the root top and next year's bud from damage by frost, which can heave it out of the ground and expose it on the soil surface. In the spring, mulch regulates soil temperature, minimizing repeated cycles of freezing and thawing and preventing early hot spells from causing plants to emerge too early and die from a severe late freeze. Finally, as mulch decomposes over time, it acts as an organic fertilizer that maintains soil humus levels, prevents soil packing, and reduces leaching of soil nutrients.

The best mulches are leaves, small-grain straw, and weathered bark/sawdust mixtures. Pure sawdust, especially if it is not well decomposed, is undesirable because it compacts too tightly, reducing aeration and evaporation, even preventing newly sprouted seeds from emerging in the spring. Hardwood leaf litter is, of course, the natural mulch of wild ginseng and is readily available to the woods grower. Leaf litter is predominantly leaves, but also bark and decaying wood. Oak leaves are an inferior mulch because they contain tannic acid, and also because they are large, tough, and slow to decompose, which makes it difficult for young sprouts to force their way up through them to the sunlight in the spring.

I use leaf litter and recommend it. It is free, handy, natural, and replaces itself each fall. I do know a pair of extremely successful growers who use wheat straw mulch in the woods. Oat straw is particularly appropriate for early fall planting because the oat seeds will sprout in the fall and then be killed by freezing weather. Although bales of straw are convenient to haul to your beds and spread, straw has two serious drawbacks: it may contain grass and weed seeds; and slugs love to live underneath it, eating seeds sown at the surface, buds that have been heaved by frost to the surface, sprouts as they emerge in the spring, as well as the foliage of older plants.

Spread mulch as evenly as possible. In western North Carolina, where I live, two or three fluffy inches of leaf litter, one inch of bark/sawdust mixture (perhaps from a chipper/shredder), or three to four loose inches of straw can be spread after planting and left on the beds year round. Farther north, twice those amounts may be needed over the winter, both to minimize frost heaves and to keep the roots from undergoing repeated, and potentially fatal, freezing/thawing cycles. Such a thick layer, if it doesn't naturally thin itself, may need to be partially removed from newly planted beds in the spring to allow seedlings to emerge.

Always keep a good layer of mulch on your beds. Make an inspection every few weeks. If the wind creates bare spots, replace the mulch and lay a little brush over the problem area. Beginning in late autumn after the leaf fall, clear any blown mulch and other leaves and debris from your trench/walkways so surface water is free to drain away.

Record Keeping

I have found that keeping written records has helped me learn from experience. Many years pass between planting and harvesting, and it is easy to forget how and when each bed was treated. So, as soon as I plant, I mulch...and as soon as I mulch, I go inside and write down things like a bed identification number and location, month and year, planting method, seeding rate or spacing, square footage, fertilization (if any), bed site characteristics, and anything else of interest. I add notations at other times if problems arise or I do anything special to the bed. Ginseng farming requires a patient long-term approach, and record keeping is an important part of that.

Pre-emergent Care

There are a few things to do over the winter: replace any mulch that blows off the beds; clear dead limbs and other debris from beds and trenches or walkways; and, if rodents have nibbled frequently on seeds or roots, set out poison baits and/or traps while their food is scarce. On nice days, the winter is a good time to begin clearing small trees and brush from future planting sites.

Depending on the climate, ginseng emerges through the mulch sometime between mid-April and early June, about the time the dogwoods are blooming. In the northern part of ginseng's range, where additional mulch is added to the beds to prevent frost heaving over the winter, that extra mulch may need to be removed a few weeks before the plants break the soil surface. Weeds may begin to sprout densely in the beds (especially in newly planted beds) well before the ginseng comes up. A few woods growers find they need to spray their beds two to four weeks before ginseng emergence with a weed killer.

Ginseng Diseases

There are at least a dozen pathogenic fungal organisms that attack ginseng, though many of them are uncommon. If conditions are conducive, the woods-cultivated grower may have to deal with one or more of three basic disease types—damping-off, blight, and root rot—each of which can be caused by several different fungi, and several of those fungi can cause more than one disease type and attack the seeds as well. Their recognition, prevention, and control are discussed below, although not in great technical detail. [Author's note: Readers interested in greater detail and current information about ginseng diseases and other pests, and their treatment, are referred to M.K. Hausbeck's "Control of Diseases, Pests, and Weeds in Cultivated Ginseng" and to the Ontario Ministry of Agriculture, Food and Rural Affairs' (OMAFRA) web site and Publication 610 (see chapter 10).] I have tried to provide the background knowledge a small-scale grower needs to cope with diseases of ginseng; however, any grower with plans to manage more than half an acre of cultivated ginseng gardens (in the field or in the woods) would be well advised to purchase Publication 610 ($20).

Damping-off

Damping-off is a generic term for a group of fungal diseases that attack the stems of young (usually one-year-old) plants. This normally occurs early in the growing season when the weather is both cold and wet, often just as seedlings are emerging through the mulch. The tender plant stem collapses, the top wilts, and the root (what little there is of it) eventually dies. Pythium is the most common cause of damping-off. It is present in a wide variety of soils, living on organic matter. It is an opportunistic fungal pathogen (technically, a "facultative saprophyte"), attacking weakened plant tissue. Thus, young seedlings just emerging are a favorite target as are already diseased, stressed, or damaged older plants. Although stems of older, established, healthy plants are undoubtedly attacked as they emerge from the soil in the spring, they rarely succumb to damping-off.

When damping-off is severe, it usually occurs in the most poorly drained areas of a planting site. Infection begins in a small area, and should wet cool weather continue, it will spread outward in all directions if left unchecked. Less severe infections sometimes attack individual or small groups of plants here and there in seedling beds. At least four different fungal organisms are known to cause damping-off: Pythium, Rhizoctonia, Fusarium, and, infrequently, Alternaria.

Damping off. Photo by Tom Konsler.

Blight

Alternaria blight, caused by the fungus *Alternaria panax*, is the most common disease of ginseng and one of the most destructive. (There are other ginseng blights and mildews—*Botrytis cinerea*, sometimes called gray mold, for instance, but they do not pose the threat Alternaria does.) *A. panax* is similar to early tomato blight (*A. solani*), but this particular parasitic fungus depends primarily on ginseng for its nutrition.

The Alternaria fungus is spread by spores (small seed-like bodies). These usually travel on the wind, in splashing rain, or on shoes and clothing, and eventually alight on a ginseng plant. This can occur any time during the growing season. The spores remain on the foliage until the weather becomes hot, wet, and humid. Then they "sprout," penetrating the stem or leaf cells. Usually only the foliage dies, stopping growth and eliminating the potential for seed production that season, but the ginseng root survives to sprout a new top the next spring. The disease sometimes attacks the senescing flower petals (as does Botrytis), eventually causing the developing seed head to abort. Only if Alternaria

Alternaria stem blight—typical collapse at point of fungal attack. Photo by Dr. C.R. Roberts, former professor of horticulture at the University of Kentucky and one of the few researchers to study ginseng growing in a woodland setting.

(or any other disease or damage) destroys a young plant early in the summer, before the bud has formed for next year's top, is there a risk of root loss. Older, larger roots will usually be able to draw upon their energy reserves to produce a new bud for the following season without the aid of any foliage.

Infection may begin here and there throughout the bed or initially be confined to a small patch. The first signs often are dark spots on the leaves or stem. Stem lesions are initially light brown and up to two inches long. In time, they will cause the stem to collapse and the foliage to wilt.

Alternaria leaf blight—typical spotting where fungus attacks the leaf tissue. Photo by C.R. Roberts.

Leaves often develop areas with a water-soaked appearance that eventually dries out, leaving a tan papery center with a darker velvety-brown border and a yellow halo. This is sometimes referred to as a target pattern.

In the right conditions, left unchecked, Alternaria blight can spread rapidly, infecting an entire hillside of beds within a few weeks. Toward the end of the growing season, as plants weaken and the foliage starts to die off, Alternaria will attack the weakening foliar tissue opportunistically. It will not cause much harm the year of initial infection, but once Alternaria is established in an area, it usually stays around—existing on dead plant material during the winter. When conditions are right during the next growing season, it will attack the ginseng again. Still-standing stems from last-year's plants are well-documented sources of

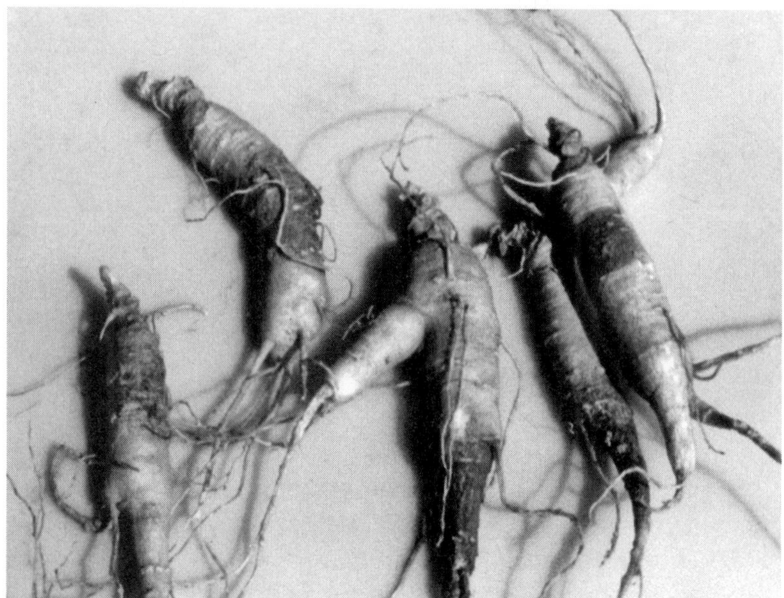

Roots that have begun to discolor and rot due to the fungus Fusarium.
Photo by C.R. Roberts.

reinfection in the spring as newly emerging tops come into contact with them.

Root Rot

In woods cultivation, root rot is the least common of the three disease types, but it is the most potentially destructive, because it attacks and ruins the roots of older, laboriously tended plants. At least five types of fungi have been associated with some form of root rot: Phytophthora, Cylindrocarpon, Pythium, Sclerotinia, and Fusarium. In addition, "rusty root"—a rust-brown discoloration of the root—may have a complex of genetic and stress-related causes.

All root rot fungi first manifest themselves when the foliage begins to droop and/or turn reddish, yellow, or bronze in spots or all over. Plants have a weak look about them. On examination, the roots usually have discolored and/or become spongy and begun to deteriorate, thus cutting off the uptake of moisture and nutrients essential to the healthy growth of the top. Most of the time, the entire root has either disintegrated or is so badly marred and discolored as to be worthless. Thick plantings in poorly drained beds are most vulnerable to root rots. In the woods, losses are frequently 20–30 percent rather than total; but, if not treated or the roots not dug, the disease often returns the next year. The following are of greatest concern to the woods grower:

Phytophthora cactorum is the most common cause of root rot. It is a water-loving, soil-borne fungus that in wet conditions will spread between the roots of

Left: Leaf infection caused by splashed spores of *Phytophthora cactorum*, a root rot fungus. Right: When *Phytophthora cactorum* attacks in force, it sometimes looks as though scalding water has been poured over the foliage, and the roots will be mushy and malodorous. Photos by C.R. Roberts (left) and Tom Konsler (right).

closely spaced plants; however, it is most dangerous when its spores are carried by the wind or splashed by water onto the surface of leaves, where it enters the plant and eventually makes its way to the root. Early symptoms of foliar Phytophthora look like boiling water has been poured onto the foliage.

Cylindrocarpon attacks only roots, usually older ones. The only symptoms are discolored leaves with red perimeters, while below ground the roots either gradually disappear until only a black stub remains or else they only become discolored (but greatly diminished in value). In the case of "disappearing root rot," *Cylindrocarpon destructans* may be present in the bed and only a few plants may succumb, or over 90 percent may disappear. Cylindrocarpon is not well understood. It is particularly troubling because it can be carried by ginseng seed (see chapter 6) and can spread fast and because no effective preventive treatment has been found.

In the case of "rusty root," the roots develop a rust-brown spot that may spread, forming a sunken, round-to-irregular lesion roughly 5 mm in diameter, usually on their shoulders near the neck, which reduces their market value (although sometimes the lesions slough off, and the root's value is preserved). The rust, by the way, is the root's way of fighting infection or adverse conditions. The rusty area is actually a "corking over" that prevents further tissue destruction.

Rust has become increasingly common and severe throughout North America, but especially in the shade gardens of British Columbia. No treatment has been found. There is good reason to believe there is a genetic component, which means that its increase is probably due to the distribution of seed from affected gardens. Various stresses on the plant—heat, cold, nutrient imbalances, damage, disease—have all been associated with its occurrence. Interestingly, in apparent contradiction to that conventional wisdom, Dr. Rick Reeleder of Canada's Southern Crop Protection and

Food Research Centre reports consistently finding a previously unknown fungus on rusty roots. Though others are skeptical, he believes it is the primary cause. In a 2007 article in *Mycologia*, Dr. Reeleder formally named the fungus *Rhexocercosporidium panacis*, but Canada's ginseng cognoscenti have dubbed it *Fungisaurus Rex*.

Disease Prevention

That old adage "An ounce of prevention is worth a pound of cure" could have been uttered first by a ginseng farmer.

Site Selection and Cultural Practices

The Koreans, who have been cultivating Asian ginseng for many decades, contend that ginseng will establish its own natural defenses against disease if conditions for good growth are maintained. My own experience has shown me that establishing healthy growing conditions certainly reduces both the frequency and the severity of disease. Proper cultural practices are cost effective over the long term for any grower, and for organic growers they are paramount. In fact, unless you are beset with weeks of unrelentingly wet weather, severe outbreaks of disease usually indicate either bad site selection or faulty cultural practices.

Many disease-preventive measures have been discussed in earlier chapters, but they are worth reviewing. First, select a good growing site: a cool, well-shaded deciduous woodlot with humus-rich soil and ginseng or many of its companion plants in evidence. If necessary, add gypsum to raise the calcium level above 2,000 pounds/acre. You must have adequate slope for external drainage, and you may need to supplement the slope with diversion ditches across the top of your beds and dig the usual trench/walkways between your beds to drain surface runoff immediately downhill below your ginseng. Do not allow your trench/walkways to get clogged with debris, but keep covered with leaf litter, or preferably a thin layer of wood chips, to reduce splashing. You must also have good internal soil drainage. If your soil is dense, then you may need to till in organic matter before seeding. Try not to compact the soil or create indentations in your beds with equipment or foot traffic. Avoid boggy areas or the base of slopes where the land flattens out, and run your beds downhill, crowning them along the center line. (The big Ontario shade growers even analyze and alter the subsurface drainage patterns in a further effort to avoid wet roots.)

Make sure your site has good air circulation. A high leaf canopy that extends beyond the garden's perimeter (especially downhill from the beds), uncrowded spacing, and control of dense weeds all facilitate air circulation. Complete weeding is inappropriate; in fact, organic growers sometimes purposely inter-plant other species, such as goldenseal or bloodroot that are thought to have antifungal properties. Avoid physically damaging the foliage in any way, as that provides an easy entry for initial infection. In particular, stay out of your beds after a frost until the foliage has thawed. A newly harvested garden is likely to have abundant damaged plant material in it, which is a potential source of infection. Do not go from a recently harvested bed to other beds, and do not seed a new bed near one that has just been harvested. If you are not producing your own seeds, know your source of supply, because seed from infected gardens can carry disease (see chapter 6). Finally, plant thinly and mulch thoroughly.

The following cultural practices will help to control disease once it has entered your beds. (1) Always work from your newest garden to your oldest, except that you should enter any diseased beds last. (2) Remove the foliage of diseased or even suspiciously weak plants from your beds and off your farm as soon as possible. This means inspections at least twice a week throughout the growing season, daily during wet or extremely humid conditions and wherever disease has been recently spotted. (3) When root rot is localized, physically isolate the disease by digging a pair of two-foot-wide trenches across the width of the bed, one above and one below the diseased area, sacrificing healthy plants where necessary. These two trenches need to be at slight angles for drainage and should run into the existing trenches along the outside length of the bed on both sides. (4) After working in an infected area, disinfect your equipment, tools, and boots with a household bleach solution, one cup per gallon of water. (5) In the fall after senescence, remove old mulch and stems from any diseased areas and immediately replace with fresh mulch. This is labor intensive, and some growers just remove the stems while others actually burn the mulch off. (6) Whenever plants appear diseased, it is wise to send plant tissue samples to a lab for specific fungus identification.

Sometimes nutritional deficiencies cause symptoms resembling diseases. Since it may take weeks to get results back from the lab, you may want to remove unhealthy looking plants and possibly apply treatment right away, if you have an educated guess as to the pathogen. State agriculture departments and soil conservation laboratories usually perform this service free or very cheaply (as they do for soil samples). When sending plant tissue to laboratories for analysis, select fresh samples plus a few ounces of soil from the infection site and follow the lab's explicit instructions for packing and shipping (see also Joe-Ann McCoy's Disease List in Appendix 5 at the end of this book).

Preventative Fungicide Application

Unfortunately, during the course of six or more growing seasons, there are often periods of constantly wet weather when diseases can establish themselves despite conscientious efforts to maintain healthy growing conditions. Many growers of woods-cultivated ginseng, especially those with plant densities greater than three per square foot, preemptively apply fungicides on a regular basis. If you do choose to employ agricultural chemicals, remember to read and follow the handling, storage, and application instructions on the labels, and be careful to keep them out of the reach of children, because some are quite toxic. Fungicides are far more effectively used as preventatives before disease develops than as controls after infection begins to spread. Pesticide solutions (and foliar fertilizers) are all best applied in the late afternoon so that the plants have time to absorb them during the cool of the night. [Author's note: Registration of pesticides for use on ginseng is often state by state, temporary, and ever changing (see Pesticide Use in chapter 3). The mention of their common application is not intended as a recommendation or endorsement. Check with your local agricultural extension agent to find out which are presently legal in your state, and then confirm their advice by reading the label, and always follow label directions.]

The most complete and up-to-date (as of this writing) list of fungicides and other pesticides approved for use on ginseng in at least one state is M. K. Hausbeck's "Control of Diseases, Pests, and Weeds in Cultivated Ginseng." Her list provides the various brand names a pesticide is sold under, what it has been proven effective against, and specifics as to how it is to be applied. Frustratingly, only if you live in Michigan or Wisconsin, will the list provide you with immediately actionable information. Otherwise, it is just a starting point. You're going to have to find out which of the pesticides on the list that are appropriate for your needs are also presently approved for use in your state.

Long before the EPA came into existence, early 'sang farmers used copper sulfate ("bluestone") mixed with hydrated lime and dissolved in cold water (three ounces each of copper sulfate and lime in three gallons of water) to prevent Alternaria blight and other foliar diseases. This mixture—called Bordeaux mix (after the French province where its effectiveness against fungi on grapes was first discovered)—is effective, and its use on ginseng does not violate federal law, if you mix it yourself. (But do not mix in a metal container because it is highly caustic and reacts with metal.) The mixing is messy and may result in blue stains; however, if you buy a premixed bag (half copper sulfate, half lime) and apply that in solution to your crop, technically you're in violation of the law, because Bordeaux mix is not labeled for use on ginseng. The copper in Bordeaux mix is generally toxic to fungi on contact, and there are accounts of its successful use against Phytophthora root rot and even damping-off, as well as Alternaria. However, if too high a level of copper builds up in the soil or on the foliage, it can damage your ginseng.

Bordeaux mix and Kocide, a copper hydroxide fungicide that comes as a wettable powder and is effective against blight, are considered natural chemicals and can be applied on ginseng that is to be certified as organically grown. Kocide and Champ Formula 2 Flowable (also copper hydroxide) are labeled for ginseng in some states. The largest, most successful producer of organically certified ginseng that I know sprays alternately with Kocide and Neem oil to prevent Alternaria. (Neem oil is extracted from the nuts of the neem tree and, in high concentrations, is a broadband fungicide and insecticide, but it is not labeled for ginseng in most states.) He uses synthetic chemicals the first three growing seasons and then switches to organic ones at least three years before harvest. A number of growers are spraying with compost teas in lieu of synthetic fungicides in order to reduce their disease problems.

Backpack sprayers are a simple, practical means of applying wettable powder foliar fungicides in the woods. However, they do get heavy with full loads on steep slopes, and it requires conscientious effort to thoroughly coat the stems and both sides of the leaves. When mixing fungicides, a commercial spreader-sticker is sometimes added to the solution to reduce the natural tendency of the spray to bead up on the surface of the leaves and to increase the adhesion of the spray to the leaves when it rains. If a "sticker" is not used, a new application may be necessary after every summer thunderstorm. It is desirable to have protection on the leaves before a period of rainy, cloudy weather. A good spreader-sticker will mix easily with the fungicide in the tank, and it will remain on the foliage even after several inches of heavy rain. Most fungicides naturally break down over time, so there is a need for a regular spraying program.

Damping-off and root rot are generally more difficult than blight to prevent with chemicals, because there are no chemical controls for several of the fungal organisms that cause them, and because most of the pathogens occur in the soil or mulch where they are hard to attack. There are preventative fungicides for some of the damping-off and root rot pathogens, but growers must rely heavily on good drainage and other proper cultural practices. Fumigation before planting will kill the pathogens responsible for these two kinds of diseases, but fumigation in the woods is not recommended, as it is hazardous to the health of the shade trees.

Natural Alternatives

A number of biological controls, including compost teas and bacteria that are antagonistic to fungal pathogens, are being investigated by researchers and tried by growers. Trichoderma, a family name for a group of naturally occurring beneficial fungi, which live in the humus layer of the soil, is the biological control that has probably had the widest use by ginseng farmers. In the soil, Trichoderma intermingle with the plant's root hairs, and an exchange of nutrients occurs to their mutual benefit. Trichoderma not only aggressively compete with other soil fungi, including pathogens, for food resources, but Trichoderma also parasitize and kill other fungi and produce metabolites that suppress the growth of competing fungi, thereby providing some protection to the ginseng roots.

QUADRIS F, which contains the toxin azoxystrobin obtained from mushrooms, is being used in Ontario and Wisconsin to prevent damping-off caused by Rhizoctonia. It is first sprayed onto newly planted and mulched beds in October and then applied onto the mulch again in the spring just before emergence. I have heard a rave review of the foliar systemic bio-fungicide, Serenade. Also, natural mycorrhizal seed inoculants (MycoApply is one brand) are reported to remain within the system of the young plants after the seeds sprout and be surprisingly effective in preventing damping-off. Theoretically, natural fungicides could not only be more sustainable but also more effective than synthetic ones. Regrettably, there is little money available to research them on ginseng.

Some 'sang growers use foliar sprays of various herbal or compost teas as natural fungal preventatives. Tea made from cloves of garlic is a mild preventative for damping-off as well as for downy and powdery mildews; chamomile tea may also be useful against damping-off; and horsetail (Equisetum) tea may be antifungal and will reduce the amount of water on leaf surfaces where fungal diseases begin.

Leaving weeds alone and interplanting other native species among your ginseng breaks up the monoculture and provides a less favorable environment for all pathogens. Several growers I know plant rows of goldenseal and/or bloodroot spaced one foot apart around the perimeter of woods-cultivated beds or simply in lines six feet apart throughout a wild-simulated garden. Both goldenseal and bloodroot seem to inhibit root rot fungi. They are forest herbs compatible with ginseng and are valuable in their own right (and covered by Jeanine Davis in the second part of this book). Although bloodroot and especially goldenseal will propagate themselves and eventually compete a bit with your ginseng, their benefits will outweigh this minor drawback. Black cohosh (another woodland perennial with economic value of its own) is also occasionally interplanted with ginseng—not as a fungal control, but to break up the monoculture and to provide a little extra shade over the ginseng where a tree has fallen or along the sunny edge of a woodlot.

Disease Control

Once a fungal disease infects a bed, it is virtually impossible to instantly kill it and stop it from destroying already infected plants, and it is not easy to stop it from spreading to uninfected plants, thus the emphasis above on prevention. When a disease shows up, review and act on your cultural practices (see numbers 1–6 in Site Selection and Cultural Practices on page 111), even as you do what you can to slow its spread. With respect to #5—removing mulch and replacing mulch—because removing the mulch (along with all the infected plant tissue) and placing new mulch on beds that have been infected with blight or damping-off is so labor intensive, some growers instead spray their infected beds (mainly their mulch) with a fungicide after about 20 percent of the plants have emerged in the spring. Any of the foliar fungicides applied against Alternaria or the damping-off pathogens are reported as somewhat useful. (Ontario growers have found that Bravo 500F is particularly effective for this early spring spraying, probably in part because its sticker, which comes already incorporated, is so effective in bonding the chemical onto the mulch and young plants despite spring thunderstorms. A natural alternative is Champ DP, a copper hydroxide that is labeled for spraying over only dormant ginseng (therefore only as a 100 percent pre-emergent), since it is both a fungicide and an herbicide. You also may want to consider initiating regular fungicide applications or changing the application regimen you're already undertaking.

When a root rot shows up, try to isolate it if practical, as described in #3, and transport the infected plant material and associated debris off your farm. Essentially, you are trying to limit your losses. A period of dry weather is the best control for damping-off and most root rots.

Dealing with Rodents, Insects, and Other Critters

I have never had my own crop threatened by a severe infestation of insects or rodents, but I know other growers who have had serious pest problems. Be alert to anything that might threaten your crop. No pests are known to specialize in ginseng. Your local agriculture extension agent, or even neighboring gardeners, probably will be familiar with, and able to help you control, whatever critters attack your green gold.

Rodents

All manner of furry mammals will roam through your ginseng beds, usually without doing more than nuisance damage. One summer, a zoology student from North Carolina State University asked to live-trap my patch, and he caught moles, voles, shrews, gray and red squirrels, flying squirrels, chipmunks, and rabbits. None of these rodent trespassers seemed to rate my roots very highly on their list of preferred food, although I suspect most of them considered the berries to be a great forest delicacy. I have had nuisance rodent problems in other years but never serious losses. Rodent predation is site specific, probably depending on population densities and the availability of more desirable alternative food sources.

Mice, in particular, go after the seeds as soon as the berries begin to ripen—deer mice sometimes even earlier while the berries are still green. Their presence is revealed by barren seed spikes and small chewings of berry leftovers on the ground. Rabbits will not eat the tops, but they will make a path across a bed, cutting down every top in their way. You will walk into the patch in the morning and see a strip littered with wilting tops and cut stems sticking up about an inch above the ground. It's aggravating, but hardly a disaster. Remove the dead tops or they may host diseases.

Mice will sometimes use mole tunnels to burrow within a ginseng bed and eat a few roots. The moles, themselves, are after worms, not ginseng. However, the woodland vole, *Microtus pinetorum*, a creature the size of a mouse but with the tunneling abilities of a mole, loves loose, loamy, tilled soil, because it is easy to dig through. He also loves to eat the roots of the herbaceous plants growing in it, including ginseng.

The gluttonous woods vole looks like a cross between a mouse and a mole. Photo courtesy of USDA.

You'll know woods voles are around when you find the thumb-sized holes in the ground that are the entrances to their tunnel systems. You'll know they're feeding on your green gold when you find the three-pronged foliage of a mature ginseng plant pressed flat against the mulch—a vole will have dug under it and pulled the stem down into his tunnel as he ate the root. The voles' tunnel system includes little cul-de-sacs in which they will cache far more than they can eat of abundant food in anticipation of lean times. Should you succeed in producing them in abundance, ginseng roots will be cached in these cul-de-sacs. Voles have small feeding ranges, but they do a good job of harvesting what's available therein.

Many growers report zinc phosphide to be the vole bait of choice, but fish-flavored Ramik Green (diphacinone), Rozol (chlorophacinone), and other common rodenticides are also known to be effective for voles and mice. (Check for what's presently approved for use in your state.) Whatever poison is used, some pellets are placed right into the tunnels near the flattened ginseng tops that show where the voles are active, and some pellets are also left above ground in bait stations. Some growers put out rodenticides all year round. The best method I've seen is to place the poison bait in the holes of a brick, cover the brick with a plastic bucket, place a cement block on top of the bucket, and then cut small holes in the rim of the bucket for access on all sides by only very small creatures at ground level. Your poison is thereby kept dry and out of the ground, dogs and cats and other larger animals can't get to it, and you can move it around to trouble spots. Pellets placed in the middle of plastic tubing, one-and-one-half inches in diameter, also works well. A few growers find they need to set up rodent barriers around the perimeter of their beds. Others set out traps. One fellow I know keeps a stable of slightly underfed cats. Personally, I have enjoyed the unsolicited assistance of screech owls, great horned owls, and Cooper's hawks, all of which nest near my beds from time to time. Wait and see what, if anything, attacks in force, and then determine your own tactics.

Browsers and Scratchers

In addition to rodents, other four-legged pests include deer, cattle, goats, and hogs, all of which will eat 'sang leaves and tear up beds. Dogs can damage plants by running over them, and they may even spread disease as they regularly trot through the beds.

Fencing is the obvious protection against most of these troublemakers (as well as rabbits). White tailed deer, however, will jump anything lower than eight feet, so fencing them off is usually impractical, and they will browse the ginseng foliage and the berries (even while they are still green). I know of no perfect solution to deer predation, but I can suggest some rather cheap, simple approaches: run a cotton rope around the perimeter of your beds and spray it with deer repellent such as Liquid Fence, Hinder, or Deer Off; treat strips of cloth with repellent and hang them either from the rope or from trees around the perimeter; stuff hair from a barbershop in stockings and hang them up; punch a few drainage holes in the bottom of one-gallon plastic milk cartons, cut part of the top off, put a bar of Irish Spring soap in, and use the handle to hang them. These will turn back most of the deer initially, but they eventually lose much of their effectiveness. It is probably most effective to regularly spray your entire planting with a taste repellent. That is rarely practical; however, you may be able to just treat the plants on the perimeter.

Recently wild turkeys have become a much more common problem for woodland 'sang growers. Turkeys have an uncanny ability to locate newly mulched ginseng plantings (either wild-simulated or woods-cultivated). The newly deposited mulch is apparently much easier to scratch aside than the normal leaf litter covering the forest floor. Sometimes the turkeys will go after seeds or tender young plants or (less often) the berries, but they are just as likely to scrape away your mulch to feed on something else. Unfortunately, not only does their feeding behavior cause damage where they scrape, but serious losses also occur immediately below the bare spots, where the mulch gets piled up and either prevents plants from emerging or smothers ones that already have emerged. Turkeys do not fly frequently, and sometimes a three-foot-high, chicken-wire fence will deter them from entering your beds. Also, for a limited period of time, almost any source of erratic movement or sound will cause them to wander elsewhere, but you have to keep changing the deterrent. String up aluminum pie pans, flagging, or just plastic grocery bags so they will move in the wind, and a radio blaring is temporarily effective as well. You can usually get permission from the wildlife authorities to shoot turkeys, even out of season, if you can show damage to your crop. After about two years, as the mulch adheres to the soil, your planting site will lose much of its attractiveness, and turkeys will become less troublesome as your ginseng ages.

Insects and Slugs

In different parts of the country, various insects occasionally do significant damage to ginseng, although serious insect problems are rare in the woods. Aphids, leafhoppers, lygus bugs, jumping plant lice, tree crickets, scale insects, cutworms, the four-lined plant bug, and slugs have all been reported as pests. Incidentally, gypsy moths and forest tent caterpillars—foliage gluttons that often are present in the woods in great numbers—have been observed to die before they would eat ginseng leaves. Growers in the Mid-Atlantic states may also have a problem with a newly recognized fly (or wasp—entomologists have been unable to identify and classify it) that lays its egg in the ginseng stem. The resulting maggot then eats and hollows out the stem, eventually causing the top to die, which shortens the plant's growing period for the year and usually means that no seeds are produced. As the top dies, growers sometimes mistake the insect's work for a disease's.

Particularly in the moist early spring, slugs and snails may attack the ginseng foliage. (Slugs are much more likely to do significant damage, especially to young plants.) Photo by Mark Haskett.

Any blanket use of an insecticide is likely to eliminate natural enemies of pests that will only then become destructive. Some insects feed heavily on fungi that cause ginseng diseases, so it is important to first identify whatever is causing damage before taking any protective action (be tolerant of a few holes in the leaves here and there). This is the one area where an ounce of prevention is not worth a pound of cure. Know that your enemy exists and is a genuine threat before taking action—perhaps in accordance with the information provided in Table 7 on the opposite page. In addition, the list of natural insect controls is endless. It includes beneficial bacteria, viruses, and other insects, as well as minerals, oils, and plant extracts. Bordeaux mix is very repellent to many insects, especially leafhoppers. If you can identify your pest, there is almost surely a natural defense.

Slugs, and occasionally snails, are much more frequently a problem than any particular insect, although they are less a problem in the woods than in shade gardens—probably because they have more natural enemies in the woods, and because most woods growers do not use the straw mulch that slugs seem to love to live under. Of course, slugs aren't insects, but mollusks—members of the land snail and sea shell phylum. They will munch the ginseng leaves, and, in late fall or early spring, will eat seeds left on the soil surface under the mulch after planting. Slugs will even destroy exposed root tops. They are most dangerous when they eat the meager foliage of young seedlings whose roots do not have the stored energy needed to survive and develop a bud for next season. You can monitor the density of your slug population by placing orange or grapefruit halves, cut-side down, in your beds and checking underneath them the next morning. If slugs are around, many of them will be under the fruit. There are many effective slug baits, both natural and synthetic. I prefer

Table 7. Suggested Ginseng Pest Control Guide*

Pest	Effective Materials	Comments
Aphids	Diazinon, Orthene, Malathion, Pirimor, (not Sevin, it kills the wasps that are aphids' natural enemies)	Examine leaf undersides. Treat as needed.
Leafhoppers	Diazinon, Sevin	As needed—watch for population build-up at alfalfa harvest.
Mice (and voles)	Zinc phosphide, treated grain or apple slices, (Ramik Green, Rozol)	Keep bait dry and inaccessible to other animals.
Lygus bugs	Diazinon, Sevin	Treat if one or more per ten feet of plant bed.
Scale	Diazinon, Orthene	When crawlers are active.
Slugs	Mesurol (preferably pellets), metaldehyde	Spread bait as needed (especially early in spring, after rains).
Cutworms	Mesurol, Sevin, Pounce	When present in numbers.
Jumping plant lice	Diazinon, Sevin	Look for white cottony material on flower spike or stem.
Tree crickets	Diazinon, Sevin	When present in numbers.
Four-lined plant bug	Pounce	When present in numbers.
Moles	Poison peanuts, mole traps	Step on a mole run, and if it has been repaired by the next day, place bait or trap there.

* Prepared by C.F. Koval, Department of Entomology, University of Wisconsin-Madison. The materials listed here have demonstrated efficacy against the pest problems indicated on ginseng and other crops. However, few are presently registered nationally for use on ginseng. The grower assumes all responsibility for such use.

Deadline M-PS (4 percent metaldehyde), which is nationally labeled for ginseng and remains effective for 45 days or six inches of rain (whichever comes first). Physical barriers of ashes, coarse sand, or diatomaceous earth are all somewhat effective in diverting slugs from your beds.

Pest Control Guide

Table 7 is a pesticide guide developed for ginseng growers in Wisconsin. I have made some additions in parentheses. Most insecticides, like the majority of fungicides, are not registered nationally with the United States Environmental Protection Agency for use on ginseng. Growers who apply insecticides often mix wettable powders in the same solution with fungicide(s), and spray both on leaf surfaces at the same time. (A natural pest control guide is listed under Other Related Literature in chapter 10.)

Other Aspects of Care During the Growing Season

Weeding

A few forest weeds are part of ginseng's natural environment, and many growers are convinced that weeding disturbs the natural balance and increases the incidence of disease. No controlled studies have been undertaken to test this hypothesis, but I think it is correct. That does not mean, however, that weeding is necessarily bad. When woodland soil is tilled, weeds sometimes respond with vigorous germination and growth. Thick weed populations are not part of ginseng's natural forest environment and may increase the likelihood of disease by reducing air circulation and by out-competing the ginseng for soil nutrients and sunlight. Of course, some competition from weeds is natural and probably contributes to the wild look of the roots.

A few weeds, then, may be desirable and certainly do not require elimination, while weeding a choking stand of undesirables is appropriate. You can greatly reduce potential rank growth of weeds by either applying a weed killer or repeatedly tilling the soil during the summer before planting and/or by mulching carefully over the entire bed surface after planting. Dr. John Proctor of the University of Guelph, Ontario, Canada, reports that Poast herbicide (Sethoxydim) will kill all narrow-leafed weeds (i.e., grasses), but not the ginseng (or the many broad-leafed forest weeds). Fusilade (fluazifop-butyl), like Poast, is a post-emergent grass killer, and it now has a national label for ginseng, though its use is prohibited in the harvest year. Dr. Konsler found that to kill broad-leafed weeds, 2,4-D (2,4-dichlorophenoxy acetic acid) could be safely sprayed over ginseng, once its leaves were fully expanded (i.e., after June). Growers also effectively spray herbicides on weeds just after the ginseng dies off in the fall.

If and when you do weed, it's best to pull the offenders up by the roots. The ground needs to be soft and wet to do this efficiently, however, and such wet conditions are hospitable to disease organisms that may be brought to the surface as the weeds are uprooted. I never weed until after the first hot and dry spell of late spring has reduced the risk of damping-off. Some growers are careful to spray their beds with a broadband fungicide immediately after weeding. One 'sang farmer I know allows the weeds to grow well above his green gold and then whacks them off with a weed eater. Although not as effective as uprooting, if timed right this procedure will prevent the trespassers from going to seed.

Weeds are the most troublesome in beds of young plants. If you do need to weed, it will probably only be during the first few growing seasons. Your mulch will hinder the establishment of new weeds, and in beds of older plants, the leaf area is often thick enough to block much of the light and greatly reduce the competition.

Irrigation

I have never irrigated my beds, and I suspect that few, if any, woods growers need to. However, there have been instances of late-summer drought when I would have liked being able to irrigate my beds. In 1988, we had the driest July and August in 30 years. Although I lost very few plants, I did find growth and berry production greatly reduced that year. The best root growth and seed production are obtained where there is ample soil moisture without wetness.

In woods cultivation, the feeder roots of shade trees permeate the soil a few years after bed preparation is completed and soon begin to compete with the ginseng for soil moisture. This is undoubtedly one of the reasons woods-grown 'sang takes longer than field-grown to attain saleable size. Thus, while irrigation would occasionally be desirable (most often in late summer when berries are developing and dry spells are common), it also might make your roots more closely resemble cultivated roots and worth much less. Consider irrigation only if your layout makes it convenient and cheap, and be careful not to completely saturate the soil.

Blossom Picking

A great deal of the energy of the ginseng plant goes towards the development of berries. If the seed spike and the blossoms are snipped in early summer, berry formation is prevented and additional energy is available for root growth. Exactly how much blossom snipping adds to root weight is known only for three- and four-year-old plants in artificial-shade gardens: 25–30 percent increase in root weight annually. I suspect that in woods-cultivated beds the annual gain would be similar. Since the benefit should be compounded as the growing seasons go by and the beneficial effect of blossom picking is widely known, it is somewhat surprising that more woods growers don't blossom snip. If you are not interested in expanding your production or in selling seeds to a neighbor, then blossom picking is a simple, natural, though highly labor intensive, way to increase root growth.

A Rough Production Budget

I hesitate to present a budget for woods-cultivated 'sang farming because each individual's circumstances are unique, and every successful grower I know does some things at least a little differently. There is certainly even more variability than for either of the other two growing methods. In particular, the labor expended varies greatly, depending on planting density, each individual grower's work rate, and his approach to the tasks required. However, I recall when I got started how interested I was in hearing experienced growers discuss their costs. (There was—and still is—precious little well-documented information on the economics of woodland ginseng culture, despite its long history in this country.)

My own experience and the observation of, and conversations with, other successful growers are the basis for the figures presented in Table 8. They include only the more essential items and assume that some basic tools and equipment are already on hand. No serious setbacks are anticipated, otherwise the projected yield is conservative, not a best-case scenario. Table 8 is based on a seven-year time to harvest, which is usually the minimum number of seasons needed to produce roots with something of a wild look. It also assumes a seeding rate of approximately eight seeds per square foot (about 48 pounds of seed per acre), which is the densest planting that I would risk. If you chose to plant less densely—probably a good choice,

Table 8. Projected Seven-year Budget for One-Half Acre of Woods-cultivated Ginseng

Seed*	
24 pounds at $120/lb.	**$2,880**
Labor	
Site preparation and planting: 300 hours × $15/hr	$4,500
Care and maintenance: 1,000 hours × $15/hr	$15,000
Harvesting seeds and roots: 650 hours × $15/hr	$9,750
Total Labor	**$29,250**
Materials and Equipment	
Chemicals (primarily fungicides but also rodenticides, herbicides, insecticides, fertilizer, gas and oil)	$1,000
Rear-tined tiller for bed preparation	$1,000
Backpack sprayers: 2 × $150	$300
Garden seeder	$75
Total Materials and Equipment	**$2,375**
Drying	
Addition of insulation and drying racks to existing room or shed	$600
Energy cost to heat (50¢/lb. of dried root)	$150
Total Drying	**$750**
Total Cost	**$35,255**
Total Expected Yield : 300 lbs. of dried root	
Gross profit: 300 lbs. × $225/lb.	**$67,500**
Net profit at end of 7 years	**$32,245**

especially for beginners—then you will reduce the cost of seed. Moreover, while you are certainly likely to harvest fewer roots from a thinner planting, the wider spacing makes delaying harvest less risky, and by growing additional years, your root weight and root quality will increase. Of course, your total costs will also increase with each additional growing season, as will your risk of loss. There is no one best practice. The rough budget projections in Table 8 are intended only as a first approximation of your actual costs. (Other woods-cultivated production budgets are presented in several of the references listed in chapter 10, under Ginseng Farming.)

Root Production

A root yield of 300 pounds from half an acre is a reasonable expectation from a dense planting, but your yield will depend on the many contributing factors discussed earlier, plus luck. Yield is extremely unpredictable, as is root quality, until you have had a harvest on your particular site. The price per pound of $225 (which is intentionally a little conservative) assumes roots that are clearly distinguishable from field-grown roots but also largely distinguishable from wild. It is as low a price as you are likely to get without dropping down to the price of field-grown roots, which can happen if you only grow for six years. If you choose to plant less densely and harvest at the end of eight or nine years, you have a good chance of harvesting a higher grade of roots.

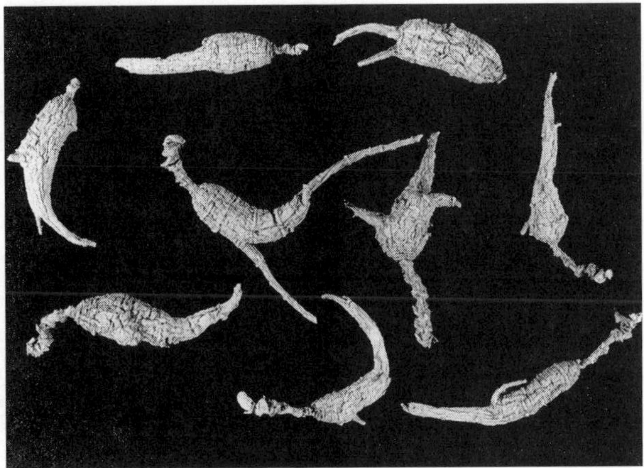

Sample of older, high-grade woods-cultivated roots from a lot that sold for approximately half the going price of wild roots. Photo by Mark Haskett.

Materials and Labor

Besides seed, you will probably need to buy various pest-control products, possibly a little fertilizer, perhaps a quantity of gypsum, and maybe some mulch if you need to supplement your leaf litter. At harvest time, you'll also need clean sand, screen or hardware cloth, a little lumber both for stratification boxes and drying racks, plus a small heater or stove of some sort for your drying area. If you decide to fence your plots, add fencing to your costs, and perhaps some security devices. You will also need a few tools, many of which may already be on hand: a shovel, an axe or pulaski, both a leaf rake and a sturdy garden rake, buckets, hammer and nails, probably a backpack sprayer for foliar spraying, perhaps a chain saw, and a small tractor or more likely a tiller (unless you favor really hard work with a mattock). You can rent or borrow the tiller, since you will need it only when you prepare the beds for planting. In addition, you may want to acquire a garden seeder.

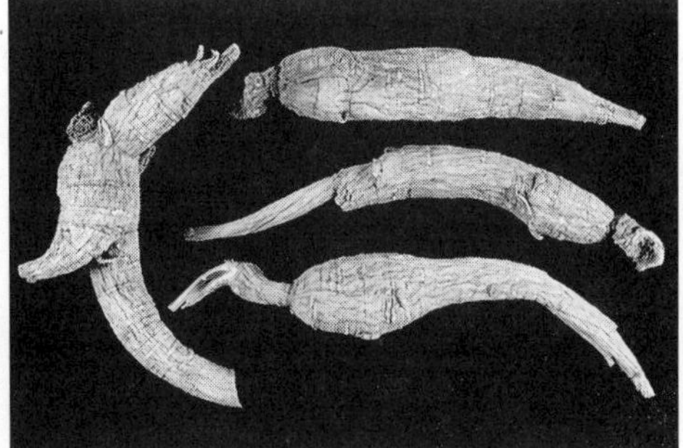

Sample of four-year-old, heavily fertilized, woods-cultivated roots from a lot that sold at the high end of the price range for field-cultivated roots. Photo by Mark Haskett.

As far as labor goes, during the growing season you should be able to manage as much as an acre of ginseng on an after-work, every-other-day basis with some extra time spent during occasional weekends. If you grow on a larger scale, you will probably need help, particularly at bed preparation and harvest time. If you do the work yourself, your labor costs will depend upon the value you assign to an hour of your own work.

A Grower's Calendar

Table 9 presents a ginseng grower's calendar of all the important tasks required for intensive woods cultivation, listing the materials and equipment needed to accomplish each activity. The best time for completing each task will vary somewhat depending on geography and climate, but the range of dates suggested is broad enough to be appropriate throughout the country. If you conscientiously complete all the work outlined in the calendar at the appropriate time, you stand a good chance of becoming a successful 'sang farmer.

Seed Sales

An acre of fourth-year or older woods-cultivated ginseng plants has the potential to produce 50 pounds or more of seed each year. This potential is not often realized, however, mainly because the grower must compete with the critters for his seed. In time, any reasonably successful woods-cultivated grower will certainly produce more than enough seeds each year for his own needs, but don't anticipate producing a large surplus of seeds for sale early in your growing career. (This is why seed sales were not included in Table 8.)

Unless you have neighbors interested in becoming 'sang farmers, selling any surplus seeds will likely prove much more difficult than selling your roots. I know from experience that advertising and other marketing expenses eat up a high percentage of the profit until you have a large quantity (at least 100 pounds) of seeds to sell. Also, marketing seeds requires careful record keeping, and a great deal of time spent talking with potential buyers. (The same requirements apply if you are interested in selling seedling roots.) The price of seed varies greatly depending both on seed quality and on the ginseng root market. To get a good price, you will have to develop a reputation for supplying healthy, stratified seeds that consistently germinate at a high rate. An alternative is to sell all your seeds wholesale (at about half the retail price) to one seed dealer who will then resell them. In sum, the seed market is not as stable or as readily available to the small grower as the root market. Aim first to produce enough seeds for your own planting needs, and consider any extra seed sales an unexpected bonus.

The Bottom Line

So what's the bottom line? Well, one way to look at it is that you can raise half an acre of ginseng with an expected value in seven years of about $67,500—not to mention the value of any seed you will harvest along the way. Your capital outlay will be about $6,000 ($2,880 of which is paid for seeds that in future years you'll probably produce yourself, with another $1,900 going to one-time disbursements for a tiller, sprayers, and drying racks). This perspective certainly suggests that there is good reason to refer to American ginseng farming as

Table 9. Ginseng Grower's Calendar for Intensive Woods Cultivation

Appropriate Month(s)	Tasks (in Order of Occurrence)	Materials and Equipment Needed
Nov–Aug	Site selection and clearing, (soil sampling for analysis)*	Axe and shovel or pulaski, (chain saw)
Jun, Jul	(Reduce perennial weeds and undergrowth)	(Tiller or herbicide and backpack sprayer)
Jul–Sep	Bed preparation (and soil amendments)	Leaf rake, tiller, axe or pulaski, shovel, and garden rake (gypsum, shredded leaves)
Sep–Nov or March	Planting, mulching, and record keeping	Seeds, garden seeder or shovel rake (mulch), paper and pencil
Oct–Apr	(Fence construction)	(Brush, or posts and wire—about 300 linear feet per 1⁄10 acre)
Year round	(Rodent control)	(Rodenticides, covered bait containers, and/or traps, barrier materials)
Nov–Apr	Replace mulch on wind-blown areas	
Apr or May	Remove all but two inches of mulch	Leaf rake
Apr or early May	Inspection (and treatment) of any seeds in stratification	(Fungicides,Trichoderma, or household bleach)
Apr or May	(Pre-emergent weed and disease control, top dressing with calcium)	(Herbicide, fungicide, backpack sprayer, and gypsum)
May–Sep	(Regular foliar spraying for disease prevention)	(Fungicides, organic fungus controls, back-pack sprayer)
May–Sep	Walking inspection of beds every 2 or 3 days	
May–Sep	Diseased plant removal (and isolation from healthy plants) for disease control; also sending soil sample and diseased tissue to labs for analysis	Garbage bags, plastic bags, (household bleach, hydrogen peroxide,watering cans)
May–Sep	(Insect control)	(Insecticide or natural controls, backpack sprayer)
Jun–Aug	(Weeding)	
Aug–Oct	Berry picking and depulping in third and succeeding years	Buckets or sacks and fine clean sand
Sep, Oct	Remove year-old seeds from stratification, water test (and treat for disease) before planting	Screen table and bucket or tub, (10% bleach solution, fungicides, or trichoderma)
Oct or Nov	Begin new seed (treatment and) stratification	Sand and stratification box (10% bleach solution, fungicides, or trichoderma)
Any time before harvest	Construct or adapt a drying room or shed before roots are dug in quantity	Wood, screen, hammer, insulation, staples, nails, etc., thermometer, heat source, (exhaust and circulating fans, dehumidifier)
Sep, Oct	Dig beds and wash roots	Shovel or spading fork, buckets, tubs or washing table, water flow
Sep–Dec	Drying	Drying room or shed with racks and ventilated trays
Oct–Apr	Storing and selling	Clean cardboard boxes or barrels

* Tasks and materials or equipment enclosed in parentheses may not be necessary.

growing green gold. On the other hand, if you look at your hours of labor as capital outlays, then the potential annualized net profit is not so outstanding, considering the investment of time and money and the uncertainties involved. Incidentally, ginseng farmers commonly deduct expenses when they are incurred, but do not pay taxes until after they harvest and sell their crop.

6

The Harvest: Picking Berries and Stratifying Seeds, Digging and Drying Roots

The Ginseng Berry

The first harvest from your garden of green gold will be the crimson kidney-shaped berries, which develop in late summer. Beginning in June, ginseng blooms and forms fruit over an extended period. During this reproductive phase, there may be flowers, green berries, and ripe berries simultaneously on different plants within the same planting. On an individual plant, berries tend to ripen in three flushes, beginning at the lower outside of the seed head and proceeding over time to the center top (see photo on the next page).

Ginseng plants generally do not flower and form berries until at least their third growing season, when most three-pronged plants will produce berries unless conditions are unfavorable. Typically, 80–90 percent of flowering plants will develop berries, although this can drop below 50 percent under prolonged hot and dry weather. In a wild-simulated setting, it may be seven or eight years before the majority of the plants are three pronged and producing seeds. In the woods, an older three-pronged plant usually forms about 15 berries; but, on a four-pronged plant growing in tilled soil, the seed head commonly contains 30 to 40 berries with two seeds (occasionally one or three or rarely four) in each berry. Thus, one healthy mature plant can yield 60 to 80 seeds annually.

Honeybees, bumblebees, and sweat bees all have been observed collecting pollen on ginseng flowers. Ginseng is, however, definitely capable of self-pollination, since blossoms enclosed in plastic bags have produced viable seeds. The wind is probably the chief pollinator, but bees should not be discouraged from helping out, since there is evidence that cross-pollination results in greater reproductive success. If the human hand does not intervene, any berries that are not eaten by the critters will shrivel and fall to the ground. In the natural setting,

As their enclosed seeds mature, green ginseng berries (dark gray in this photo) ripen to a bright red (light gray). Typically, this ripening begins at the bottom of the berry cluster and progresses to the top during about a three-week period starting as early as late July. The large triangular red (light gray) berry near the center of the higher cluster probably contains three seeds rather than the usual two. Photo by Tom Konsler.

less than two percent of ginseng seeds survive, germinate, grow to maturity, and in turn reproduce.

Picking the Berries

Ginseng berries are harvested by hand. There is no effective mechanical device to help. Only red berries should be picked. Green berries are much less likely to be viable because their seeds are not sufficiently developed. In the woods, failure to pick berries as they redden will allow rodents and occasionally birds, deer, and even porcupines to make off with more than their share of seeds.

An acre of ginseng under artificial shade often produces more than 200 pounds of seed each year, but you would do well to harvest 20 pounds of seed from an acre of mature wild-simulated plants. Indeed, on a wild-simulated site, it may not be cost effective to pick berries at all. Each plant's production is usually modest, and critter predation can be extensive. The rate at which berries mature differs from year to year depending on the weather, but the majority usually turn crimson over a three-week period. A single picking, after nearly all the berries have had a chance to ripen, is usually all that is worthwhile. In woods-cultivated gardens, on the other hand, some berry picking is almost always worth the effort. Since the woodland critters will often start dining as soon as the first berries ripen, three separate pickings may be called for. Extracting only ripe fruits from the cluster requires care and delicacy to avoid breaking the seed spike and thereby losing the remaining seeds. Small hands are an asset, and women often do a better job than men. Alternatively, you can take the wild-simulated approach and just wait until nearly all the berries have ripened and then pick what the critters have left for you.

If you are a beginner who is starting out small, your plants probably will not produce a great many seeds. Nevertheless, at the end of your third or fourth growing season, take the time to pick whatever berries

Berry picking in a woodland garden of mature plants at Harding's Ginseng Farm. Photo courtesy of Larry Harding.

are produced so that you can go through the stratification procedure (see below) and learn to do it properly before you begin to harvest seeds in any quantity. Incidentally, you have to pick roughly one gallon of berries to obtain one pound (just over one pint) of seeds. There are exceptions, but a pound of seeds commonly contains between 6,400 and 8,000 seeds. (The seeds from older, larger plants tend to be bigger and result in fewer seeds per pound.)

Depulping the Berries

Although not all growers do it, extracting the seed from the berry's flesh before stratification is a good idea. The fruit will take up a great deal more room than just seeds in whatever storage container you use or construct—a bushel basket of berries contains only about one gallon of seeds (about 60,000, or about eight pounds). Furthermore, putting seeds in moist storage surrounded by their pulp strikes me as an ideal methodology for the attraction and growth of all kinds of potentially harmful micro-organisms.

There are several methods of depulping freshly picked berries. The old-fashioned traditional way, which I have used for many years, is to fill a burlap bag or a bucket almost one-half full of berries and mash and churn them thoroughly. If they are in a burlap bag, you can mash them quickly by trampling on them. In a bucket, you can mash the berries with your hands or with a hand-held implement. I repeat the mashing and churning several times over the period of a week to ten days, but some growers skip that. During this time, keep the mush of seeds and pulp protected in a cool shaded place, such as a shed or basement. Fermentation will soon begin, and the pulp will disintegrate.

After most of the pulp is gone, to remove the remaining fleshy detritus, dump the smelly seeds in a bucket of water. The last bits of pulp will float to the surface along with, hopefully, only a few seeds (which are traditionally referred to as "floaters"). About 95 percent of the floaters are partially hollow and not viable because the kernel did not develop fully or was at least partially destroyed by disease or some other natural cause and does not completely fill the seed husk. Skim the floaters and the bits of buoyant pulp off the water surface and throw them away. Before the sunken seeds absorb too much water and are ruined by bloating, carefully dump most of the water on the ground (not in a stream or pond, where it may create a biological oxygen deficit), and then lay the seeds that have sunk to the bottom of the bucket out on a hardware-cloth table (a window screen supported by two sawhorses will do). Drain and air dry the seeds for several hours in the shade or other cool location with good air circulation. Sometimes it is helpful to briefly repeat water immersion several times to get only clean seeds remaining in the bottom of the bucket.

When I float my seeds to complete the depulping process, I usually use a 10 percent household bleach solution (1 part chlorine bleach to 9 parts water), instead of plain water, and I am careful to let the seed stay in the treatment solution only between 10 and 12 minutes. Then I immediately dump the seeds on a hardware-cloth table in the shade and hose the bleach off. This treatment does an excellent job of killing any fungus or bacteria on the seed coat. However, research has shown that a very few fungal spores can survive on the seed coat or husk by being trapped in microscopic air bubbles within little microcaves on the surface of the husk. Also, a bleach treatment does not affect the inside of the seed, where pathogens may be lurking in complete safety.

Finally, I leave the wet seeds out in the shade to drain and air dry for a few hours to the point that I can run my fingers through them and not have to dry the excess water off my hand. Then I pour them into burlap bags and store them temporarily in my air-conditioned shed. Any cool, secure spot will do. If you allow seeds to stay wet, they will rot or mold. The burlap bag wicks remaining excess moisture from the seed husks, which is good; but ginseng seeds can dry out and lose their viability, which is bad. I only keep them in the bags for a day or two (much longer, and they will begin to desiccate), until I can pick up a handful of seeds, squeeze them, and find—on relaxing my grip—that they don't tend to stick together or to my fingers. At that point, I put my seed into stratification (see below) as soon as practical. Incidentally, freshly depulped seeds are often referred to as unstratified or "green" seeds, though in fact the husks are a light beige or cream in color.

Another simple depulping method takes longer, but is a bit less trouble. I have used it successfully when I've had small quantities of seed to process. I put my red berries in buckets or tubs in a shed, alternating one-inch layers of fine clean sand with half-inch layers of berries, beginning and ending with a layer of sand. I lay a damp towel on top of all the layers. In six weeks or so, the berries decay, exposing their seeds. Until then, the pulp keeps the seeds sufficiently moist. When the flesh has nearly completely deteriorated, the seeds will begin to dry out and lose their viability. I once forgot for several months about a bucket of sand and berries from one day's picking and lost most of those seeds. So, periodically, I brush away the covering layer of sand to check on the decay process.

When nearly all the pulp is gone, I dump the sand-and-seed mixture onto my hardware-cloth table and hose away the sand through the mesh. I then water-test the seeds as above by briefly dumping them into a 10 percent bleach solution. I throw away the floaters, wash off the bleach, drain them, and proceed with stratification.

Finally, Canadian growers have developed a machine that separates seed from pulp. Increasing numbers of United States growers have used it in recent years. Machine-separated seed is thought to be a little healthier, because the shortened separation process reduces problems associated with fly larvae, bacteria, and fungi growth. The depulping machine (see chapter 9 for equipment sellers) is a modified grater that uses water to rinse away the berry pulp as it is scraped off the seed.

Canadian growers usually store their depulped seed in burlap bags in a cool

Depulped seeds ready for stratification. Photo courtesy of Tom Condon.

(45°–50°F) location, occasionally adding moisture, until the soil temperature gets below 60°F. Only then do they treat their seeds, drain them again, and put them into stratification. If soil temperatures are too high when the seed is buried, the Canadians think there is a risk that a percentage of the seed will germinate the first spring, and there is no good way to separate the early sprouters. I have never had a problem with early sprouting, but two other southern growers have told me that, on one or two occasions, when their berry harvest has been early, some of the seeds from their first picking did sprout the first spring and were lost. It makes sense to separate and plant the early ripening seeds from your first couple of harvests to find out whether a significant number will germinate a year ahead of schedule. If you do get much early sprouting, you can mimic the Canadians and hold your seed in a cool location until the weather and your soil cools, or you can plant your first picking right away.

A limited market for berry juice has developed very recently, no doubt accelerated by new research showing that ginseng berry pulp has antidiabetic properties. One grower I know depulped his berries with a cider press and then canned and sold 400 quarts of the berry juice (and, of course, also stratified the seed).

Immediate Planting

After seed is depulped and air-dried, it is usually stored in a stratification box for a year, although seeds can be planted immediately. I think it unlikely, but you may find that you want to keep the first batch of seeds you pick each summer separate, and plant them right after depulping. There are other legitimate reasons to plant immediately and risk leaving the seed scattered in the ground the necessary 16 to 22 months before germination. You avoid putting all your eggs in one basket—that is, all your seeds in one stratification box—where they might all be lost to some unforeseen disaster, such as disease or even theft. By planting the same fall that you harvest, you can spray your newly planted beds with herbicide once or twice the next summer (while the seeds are in the ground, still developing), thereby greatly reducing future weed problems. This is especially helpful if you mulch with straw, which usually contains many weed seeds. Finally, you can mulch heavily after planting and allow that mulch to decompose into organic nutrients for the extra year before your seedlings emerge and utilize that additional nutrition.

Nevertheless, nearly all experienced growers store their seeds for a year in a stratification box before planting. Consistent losses of seeds planted immediately are probably due to rapid temperature fluctuations on the soil surface as well as varied other natural causes, including critters foraging under the mulch or leaf litter. This has convinced most 'sang farmers that storage for a year in a protected, relatively ideal environment is well worth the trouble. If you start producing more seeds than you have time or space to grow, you will almost certainly have to stratify them to sell them, although you'll probably need a considerable surplus to make this very profitable. (See Seed Sales at the end of chapter 5.) Folks simply do not want to buy seeds that will take a year and a half to sprout!

Stratifying Your Seeds

Ginseng seeds do not, for the most part, sprout the spring following picking, but remain dormant for 16 to 22 months until their second spring arrives. They must be "aged" for a year before they will germinate in the next spring following a fall planting. The common method of aging seeds is called stratification and involves mixing the seeds with fine, moist sand and storing the mixture underground for a year in either a plastic or a wooden container. If the traditional wood-frame box is used, both the top (or lid) and the bottom of the box are covered with a screen or hardware-cloth mesh (⅛"), which allows rainwater to seep down and moisten the sand and seeds, but also allows good drainage through the bottom, and keeps the critters out. If a plastic bucket or tub is used, then sufficient holes (smaller than the seed) must be made in both the top and the bottom to allow for good drainage. For small quantities of seed, simply fold a piece of aluminum screen over on itself and staple or wire the two opposite sides together to form a pouch, then fill it with the seed/sand mixture, and close the open side securely.

At least twice the volume of clean sand as seed should be mixed thoroughly with the seed, and the mixture then poured into the stratification container. A top layer, several inches thick, of only sand will insure that seeds near the surface do not dry out. The sand is crucial, as it has the fascinating ability to drain quickly but retain just the amount of moisture that 'sang seeds need to stay viable and continue to develop.

It is possible to stratify seed in sand above ground in a well-controlled environment, where the temperature is kept at about 35°F for nine months, followed by 60°F for three-and-a-half months, before fall planting. Moisture content must be monitored as well as temperature, since there will be no rain to add moisture. A 15 percent moisture level is ideal, but 25 percent or more can kill the seeds. A $200

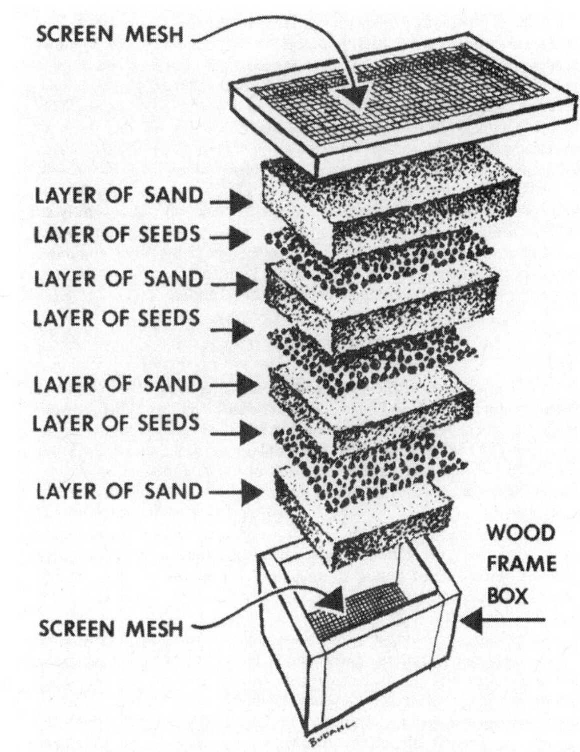

Stratification of seeds in the traditional layered manner. Alternatively, seed and sand may be thoroughly mixed together.

tensiometer can be used to monitor moisture content; but, since sand won't retain too much moisture if drainage is provided, experienced growers just regularly check the top layer of sand to make sure the entire medium remains moist, and the seed doesn't dry out. Still, there is some inherent risk involved, and, not surprisingly, most growers bury their stratification box outside in the shade, relying on Mother Nature and the proven properties of sand.

I always select a burial site in well-drained soil on a slope. Robert Beyfuss suggests that you verify the good drainage

Wood stratification box with year-old seed ready for planting. Note that some of the seeds are "smiling" encouragement to the grower. Photo by Mark Haskett.

of your prospective burial site by digging a one-foot square and one-foot deep hole, filling it with water, letting the water drain out, then immediately filling it with water again. If the second filling drains out in less than 30 minutes, then the location has good drainage and it is safe to store your seed there. Stratification boxes, buckets, or tubs should be buried deep enough so that the top layer of seeds (not the container, itself) is at least eight inches below the soil surface. Shovel dirt on top of the box until the hole is filled level with the soil surface. At this depth, soil temperature is usually close to ideal and does not vary much. Ginseng seeds are subject to both premature and delayed germination as a result of hot or cold temperatures (see Germination Requirements below). If you are to the point of burying several boxes, it is prudent to place them in different locations in case problems occur. Don't forget to throw on some leaf litter to hide your work and then mark the location(s) inconspicuously.

It is a good idea, especially for beginners, to inspect the seeds occasionally. Stratified seeds are most vulnerable during extended periods of extremely wet or extremely dry weather. After such periods is a good time to examine your buried treasure. The beginning grower should definitely look at his seeds after the first wet spell in the early spring. Feel the sand for moisture, and cut open a few seeds with a pocketknife. If some are spoiling—that is, if they are soft, mushy, and foul smelling, or have mold or fruiting bodies (sporodochia) growing on the husks—then either you have not stratified them properly, or they were infected with disease when you buried them. If you check your seeds and find no problems, it is okay to rebury the box at once. However, it is easy, once the container is excavated, to stir the sand and seed—which will aerate the mixture—before reburying. Since I have had no history of seed spoilage, I now leave my seeds unattended for the entire stratification period.

The next fall, when you are ready to plant (or sell), uncover the stratification container and separate seed from sand. I spread my seed/sand mix on a hardware-cloth table and hose the sand away through the mesh. Next, I repeat the same process that I use when I finish depulping. I treat the seed in a 10 percent solution of bleach for ten to 12 minutes, skim the floaters off the surface, pour the seed back on the table and thoroughly hose the bleach off. I let the seeds drain in the shade until they dry to the point that I can run my fingers through them without getting my hand soaking wet, and then scoop them into burlap bags.

A rule of thumb is that for every seed that floats, there is another one in the process of going bad that didn't float; thus, if more than five percent of your seed are floaters, you may want to throw all your seeds away and not plant bad, quite possibly diseased, seed. I have never had many floaters; but if I did, before throwing the seed away, I would first cut a sample of the floaters in half with a pocketknife to make sure that most were indeed bad seeds—that is, the kernel was shrunken or rotted. If most of the kernels were firm and white and filled the entire husk, and there were no reddish sporodochia growing on the husks (see Seed Diseases below), then I'd certainly refloat the seed that floated and see if they sank on a second try, indicating that the problem was slight temporary dehydration, not disease or incomplete development.

Plant the seeds as soon as possible, and let Mother Nature take care of them. Inevitably, on occasion, you will need to hold seeds for weeks before planting. Hold them in a cool (below 60°F), shady, secure location (even in a refrigerator, but not in the freezer) and mix them up every few days to aerate and maintain an even level of moisture throughout. Add moisture when the husks begin to lighten in color. Seed can dry out rather quickly if stored in a burlap bag, because the burlap will wick away moisture from the seed and then the burlap itself will dry out. The outside of the bag may need to be sprayed with water every few days to prevent too much moisture from being wicked out of the seed. If you are holding only a small quantity of seed, it may be convenient to place them in ventilated plastic bags. (I have used Ziploc bags with small holes punched in them.) Before you place them in plastic, be sure seeds do not stick to your fingers when you squeeze them, as wet seed will mold.

During their year in storage, the seeds will darken, increase slightly in size, and some of them will split—that is, the husk will open slightly along a seam on one edge of the seed so that the white kernel is visible. The seed is then said to "smile." Smiling, or split, seeds are a good sign that storage conditions were suitable and that the seeds are developing normally. It is common, however, for only a few seeds to be splitting when the stratification container is emptied in the fall, and many viable seeds often remain closed until early spring. (A very few will not split and germinate until their third spring.) If your seeds are not splitting after a year in stratification, look for a thin black line along one edge of the seed husk. If this line is clearly visible on most of your seeds, then the embryos are probably continuing to grow satisfactorily and your seed will germinate on time. Only if no seeds are smiling on Thanksgiving would I begin to be concerned.

Germination Requirements

New, or green, seeds usually require a year and a half spent in a moist (but not wet) medium with exposure to a cold/warm/cold sequence of temperature changes (corresponding to winter/summer/winter seasons) before they will germinate. The temperature changes stimulate embryo growth. These requirements are met naturally by seeds that fall from the plant, work their way down into the soil, and remain there for 20 months until sprouting in the spring. Growers usually pick and depulp their berries, extracting the seeds, in the fall, store them for a year in moist sand (stratification), remove the seeds from stratification and plant them the next fall to sprout the following spring. The Ontario Ministry of Agriculture, Food and Rural Affairs booklet reports that if seed is subjected to temperatures near freezing after picking but before stratification (such as during storage in a too-cold refrigerator), there is a risk of delayed germination—that is, some seed may not emerge on schedule in the spring 20 months later but may delay germination another 12 months. In addition to time and temperature requirements, I am convinced that the percentage of seed germination also is affected by the amount of moisture in the soil during the early spring. Fairly wet, but not soggy, spring weather seems to produce the highest germination percentages.

Dr. Tom Konsler studied the effects of cold temperatures on the sprouting rate of stored, one-year-old, seedling roots, and I suspect his results are relevant to seeds as well. Dr. Konsler found that most seedlings will sprout in the spring, if they have been exposed to at least 45 days of temperatures at 36°F or to at least 60 days at 48°F. Without cold stimulation below 49°F, the roots did not sprout regardless of exposure time, although they did not deteriorate either. These findings probably explain why ginseng does not grow in subtropical climates and why stratification at room temperatures has been unsuccessful.

Dr. John Proctor of the University of Guelph has found that *P. quinquefolius* seeds can survive at least 24 hours of exposure to temperatures as low as 5°F, if the seeds do not have a high moisture content. This is not true of roots, however. They begin to suffer damage as the temperature drops below 14°F (or if subjected to repeated cycles of freezing and thawing). These findings probably explain why ginseng does not grow north of southern Ontario.

Many attempts have been made to encourage American ginseng seeds to sprout in just eight to ten months. Most efforts have involved simulating nature's year-and-a-half seasonal temperature changes within a ten-month, or shorter, period. To the best of my knowledge, while occasional limited success is not uncommon (especially in controlled environments), a method for consistently causing a high percentage of green seeds to sprout the first spring in field conditions has not been developed. (Asian ginseng seeds are regularly being germinated and induced to grow in just eight to ten months.)

Seed Diseases

There are at least five known fungi that infect American ginseng seeds: Rhizoctonia, Botrytis, Fusarium, Cylindrocarpon, and Pythium. It is not fully understood how and when these pathogens actually infect the seeds, and they are difficult to detect until infected seed begin to decay. When decay occurs during stratification, growers are able to eliminate bad seed by water testing them and removing those that float. Unfortunately, if the disease is not advanced, the kernel remains largely intact, and the seed will not float.

Rhizoctonia and Cylindrocarpon can be detected, even in seeds that sink, by cutting the seed in half and looking for any orange areas in the otherwise white interior. When seed is infected with Pythium, its interior remains white, but it looks distinctly crumbly and cheesy. If you cut a seed and a spray of foul-smelling, milky or black liquid comes out, then bacteria has probably rotted that seed, but it does not threaten to spread disease to a newly planted garden. Seed that is hollow when cut may have been the victim of Botrytis, which thoroughly rots the interior, causing most of them to float.

Seed infected with Fusarium and Cylindrocarpon produce fruiting bodies (called "sporodochia") on the external husk. These are usually colored in the pink-to-red range and can be detected by close eye inspection. Each fruiting body contains vast numbers of spores, which will be introduced into the site if the seed is planted. Those that do not float need to be removed by hand to the extent practical. If your floated seed has more than scattered individual seeds with fruiting bodies, you may be better off not planting the seed.

Large-scale commercial shade growers often have some seed problems and will typically plant their seed if less than 15 percent float. I would be very concerned if more than five percent floated, because fungi reproduce themselves in massive numbers and a small percentage of infected seed could soon result in widespread disease. A recent Ontario study of arbitrarily collected seed from commercial sources found a high level of infection by fungal pathogens, even in seed that had been surface sterilized. Plant stands six weeks after seeding were only 17 percent of the seeding rate.

Seed Treatments

Infection and spoilage can be the result of improper stratification or of a poorly chosen burial site, but fungal diseases are certainly also found in properly stored seeds. To protect seeds from fungi, many growers utilize household bleach, as described above. Another widely applied seed treatment is to mix one ounce of formaldehyde per gallon of water and soak seeds in this solution for 20 minutes before and after stratification (see the OMAFRA Publication 601 for more details). If undertaken, this soaking should also serve as water testing, because repeatedly soaking seeds is not a healthy practice, especially for 20 minutes, and most especially if many of them are already split.

After seeds are brought up from stratification in late summer or early fall, it is a common practice among shade growers to first soak them in a formaldehyde or bleach solution. After soaking, they wash off the solution, let the seeds dry off a bit, and then cover them with a fungicide (in powder form): Thiram (tetramethylthiuram disulfide), Captan (a multi-chemical formulation), or Trichoderma, using only enough to lightly coat the seed (one ounce of wettable powder with 50 percent active ingredients per 25 pounds of seed). The British Columbia Ministry of Agriculture's online "Ginseng Production Guide for Commercial Growers" recommends this procedure (see chapter 10, Websites of Interest for web address). It is intended not only to help prevent contaminating beds with disease-carrying seeds but also to prevent good seeds from being attacked by fungi between planting and emergence. Also, natural mycorrhizal seed inoculants (such as MycoApply) are reported to be effective and to even continue protection against damping-off during the plant's first growing season. [Author's note: The same labeling-for-use-on-ginseng issues apply to the use of seed treatments as to pesticides—see Pesticide Use in chapter 3]

Digging the Roots

Many farmers develop an appreciation, even affection for their crop. This is especially true of ginseng farmers, because they protect and nurture their plants for many years before they finally are able to dig their green gold. When I harvest a particularly nice root, the greedy part of me says: "That's worth two dollars. Throw it in the bucket and grab the next one!" Another aspect of my personality wants to hold the root a moment, admire its nature, and take satisfaction in what I have grown. Digging ginseng roots is hard work that puts you on your hands and knees grubbing in the dirt, but it is also an enjoyable, exciting time—especially that first harvest.

The best time to dig ginseng is in the fall after the tops have at least begun to die down, indicating that growth has stopped for the season. As the foliage senesces, it adds about 2 percent additional weight to the roots. To harvest early in the summer sacrifices the plants' last seed crop as well as additional root weight. Also, roots definitely dry with a better look when dug soon after leaf senescence. Roots dug and dried in the spring or early summer, tend to have undesirable vertical creases and are recognized and disparaged by buyers as "spring roots." Even during senescence, the root's

In his woodland ginseng gardens, Scott holds a mature American ginseng plant. Photo by Mark Haskett.

chemistry is changing. There is evidence that the medicinally active compounds, the ginsenosides, are at their highest levels in the fall.

If you have grown your ginseng in beds, rake aside the mulch and dead or dying tops before digging. It is a good practice to completely remove this residue from your ginseng farm, if practical. Whether you have grown in beds or in a wild-simulated manner, begin at the bottom of the planted area and dig the roots, being careful to do as little damage as possible without taking endless time. A noticeable number of badly mutilated roots will reduce the value of your entire lot of 'sang. A few undersized roots are acceptable, but an abundance of them is not. If runts are common, pick them out and consider transplanting them, though the labor required makes that a marginal proposition. On a wild-simulated site, just leave the runts in the ground to continue growing. It is unknown whether undersized roots are stunted primarily because of genetic inferiority or primarily due to competition.

Artificial-shade growers with large operations harvest their crops with a machine similar to a potato digger. A tractor pulls it through the beds and deposits most of the roots on the soil surface where they are easily gathered by hand. A second pass through the bed completes the task without a back muscle being strained. Unfortunately, the slope, trees, stumps, and large tree roots in the woods usually make such machines impractical, although a few woods-cultivated growers have modified their equipment to do it. I have even seen a backhoe used effectively in the woods, but you will probably have to dig the hard, slow way by hand with a shovel or spade fork or even a large flat-headed screwdriver. The best I am able to do to reduce wear on my own back is to hire college students (a nonresident population that is less likely to return and poach) to help me with the digging. Their music blares across the usual tranquility of my valley, but they've gotten my roots dug. Many woods growers enlist family members or day laborers (sometimes risky) to help at harvest time. It is easier to dig the day after a rain, when the soil is moist and loose, and you may need to till or rake the site after digging to reduce erosion.

Freshly washed roots draining on a screen-bottomed table. Photo courtesy of Tom Condon.

Washing

There is a growing market for fresh roots, but unless you have adequate cold storage space or a delivery contract in hand, it is usually prudent to wash and dry your harvest. Once roots are dug, toss them into buckets and keep them in a shady, cool, preferably breezy spot until they are washed later the same day. Although shade growers now routinely place their roots in refrigerated storage for two to six weeks before drying them, this is tricky business and not necessary for high-grade, woods-grown roots. If you need to wait a few days after digging before washing your roots, put them either in open containers with a moist towel laid on top or in burlap bags that you have sprayed briefly on the outside with a hose. Keep the roots cool (preferably below 50°F) and secure. Monitor them daily to make sure they do not start to dehydrate, and move them around gently for aeration. At the first hint of foul odor or the first sign of mold, wash and dry the roots. One bad root, like one bad apple, can quickly spoil the entire lot, so it is wise to keep separate any roots damaged in digging.

To wash, lay your harvest out on screens or dump it into tubs and clean the loose dirt off with running water. A pressure hose is commonly used, but make sure the force of the water is not strong enough to abrade the skin of the roots. More water under less pressure is much better than less water under more pressure. Abraded skin reduces the value and makes spoilage more likely during the early stages of drying. Don't scrub and don't try to thoroughly clean every speck of dirt off—soil left in the ridges and crevices, especially in the horizontal rings, is desirable. If your roots have a sheen to them, they've been over-washed. Do remove the debris, such as mulch, bits

A cement mixer makes an excellent ginseng root washer. Photo by Al Oliver.

of tree roots, bark, etc., that has a tendency to get caught in the small feeder roots. Buyers will pay a high per pound price for roots, not for dirt and debris. After washing, allow your roots to drain in the shade for an hour or less before you start to dry them. (A small cement mixer makes a good root washer, and there are mechanical root washers specifically designed for ginseng on the market—see chapter 9.)

Drying

Wet roots will soon begin to rot if air drying is not started immediately after washing and draining and then continued without interruption until dehydration is nearly complete. Small roots dry in a few days, but large ones may take up to six weeks, depending on the heat and efficiency of the drying method used.

Improper drying can greatly reduce the value of ginseng. Large-scale growers commonly use modified tobacco kilns for drying. On a small scale, simpler, more primitive procedures can get the job done well. Successful drying operations vary widely among growers. What follows here are general guidelines and requirements.

The Drying Room or Shed

You can dry a few pounds of fresh roots in any area of your house that stays warm at night and is protected from drafts of cool wet air on rainy days. Most successful ginseng farmers find that a room or shed, which is designed or adapted to dry the harvest each fall, is needed. The drying area should be well insulated, and most growers provide a dependable source of heat, a means of circulating air, and a way of controlling

Drying rack with moveable screen-bottomed trays. Photo by Kim Fadiman.

humidity. It is furnished with drying racks—that is, frameworks into which large shallow trays filled with fresh roots can be slid in and out. The frames usually are built against the walls like bookcases, with extremely broad removable shelves or trays spaced at least 8" apart from floor to ceiling. The trays are sometimes as large as 4' × 12' and 3" deep, always with a screen or lath strip bottom for aeration. Since drying may only take a few weeks, you will want to be able to utilize the area for other purposes during the rest of the year.

Expose to Moving Air

During the dehydration process, moisture evaporates from the roots into the surrounding air. To facilitate this moisture exchange, roots are spread thinly on shallow drying trays to maximize surface area exposed to air. Newly washed roots can be piled on top of each other two or three inches thick only in a high tech, forced-air drying operation with ideal conditions for quick dehydration at temperatures above 90°F. (The British Columbia "Ginseng Production Guide for Commercial Growers" has a design for such a system.) In a more primitive operation, roots should be laid out on the trays one layer thick with only the fine feeder roots touching.

If only a few pounds of roots are being dried in a large volume of air space, then speeding up air circulation is not a concern. A few roots can simply be laid out to dry on newspapers in a warm room or on a northeast-facing window ledge for exposure to morning sunlight in good weather. If, however, a large volume of roots is dried in a small area, humidity will be high, and the air must constantly be moved over the roots as they are drying. A large electric fan is the usual method of rapidly circulating the air.

Regulate the Temperature

The air temperature at which roots are dried is important. Everything else being equal, hotter air is drier air, and the drier the surrounding air, the more rapidly moisture in the roots will evaporate. At temperatures below 60°F, it is very unlikely that dehydration will be completed before the roots begin to spoil. Thus, you will need not only a well-insulated drying area that retains heat, but also a source of heat for regular use or at least available in an emergency such as a spell of cold, wet weather.

Any source of dry heat—wood, gas, oil, electricity—can be used. If possible, do not locate the heat source right next to your ginseng or you may dry the outside of the near roots too rapidly, discoloring (darkening) the skin and hardening the exterior before the interior is dehydrated, thereby greatly reducing the value. Placing the heat source in front of a fan that blows the heated air over the roots works well. Heating costs are reduced if you dig and dry your roots early in the fall before the nights get cold, and if you locate the drying area so that the exterior is exposed to as much sunshine as possible during the day.

Once drying has begun, monitor the air temperature with a thermometer, and, if practical, either maintain a constant temperature of at least 70°F or gradually raise the temperatures to as high as 90° or 95°F over a period of days or weeks as the roots dehydrate. Lowering the temperature may cause moisture in the air to be taken up again by the roots and that, in turn, will not only slow down the process but also may cause mold or mildew. Partially dehydrated roots have lost most of the moisture near their surface, and increasing the air temperature, and/or the air flow, will help volatilize the moisture deep inside large roots. An inexperienced grower should not allow the air to get above 95°F, as higher temperatures can darken the outside of the root and caramelize the interior, markedly reducing value.

In lieu of, or in addition to, gradually raising the air temperature as drying progresses, growers raise their roots. That is, when drying begins, a tray of freshly dug and washed roots is slid into a slot near the floor (where any last dripping will splatter harmlessly). As time goes on, the tray is slid out and moved up to progressively higher levels in the drying rack, until it arrives on top where dehydration is completed. This raising-the-roots method works because the air near the floor of a drying area usually is about 10°F cooler than the air near the ceiling. Raising the roots has the advantage of allowing the grower to dig his roots over an extended period, sliding a day's harvest into bottom racks after moving up trays of roots dug during preceding days. Thus, freshly washed roots do not drip on already partially dry trays.

Roots of roughly the same size can be grouped together on trays, and columns of similar trays moved up at a rate appropriate to their size, with columns of trays of larger roots progressing upwards more slowly. If trays are progressively raised in this manner over several weeks, once roots have reached the top rack or tray, it's usually safe to dump them on top of each other two or three inches thick and let them finish drying. After a maximum of six weeks in the drying area, almost any root should be dry and ready for sale or storage.

Replace Wet Air with Dry Air

When many pounds of roots are dried in limited space, the air becomes saturated with moisture before the roots are completely dehydrated, and they can begin to rot or mold. Either fresh air must be exchanged for the saturated air through intake ventilation and an exhaust fan, or the moisture must be removed from the old air by a dehumidifier. Either technique works well, and they can be used together.

When roots are dried during cool rainy weather with no dehumidifier present, the humidity in the drying room may rise, and the air may become wetter per unit of volume than the interior of nearly dried roots. When that happens, the roots may reabsorb moisture and begin to mold or mildew. Thus, in cool, rainy weather, it's a wise precaution to crank up your heat source and raise the temperature in your drying area as much as 10° (but not over 95°F), which lowers the humidity in the air and helps volatilize the moisture remaining in nearly dried roots. If mold is ever detected, expose the affected roots through a closed window to the first available bright sunlight for several hours on both sides.

Keep an Eye on the Process

Once your roots are safely laid out in your drying rack, you still cannot relax. Cool temperatures, excess humidity, mold, mildew, malfunctioning equipment, overheating, and even the possibility of a fire in such a dry situation are all reasons to keep a close eye on your green gold.

You should inspect your roots twice over the first 36 hours of drying, and turn them at least once to prevent water from accumulating where the root touches the tray bottom. The first noticeable change is the roots shrinking and shriveling as moisture leaves their surface. The roots look wilted and pathetic, and they feel limp and rubbery during this period. As long as there is no sign of mold or mildew and no pungent odor of spoilage, do not be alarmed. As dehydration continues, the smaller roots harden quickly. The larger ones continue to shrink, and very gradually become less rubbery until they, too, eventually harden.

Your roots are ready to sell when you cannot bend them, and they break with a snap. The interior should be white and have a slightly waxy texture. Before storing, break and carefully inspect a small sample of your largest dried roots to make sure that their central core is properly dehydrated. (Actually, there should be about eight percent moisture remaining in the root, but you wouldn't know it.) If the broken cross section shows a thick brown circle around

Tony Hayes (center) of Ridge Runner Trading Company in Boone, North Carolina, showing dried roots to a buyer for a Hong Kong auction house (right) and his Canadian associate (left), before packing the roots in cardboard barrels for shipment and resale in Hong Kong. Photo by Sandy Hayes.

a white center, then the roots were exposed to too much heat too rapidly. Between 3 and 3½ pounds of fresh roots eventually make one pound of dried roots. There are typically anywhere from 80 cultivated to 270 wild-simulated roots in a dried pound.

Feedback

Buyers do not hesitate to give you feedback (and a lower price) if your roots are poorly dried. The more knowledgeable buyers will have a good idea what you must have done wrong and will suggest how you can make improvements. Try to get some experience with a small lot or two before the year when you have a large valuable harvest to dry.

Storage

Store your dried roots in rodent-proof containers, away from all dampness, until you are ready to sell them. Clean cardboard boxes or barrels are commonly used for both storage and shipment. Don't put your roots in plastic bags or other airtight containers for an extended length of time, because, if even a few of the roots are not sufficiently dehydrated, without ventilation they may spoil and jeopardize the rest of the lot as well. However, when roots are shipped, most growers do seal them in large plastic bags inside cardboard barrels for protection from humidity or excessive dehydration and weight loss while in transit.

Dried 'sang keeps fine at normal room temperatures or slightly cooler, as long as the atmosphere is not humid. If placed in a damp basement or cellar or beside an open window in rainy weather, the roots will absorb water from the air and will mold.

Since a barrel of green gold has considerable value, it is wise to hold your harvest inside your house or in some other secured building. Some insurance companies will insure your roots against fire and theft. Ginseng has a long shelf life, but hoarding for more than a few months runs the risk of loss; plus, unless you have good reason to think the price will rise, it's like keeping money under your mattress.

7

Business Decisions and the Future Market Outlook

Raising ginseng is an undertaking that a patient person who likes to grow things will find not only interesting and challenging but also potentially profitable; working against you will be the need to make early business decisions before you've had a chance to gain experience. Some of those decisions will not be obvious. This chapter provides a little guidance to help you make those choices.

Which Method

American ginseng is one of the world's most valued herbs, and growing green gold is occasionally advertised as a get-rich-quick scheme. Be assured that it is not. In the words of Donan Jenkins, a highly successful, full-time Kentucky 'sang farmer, now retired:

> There are 98 ways I know of to grow ginseng, and only two ways out of 100 not to. The two ways not to are FAST and EASY. There have been thousands of pounds of seed scattered over the countryside by scheming individuals who intended to get rich quick and easy. This idea simply has never worked in the past, is not working now, and will not work in the future.

Most of the information you need in order to decide how you want to grow ginseng is contained in the preceding chapters. To summarize the most important points: The great majority of the ginseng produced in North America is cultivated in open fields under artificial shade in large-scale, capital-intensive, highly mechanized farming operations, sometimes under corporate ownership. Shade farming on a small scale is a risky economic proposition since rapidly expanding artificial-shade plantings have exponentially increased production and until quite recently dramatically reduced the profitability of that method of ginseng farming. I am only comfortable encouraging growers to employ the artificial-shade method for seed production, for personal consumption, to supply some niche market that they develop, or simply as a horticultural challenge. If you do decide to try it on a small scale, then much of the following advice regarding woods cultivation should be relevant. If growing under artificial shade on a large scale remains your choice, then go to References for some excellent information detailing proven, commercial, shade-growing practices.

The alternative advocated here to the artificial-shade method is to grow this native perennial in its natural woodland environment. Woods growing requires only a small initial capital investment, it utilizes uncleared land that would probably otherwise be idle, it does not destroy the forest, and it can be undertaken profitably on a small scale. Moreover, the roots produced in the woods differ from the field-grown variety and do not directly compete with them in the marketplace.

The price of wild and wild-simulated roots has resisted any downward pull caused by the oversupply of field-grown ginseng. Since top-grade (wild-looking) woods-cultivated roots usually bring between 30 and 50 percent of the price paid for wild roots, their price has been rising as the price of wild roots has been rising. Low-grade woods-cultivated roots usually bring the same price as the best artificial-shade roots.

If your woodlot holds wild 'sang or an abundance of its companion plants, then all options are available to you. If your site is not one where wild ginseng is likely to thrive, then neither my wild-simulated method nor a large scale woods-cultivated planting is a good option for a commercial operation (though I'd never discourage someone from sowing a couple of ounces of seed as an experiment in almost any woodlot). The Hankins Method will give you a chance for success on a less than ideal site. If you have over five acres of prime potential planting ground, consider a combination: bedded gardens near your residence to insure a seed crop, with more extensive wild-simulated plantings eventually covering all your suitable sites. However, do not get involved with cultivated beds (in the woods or under shade) if you do not want to regularly spend quality time with your plants. If you anticipate extended stays (over ten days at most) away from your ginseng during the growing season, you will have to arrange for someone to care for your crop while you are absent. Understand that all farming is a risky proposition and ginseng farming particularly so because your crop has to survive multiple growing seasons.

Where to Obtain Planting Stock

It is still possible to get started in the ginseng business by transplanting legally dug wild roots. You would then have to pick, depulp, stratify, and eventually sow the seeds of your transplants, thus delaying your first harvest several more years. Moreover, since there is not an abundance of wild 'sang left, and you'd have to spend a great deal of time searching it out, this is not a practical option for most people. The usual approach is to purchase seed (or possibly transplanting roots) from an established grower.

It is not difficult to search the Internet or look in the classified ads of alternative lifestyle, sustainable agriculture, or various other farm-related publications and find ginseng seed for sale, offered directly by either growers or resellers. It is difficult, however, to be sure that the seed you buy is quality seed that not only has been depulped, stratified, stored, packaged, and shipped properly so as to insure that it remains viable, but also is largely free of the diseases described in chapter 6 that are too often carried in and on ginseng seed. Advertised representations as to quality and viability should be read with a healthy skepticism. Caveat emptor, or "buyer beware," applies when purchasing ginseng seed.

Fortunately, good seed is obtainable if you do a little homework and check out the source. Keep in mind that any growers are going to plant their best seed themselves and sell their surplus. Thus, you have to find a supplier who has mastered the techniques of stratifying and shipping seed and who either has nearly disease-free gardens (not common, especially among artificial-shade growers) or has the integrity not to pick berries from plants growing in beds with disease present and to discard seed if there is evidence of disease after stratification. Ideally, you would like to find growers who have established an excellent reputation among folks who have planted their seed and whose gardens are close enough that you can personally inspect them in mid-summer, when the berries are forming, to confirm that there is little disease. Such a local grower could supply you not only with good seed but also with local expertise and advice. In addition, other things being equal, there is probably a slight benefit from using seed that is harvested from plants that have adapted over a few generations to the local growing conditions. Indeed, if there is a local supplier whose stock has been entirely propagated from wild roots transplanted years ago, there may be a great benefit from planting those seed, which have evolved for many millennia to survive in your environment.

Since it is uncommon to have a well-established 'sang farmer as a local resource, seed is usually bought by mail order based primarily on the basis of word-of-mouth reputation. Actually, price is too often given the primary consideration, but buying disease-carrying seed at a cheap price is no bargain and a common cause of failure. The profitability of your ginseng planting does not depend on whether you pay one cent or two cents for a seed that grows into a two-dollar root. Consider price only when all else seems about equal. Contact your local agricultural extension agent. He or she may very well be aware of growers in the area who, while they may not be producing surplus seed for sale, have found a reliable seed source. Although ginseng conferences and workshops are not common, a few are usually held each year in the eastern United States, and one may be within reasonable driving distance. In short, evaluate potential seed suppliers by consulting folks who have planted the seed, not those who are trying to sell the seed.

Most reputable sources of seed will allow you to immediately return seed for a refund if they arrive spoiled or otherwise unacceptable. However, once you plant the seed, they are yours. There are no money-back guarantees should they fail to germinate or all your young plants succumb to disease during the first few growing seasons. It used to be possible to buy a sample of seed and have them tested for viability and/or diseases before purchasing a larger lot. Unfortunately, the only private lab I know of that presently does such testing of ginseng seeds asked not to be mentioned in this book, because they are phasing out of that business. Some land-grant colleges still may offer seed testing services for a fee, when they have qualified personnel on staff. (Contact your county extension agent to determine if your state's school offers this service.) Some seed suppliers with good reputations are listed in chapter 9, but that is very incomplete list, which will no doubt soon become outdated. It is intended as a backup to your own research.

As soon as you receive an order of seeds, inspect them. If they smell foul, reveal a milky or cheesy center when cut open, or have sporodochia growing on their husks (see Seed Diseases in chapter 6), return them for a refund, not for replacement. When I first started, I twice had to return bad seeds. If you order seeds by mail and do not know that the grower has already treated your order before shipment for fungal contamination on the seed husk, it's wise to do so yourself. There are several treatments to choose from, as discussed under "Seed Diseases." The simplest is probably to soak seeds for no more than 12 minutes in 1 part chlorine bleach to 9 parts water, and flush with fresh water.

New planting stock should be put in the ground as soon as possible. Have your site prepared and be ready to plant when your stock arrives. If you need to store seeds temporarily, do so in your refrigerator (not in the freezer), or in another place where the temperature remains between 36° and 50°F. Stock usually is shipped in plastic or burlap bags. Open and inspect these every day or two. If the seed husks are beginning to lighten in color, spritz them with water to maintain moisture, and either carefully shake or gently pour the contents into a second bag to aerate and prevent molding. Inexperienced growers should plant purchased seed as soon as possible due to its sensitivity to moisture level.

If you receive your seeds in the spring, a few may begin to germinate as early as February 1. (Spring planting is rarely attempted north of Virginia and Kentucky.) After sprouting, it is difficult (but possible with care) to plant seeds without some damage. The radicle (which becomes the root) that first sprouts from the seed is easily broken and especially subject to drying out. Storage in a refrigerator does not effectively slow germination.

What Kind of Initial Planting

In considering how you want to get started, two time-honored clichés apply: "Don't bite off more than you can chew" and "Don't risk more than you can afford to lose." I have been selling planting stock to prospective growers for 25 years now, and I have often seen those nuggets of common sense ignored, sometimes despite my clearly expressed concerns. There have been a very few instances when I thought my conservative advice was foolishly ignored, only to see the new grower quickly have large-scale success. Much more often, though, the inexperienced pay a price in time, labor, and capital for their impatient overreaching.

Before sending off your initial order for planting stock, you'll have to decide whether you want to either just stick a tentative, experimental toe into ginseng farming or take a modest risk and go large enough initially to have a chance for a little supplementary income. I trust that no one reading this book will immediately quit their job to become a 'sang farmer. Though ginseng economics are attractive, wait until you have had some success and know what you are getting into before you consider changing your vocation.

An Experiment

With a seed order of just two ounces (at least 800 seeds), you can plant several test plots, experimenting with wild simulation and/or woods cultivation on different areas of your woodlot and with different seeding methods. Plant the seed on your sites, see how it performs, evaluate your prospects, and then proceed from there.

On a wild-simulated seeding, a conservative approach would be to wait as many years as it takes for the majority of your surviving plants to have two (or more) prongs. From that stage, chances are good that most of them will last to maturity and harvest. If you are averaging one or more plants per square foot at this point, consider the experiment as a success and a more substantial investment as a good gamble. While you wait for your experiment to run its course, continue to plant a few ounces of seed each year (assuming your initial seeding looks encouraging). Learn to improve your seeding technique or experiment with variations on it, and test the marginal areas of your woodlot.

If you are running an experiment in woods cultivation, then an even more conservative evaluation is needed, because problems often do arise even after the plants reach the two-pronged stage. You have to get your experimental planting through to at least the sixth year and at least the three-pronged stage before you can be confident that preparing 5,000 square feet or more of beds each year (which I believe is about the minimum area needed to produce a net profit) is a truly wise investment. Of course, few folks have this much patience. I know I didn't. Just be aware that the failure rate is high and, until you have gained the experience of growing in tilled woodland beds from seed to sale, the odds are probably not weighted in your favor.

Do not be discouraged too easily, however. Even if your survival rate is disappointing, if your plants that do survive have grown vigorously and have desirable roots, then your site may be suitable. You will have to learn how to bring a higher percentage through to maturity.

Aiming for a Little Supplemental Income

To make at least a little supplementary income, you'll need to harvest enough pounds of root to compensate yourself not only for the cost of planting stock and for the labor of site preparation, planting, care, protection, harvesting, drying, maintaining equipment, and selling; but also enough to compensate for the time spent going and coming from house to woods, obtaining supplies, and learning from this book and other sources. To get past the break-even point requires investing in about six pounds of seed (not necessarily in the same year). If you choose not to try an experimental planting first, your effort and investment will be at considerable risk, even more so if your site is not ideal or not secure. Of course, if you place little or no dollar value on your own time and labor—perhaps regarding it as exercise and recreation—then these guidelines are not appropriate for you.

Andy Hankins believed that a grower using his wild-simulated seeding method (see end of chapter 4) can be profitable planting only one pound of seed and then harvesting the results in nine years. I'm not convinced Andy was right about that, but I'm certain he's right if a pound is planted each year for six years. I've found that my own wild-simulated method of seeding becomes potentially profitable once a quarter of an acre (over 10,000 square feet) is planted, which requires six pounds of seeds sown either all in one year or over several years. The loosely estimated harvest from six pounds of seed planted in a wild-simulated manner and harvested after nine years of growth is just under 40 pounds of dried roots (worth about $27,000 in today's market), which projects to a net profit of better than $20,000. Obviously, larger plantings increase the potential profit, but they increase the risk as well.

For a woods-cultivated operation, the economics are highly variable and depend on how thickly you sow seed, how many years you grow before harvesting, and the quality of the roots that your site and your techniques produce. The best general guideline I can give is that sowing six pounds of seed into 5,000 square feet or more of actual bed space should be enough to generate a profit. If you hire labor, your estimated net profit is roughly $7,000; but, if you pay yourself for the necessary labor, you'll in effect earn perhaps an additional $7,500 dollars over six plus years. Because regular trips for care and protection are required throughout the growing season for a woods-cultivated garden, and because it does not take, say, four times as many hours to look after 20,000 square feet of beds as it does to look after 5,000, there is significant economy of scale to be gained by larger woods-cultivated plantings. However,

The late Kenneth Harding (right), founder of Harding's Ginseng Farm, and Glenn Hutzel harvesting another berry crop. Kenneth's son Larry now directs the Hardings' 40-year-old woodland operation. Photo courtesy of Larry Harding.

while one person can maintain many acres of wild-simulated ginseng, once your total woods-cultivated bed space begins to approach an acre, then the demands on your time will probably start to exceed the hours you have available on a part-time basis.

Whether to Produce Your Own Seed

Anyone growing ginseng to maturity will be able to harvest some seed, but woods growers are likely to experience limited seed production. The woodland critters will dine on your berries, only the oldest plants will yield many berries, and you will often be digging your roots the first season they are fully mature and producing a large berry cluster. Wild-simulated growers normally have to purchase seed each year for new plantings, and this is also true of some woods-cultivated growers. Thus, if they want to consistently obtain significant seed production, woods growers may have to set aside some of their plants from harvest and give them special long-term care and protection.

When I began growing green gold, I initially ordered seed from growers all over the country, even from Canada, searching for the best planting stock. The packages that arrived on my doorstep contained seed of dramatically varied quality, but I soon found a couple of sources of good seed. Nonetheless, I determined early on in my ginseng-growing career that I wanted to start producing my own seed, for several reasons: to avoid the annual expense of purchasing seed; to insure a dependable supply of seed (my original sources of quality seed soon either died or retired); to have seed to sell to other growers; and to improve my planting stock by transplanting selected roots into permanent seed-producing beds. All of these reasons proved in time to have merit, and for many years I strongly encouraged woodland 'sang farmers to set up their own permanent seed-producing beds. However, I have come to realize that many growers are probably better off simply buying whatever seed they need each year.

When Seed Production Is of Questionable Value

To maximize berry harvest, plant seed-producing beds in well-tilled rich soil (perhaps amended with gypsum) on an excellent site and then weed, care for, and protect it conscientiously. Even in a woodland garden, growers will need to inspect their plants frequently, remove plants at the first sign of disease, and prudently undertake a disease-preventative program (see Disease Prevention in chapter 5). This can be a time-consuming regimen, which requires learning the proper disease-prevention techniques, and it may all be wasted effort if the grower is away for an extended period of time when disease attacks prized plants.

Growers using the woods-cultivated method for larger planting operations will need to learn and exercise proper disease prevention. They will already be tending their beds during the growing season. For the wild-simulated grower, on the other hand, constant care and protection of the plants (other than security) are not required, and maintaining seed-producing beds represents considerable additional time, effort, and learning. Most wild-simulated growers who do not plan to eventually plant at least four or five acres would seem well advised to go ahead and buy seed as they need it. For large-scale wild-simulated growers the choice is not so obvious. If you have the potential to plant large acreage, and you plan to do this over an extended period, then it may indeed be worth your while to set out permanent long-term seed-producing beds to supply yourself with cheap, dependable, quality seed (see photos of Scott's permanent seed-producing beds in the color photo section).

Seed-producing beds are not useful to every woods-cultivated grower either. Growers with limited land area suitable for ginseng may prefer to go ahead and buy seed and get the entire site planted, rather than wait the several years it will take to start producing their own seed. They may never have any more need for ginseng seed because replanting in beds is so hazardous. (See Replanting in chapter 5.) Most successful large-scale woods-cultivated growers do produce their own seed, but many of them find they do not need to set aside special beds of older plants for seed production. Once their operation has been ongoing for six or seven years, they may harvest enough seed from their mature plants each year to meet their planting needs the following year. Regardless, a grower with a large-scale woods-cultivated operation may still choose to have permanent seed-producing beds in order to have surplus seed for sale and/or to improve planting stock. Also, there is the new (but certainly not well established) berry juice market.

Setting Out a Permanent Seed-producing Bed

Permanent seed-producing beds are usually started with roots purchased from an experienced grower and spaced well apart (no closer than 9" × 9"). Seeds, rather than roots, may be used, but roots save years. Eventually, you will be able to start seed-producing beds by transplanting roots selected from your own gardens for their size, shape, and vigor.

Within three years, the value of the seeds you have harvested should at least equal the initial cost of transplanting the roots. If the beds are well tended and healthy, the grower can expect to harvest at least 30 seeds per plant in a short time. A few thousand mature plants can thus provide the seed needed for each year's new planting. About 2,000 plants, spaced out over at least 1,000 square feet of actual bed space, is about the smallest area that's worth a businesslike effort to prepare and monitor conscientiously as a seed-producing bed.

For many years, I had a permanent seed-producing bed under artificial shade in my back yard. This not only reduced critter predation of the berries but also added security from human poachers, who are sorely tempted by beds of large, lush plants. Of course, a shade structure in the backyard does not contribute to the aesthetics of your lawn, and it is expensive. My permanent seed-producing beds are now all in the woods. By concentrating the bulk of seed production in such beds, it becomes more cost effective and efficient to take counter measures against both berry predation and poaching.

As far as I know, there are no mechanical devices that set out ginseng rootlets. You have to do it by hand. When I plant, I start at the top of a well-tilled bed, dig a narrow trench four or five inches deep across the width of the bed with a garden hoe, set out the roots at least nine inches apart in the trench, and firm the dirt over the roots with the flat side of my hoe. While I am doing this, I am standing on a flat board at least nine inches wide, which I have laid across the bed just below the trench into which I am setting the roots; standing on the board while planting reduces compaction of the soil. After I have planted the row, I step off the board to the side of the bed and roll the board over once downhill. Then I step back onto the board and dig another shallow trench across the bed, which will be at least the width of the board below—and parallel to—the first one. I repeat this procedure all the way down the bed, using the wide board as a movable guide for my rows. A rootlet can be set in at any angle—or even flat—as long as the bud doesn't face down and is level with or above the trunk of the root. The bud should lie one to two inches below the surface. Finally, it's a good practice to only begin placing roots several inches in from the outside edges of your bed, because the sides of the trench/walkways can get very wet during heavy rains. After planting seedlings, the bed surface is usually pitted and uneven, and it should be raked smooth before mulching.

Studies of seed production at various plant spacings clearly show that wide spacing is the only way to go. This may be 9" × 9" (two plants per square foot) or even 12" × 12" (one plant per square foot). Rootlets are far more expensive than seeds, so getting the most out of each one, maximizing its seed production and its life expectancy, is important. Wider spacing not only reduces competition, stress, and susceptibility to disease, but also slows the spread of disease so that you will have a better chance to spot it and deal with it before many plants are affected.

When, Where, and How to Sell Your Roots

As soon as the ginseng plant produces a sizeable berry cluster, its root is likely to be big enough to sell. Therefore, most 'sang farmers choose to avoid risk and to dig their green gold soon after it reaches reproductive maturity and begins producing berries. Once ginseng gets big enough to produce large clusters of berries each year (usually on four-pronged plants), the annual growth of the root slows down and makes the decision to dig easier. Many factors, some of them beyond the grower's control, influence the rate of growth, so it is difficult to state exactly how old plants will be before they are ready to harvest. Artificial-shade growers usually are able to dig marketable roots after only three years; woods-cultivated growers generally harvest after six to eight years; and wild-simulated growers after anywhere from seven to eleven years. I recently spoke to an experienced wild-simulated grower who says he intends to wait 20 years to harvest some of his plants. I have seen too many things go wrong to recommend that approach. Roots usually do continue to get heavier with age, and a significantly higher per-pound price comes with an age of 20 years; on the other hand, delaying harvest means risking disease, poaching, flood, or some other serious problem that could wipe out all of your profit.

The Traditional Route: Selling Wholesale to Exporters

Because the consumer market for ginseng has been almost entirely in the Orient, until recently growers have had little choice (or desire, given the prices offered) but to sell their roots wholesale to exporters who then resell into Hong Kong and the Pacific Rim countries. Selling to these exporters may still be your best option. Buyers are not hard to locate. They are usually desirous and capable of buying all the roots a grower has, and the transaction is quick and simple with no paperwork for the grower. Payment is immediate upon delivery, often in cash for small sales if the buyer and seller meet.

Major ginseng exporters have networks of agent-buyers who are located in the ginseng-growing regions of North America and who will quote the ginseng farmer a price. Unless there are no other buyers in the area, these local dealers (often shrewd old-timers) will quote a price for wild (or wild-simulated) 'sang that is in line with digger prices throughout your region of the country. However, most small dealers are somewhat unsure of themselves in grading the value of cultivated roots (either field or woods grown), and they are likely to quote a low price in order to be sure that they can resell at a profit. Growers of woods-cultivated or wild-simulated roots who have more than ten pounds to sell are probably well advised to sell to a large well-established and knowledgeable buyer. (A list of reputable, knowledgeable buyers is provided in chapter 9.)

Obtain bids from at least two buyers before making a large sale of woods-cultivated roots, and be aware that the price sometimes does fluctuate over even short periods. Buy bids can be obtained by mail from dealers on the basis of a sample. The best procedure is to write a letter (enclose a self-addressed stamped envelope for a reply) or send an email stating the number of dried pounds that you have for sale and your growing method. The buyer's reply will indicate the size of the sample required for grading, the preferred manner of shipping, and usually an offer to pay for the sample or at least to pay the cost of return shipping. When you finally send your crop to the highest bidder, insure the package(s) and keep the receipt. A dealer's continuing operation depends on reputation, but there are short-term unethical operators out there. Until you have developed an ongoing relationship with a buyer, do not extend credit. Insist on payment upon delivery. Most years it is a seller's market, and it is often best and certainly most active between October and December.

Once you have established yourself as a producer of quality roots, buyers will aggressively solicit a sale, pestering you with phone calls (especially when prices are rising) or even coming to your farm in the early fall to grade your roots and offer a bid before they are dried. Growers who can learn to grade their own roots with some accuracy stand a more certain chance of getting a fair price. However, I have found fine grading to be as much art as science, and I recommend relying on competitive bidding. You can help yourself by making sure your roots are kept clean of dust and debris, by drying properly so they have a lighter color (a slightly golden tint is optimal), and by separating out blemished, damaged, and diseased roots so that they do not downgrade your entire harvest. Some dealers look to buy roots fresh out of the ground and oversee the dehydrating themselves.

The Nontraditional Route: Direct Marketing

Because each link in the export chain receives a commission and then there is a large retail markup (very large, if fancy packaging is added), the consumer normally pays much more for ginseng roots than the grower's wholesale price. Thus, there is a potential of additional profit for the grower by selling roots directly to the retailer, and considerable additional profit in selling directly to the consumer. For example, a number of small-woods growers with family ties to local Chinese or Korean communities have told me that they have been able to get at least twice the price exporters have offered them by selling directly into those communities door-to-door and by word-of-mouth. There are many possible targets for direct marketing, not all of them of Asian ethnicity: Chinese and Korean restaurants, acupuncturists, naturopaths, homeopaths, herbalists, health food stores, and other retail outlets that cater to upscale, health conscious consumers. Some of these customers will pay an even greater premium for fresh or organically certified roots, but understand, there is no premium paid by the traditional Asian market for organically certified roots. These roots must be marketed in a nontraditional manner to garner additional value. (Chapter 9 has a short list of buyers usually willing to pay a premium for small organic lots.)

Identifying these markets and then efficiently distributing your roots to them on a timely basis is a challenge; but, then, that's why folks like me are not doing it and why the opportunity may be there for you, especially if you have a personality that enjoys and excels at marketing. Growers of the herbs covered by Jeanine will have a difficult time being profitable if they don't take a direct approach (at least direct to the retailer). A more thorough and knowledgeable exploration of direct marketing techniques—most of which can also be applied to ginseng—can be gleaned from her chapters. The North American demand for high-grade ginseng is limited, however, and growers with hundreds of pounds of root will probably need to sell some of their crop to a traditional exporter.

Another marketing approach is to try the Internet, through which woodland ginseng farmers, even in places like Tuckasegee, North Carolina, can directly reach potential customers spread out on the other side of the world. I have not taken much advantage of the Internet, myself; but, based on conversations with several Internet-savvy growers, it is apparent that a few points of caution are appropriate here. First, if you don't have at least 100 pounds of high-grade roots to sell, the cost and time required for doing business over the Internet will eat up any additional profit you might gain by selling your crop at closer to the retail price. Second, selling ginseng internationally requires more than a website, online order forms, etc. You will also have to learn to deal with the regulations, requirements, permits, and certifications required for international transport of an agricultural commodity that is subject to an international trade agreement (see CITES in chapter 1). Third, and perhaps most difficult, you will want to learn how to grade your roots. Otherwise, not only may you have difficulty getting your best price, but you also run the risk of misunderstandings, returned orders, and a spoiled reputation. Despite these difficulties, more and more of the trade in American ginseng will probably be conducted over the Internet, directly between grower and retail dealer or end consumer. This is not necessarily a good thing, since the traditional buyers have maintained a stable and dependable export market for decades.

The Future of the Market

Demand

As described in chapter 2, except for interruptions during times of war and trade embargoes, the North American ginseng trade with the Orient has been continuous for about 275 years. The price paid per dried pound of wild root has risen from the two or three francs per pound paid by the French Canadian fur traders in the early 1700s to typically around $675 per pound in 2012 (see Recent Prices at the end of chapter 2). Most wild American 'sang is now exported to Hong Kong for distribution on the island itself as well as to Taiwan, Singapore, Japan, the rest of Southeast Asia, and into Mainland China (especially the more affluent sections), which is by far the world's largest consumer of *P. quinquefolius*. The Mainland Chinese have been known to pay two months' earnings for a few ounces of dried root. Like many generations of their ancestors, they believe it is a regenerative tonic with broad and subtle efficacy.

As economic reforms continue in China, per capita income should also continue to grow and, with it, per capita consumption of ginseng. Recent statistics show the average Chinese consumes two grams of ginseng per year, as compared to the average Taiwanese who consumes over nine grams, and consumption is 74.6 grams per person annually in affluent Hong Kong. Thus, there is considerable room for growth and, since the Chinese Communist Party is softening its view of American ginseng as a decadent luxury item, growth in consumption should occur as rapidly as economic growth. The Mainland Chinese now buy primarily the lowest grades of ginseng, but China represents a huge future market for the woods grower, if per capita income continues to rise.

Our wild roots are being heavily hunted because their value has continued to rise. The wild harvest is diminishing. The trend is fewer pounds of wild roots harvested annually, with more (and therefore smaller) roots in each pound, which could eventually lead to more restrictions on digging. (Indeed, it's worth repeating here that, because there is some future possibility that foraging for wild 'sang will be curtailed by law to protect the species, growers should have a government employee, such as a county extension agent, document their woodland plantings to insure that their roots will be saleable if such a ban occurs.) As wild ginseng becomes increasingly scarce, demand for wild-simulated and older woods-cultivated roots should increase to help fill that high-priced sector of the affluent Southeast Asian market.

Demand for green gold has risen significantly in North America and Europe over the last decade. Our health food industry is now selling more American ginseng, which, until recently, had been looked down on here while being highly thought of in the Orient. There is even a slowly growing appreciation in the West

Table 10. Estimated Woods-grown Ginseng Farming by State in 2000

	Woods-cultivated		Wild-simulated	
State	Growers	Acres	Growers	Acres
Alabama	15	4	150	50
Arkansas	3	4	25	5
Georgia	10	8	50	150
Illinois	80	150	250	170
Indiana	8	7	35	12
Iowa	10	5	300	60
Kentucky	135	31	200	30
Maine	0	0	40	40
Maryland	26	35	80	10
Michigan	25	150	5	10
Minnesota	43	23	200	60
Missouri	7	10	30	10
New Hampshire	0	0	3	10
New York	25	20	250	30
North Carolina	40	17	130	200
North Dakota	0	0	2	1
Ohio	28	26	350	65
Oregon	1	1	0	0
Pennsylvania	75	110	400	115
Tennessee	175	50	650	250
Vermont	0	0	4	4
Virginia	30	12	170	45
West Virginia	9	14	35	10
Washington	0	0	6	2
Wisconsin	6	6	50	30
Total of all States	751	683	3,415	1,369

for high-grade American ginseng roots. I recently helped a grower sell 100 pounds of his fresh, high-grade, organically certified woods-cultivated roots for processing into a rather expensive extract that is being successfully retailed to up-scale Americans on a trial basis. Increasing concern for the chemical residues found in field-grown roots is also spurring interest in wild and woods-grown 'sang. There is a very large potential market in Europe for organically certified roots, but regulations have so far prevented this market from opening to the grower.

Supply

Throughout eastern North America, wherever there is a mature hardwood forest, humus-rich soil, and slope for drainage, a few patient hard-working folks are likely growing ginseng. In the United States, woods growing is now practiced by many more people than field growing. Nevertheless, although recent surveys indicate there is a modest trend toward increased acreage planted in the woods, woodland production remains quite limited in terms of both land planted and dried pounds harvested.

I first conducted a survey of woodland ginseng production back in 1994 for the International Ginseng Conference in Vancouver and then updated it in 2000 for the Ginseng Production in the 21st Century Conference in New York. Table 10 shows the 2000 survey results. These figures are estimates, obtained by asking knowledgeable people how much ginseng was being

Table 11. Estimated U.S. Woodland Ginseng Farming in 1994 and 2000

Production Method	Number of Growers		Number of Acres		Potential Increase in Annual Production
	1994	2000	1994	2000	1994–2000
Woods-cultivated	814	750	401	682	19,100 lbs.
Wild-simulated	3,334	3,416	905	1,339	7,715 lbs.

raised on how many woodland acres in their area. Woods-growers are understandably secretive about their operations, and there is not (and never has been) any hard documentation of the extent of woodland plantings or of the amount of woods-grown ginseng that is harvested annually. According to the survey results, just over 4,000 growers were estimated to have just over 2,000 acres planted in the forest. That is an average of one-half acre per grower. Clearly, woods growing as practiced in the United States is predominantly a small-scale enterprise with the potential to provide only a little supplementary income.

Table 11 is a combined summary comparing the 1994 and 2000 survey results. It appears that the total number of woods growers remained almost constant at around 4,150 over the six-year period between surveys. Certainly some turnover in growers has occurred, but how much is not known. While the number of growers has remained constant, the number of acres they are planting in the woods has increased meaningfully—probably because, with rising prices paid for high quality roots, potential profit has increased. Woods cultivation has grown roughly 39 percent (from 491 acres in production to 682), and the number of acres planted in a wild-simulated manner has increased approximately 48 percent (from 905 to 1,337 acres).

Based on the estimates of acres planted presented in Table 10, it is possible to speculate on how many more pounds of ginseng were being produced in the woods in 2000 than in 1994. If woods-cultivated growers are harvesting an average of 600 pounds of dried roots per acre after an average of six years' growth, then the maximum potential increase in poundage of woods-cultivated root that might be produced each year (assuming, with great optimism, no failures) would be 19,100 pounds (682–491 = 191 acres × 600 pounds divided by six years). Since the price of high-grade woods-cultivated roots rose slightly between 1994 and 2000, one is drawn to conclude that either nothing like maximum potential production is being achieved or that so far the market has had no trouble absorbing thousands of pounds of additional high-grade roots. (Either conclusion bodes well for the future market.) Some of this volume is almost certainly substituting in the marketplace for the diminishing supply of wild roots. There is general agreement among buyers that, during the last 50 years, there has never been enough woods-cultivated 'sang available to test the limits of demand for this kind of root.

Similarly, if wild-simulated growers are harvesting an average of 160 pounds of dried roots per acre after an average of nine years of growth, then the maximum potential increased annual production of wild-simulated roots (assuming, again, no failures) would be an estimated 7,715 pounds (1,339–905 = 434 acres × 160 ÷ 9). Due primarily to the increasing scarcity of wild root, the number of pounds of wild root exported dropped from a peak of 132,000 dried pounds in 1994 to roughly 60,000 pounds in 2000. Even if the maximum of 7,715 additional pounds of wild-simulated root were being added annually to the market (and there is a consensus among buyers that some significant portion of it has been), it is not surprising that it has had no noticeable effect on the total amount of wild root exported and certainly not on the price of wild ginseng. Actually, with diminishing supply, one would have expected the price of wild roots to have gone up markedly during the period between surveys. At least a partial explanation is that the economies of the Pacific Rim countries were troubled during those years, and there was stiff resistance to higher retail

Stuart Mirfin and his son Scott (in foreground) prepare to plant American ginseng underneath the natural shade of their radiata pine plantation near Richmond, at the north end of New Zealand's South Island. The tents of netting have proven absolutely necessary to prevent several species of native birds, plus imported California quail, from destroying young plants. Though grown a world away, the Mirfins' roots look surprisingly like those produced on the hardwood hillsides of Tuckasegee, North Carolina. Photo courtesy of Stuart and Scott Mirfin.

ginseng prices among the end consumers. On balance, it seems very probable that the demand for wild root is, and will remain, much greater than the supply (historically at least 100,000 pounds annually, but now only half that).

The production of woods-grown 'sang is very labor-intensive; the time to harvest is extremely long; the risk of loss over such a long period is great; and there is no standard accepted growing method. The kind of large-scale, corporate-financed operations that have resulted in the overproduction of shade-grown roots have not been established in the woods. There is small-scale woodland production of American ginseng in many countries outside the United States, even in the Southern hemisphere—in southern Australia, on New Zealand's South Island, and in Tierra del Fuego, Argentina; but I have visited many of those places and found the number of growers and the scale of their operations to be small and not likely to significantly affect the available supply of roots any time soon. In addition, due to centuries of deforestation, there are not many acres of mature hardwood forest in China suitable for woods growing. Thus, at least the short- and medium-term market outlook is good. For the foreseeable future, there remains a golden opportunity for the patient, persevering, woods grower.

8

A Grower Tells His Own Story: Oscar Wood

The folks who grow American ginseng in the woods tend to be an optimistic, self-confident, diverse, and independent group. Most of the successful ones utilize at least some of the methods suggested in the preceding five chapters, but almost all of them have developed their own particular practices—sometimes out of necessity, perhaps occasionally out of ignorance, but most often in response to their unique circumstances. There is no best way to farm ginseng in the woods. You too may be faced with unusual conditions or have a strong inclination to try a different approach. Diverging from my guidance provided in this book is certainly not a formula for sure failure.

My friend Oscar Wood grew ginseng for 20 years on a steep wooded hillside near the top of Balsam Mountain in western North Carolina. Oscar was growing 'sang long before I got interested in it and before there was much written information on how to cultivate it. He developed his own methods based on common sense and a lifetime of experience hunting the plant in the wild. Oscar was successful and, in turn, helped me to get started.

He used to pluck some weeds from his beds once in a while, but, basically, Oscar just sat back and let his 'sang grow naturally. He wanted to avoid the pollution and the expense of synthetic chemical fertilizers, fungicides, and insecticides. He told me that his best beds produced about 30 pounds of roots (dried weight) per 1,000 square feet and that he got premium woods-cultivated prices for his roots.

Oscar is dead now, but years ago, when I asked him to tell his story, he settled himself deep into his favorite recliner and freely shared his experiences as a 'sang grower. Some of the techniques Oscar used, or did not use, I do not consider to be the best practices, but they worked for him.

This is Oscar's story as he told it to me.

Oscar Wood inspects a ginseng root. Photo courtesy of Dennis McClintic.

Oscar's Start

After I got crippled up and couldn't work reg'lar, I had to find somethin' to do. I was layin' in bed one morning and, like a vision, I pictured the whole hilltop back of the house covered up with ginseng! I got up, and my wife asked me what I was going to do. I said I was fixin' to go 'sang diggin', replant it, and start growing it on our mountain.

I ate my breakfast and took off. Hunted 'til noon and hadn't found one bunch. I was restin' on top of a ridge when I spied a black snake coiled by the side of a stump. I reached down with my hoe to move the critter...and right there stood a four-pronged bunch of ginseng. I looked again, and there stood another big bunch.

I dug 118 roots off that ridge in two hours. I 'bout give out, so I gathered it up and brought it home, rested awhile, and then set it out. Went back there two more days, and then hunted all over the area that fall, which was in 1974. In six weeks I dug enough bunches to set out a bed 113' long by 4' wide with the roots spaced 6" apart each way...it come to about 2,100 plants altogether.

Then I began pickin' the berries from those plants every year and sowin' the seeds. I got half-a-pound of seeds off 'em the first year, a pound the next year, and two-and-a-half pounds the third. The first several years, when we'd get a little ahead, I'd buy a few thousand seeds too. This past year I planted 20 pounds of my own seed.

Bed Preparation and Planting

Before I plant, I cut out the undergrowth, remove the stumps, and drag off the brush. That first year, when I transplanted, I used a mattock to break up the ground and get the roots out. Let me tell you, that's hard work, brother! Now I use a tiller. I dig the leaves right into the ground, and that's the only fertilizer I use. We plant before the leaves fall—about mid-October here—so that the new-falling leaves are my mulch.

I set up the beds 6' wide and maybe 200' long down the face of the slope, so they'll drain well, and leave two-foot wide walk-

The hillside (in late fall) where Oscar Wood grew his green gold. Photo by Kim Fadiman.

ways between beds for weeding. I lay each bed out with string stretched between four corner stakes. If a large tree is in the bed, I just let it be... plant right next to it. Grows fine like that.

My wife does all the planting now, during the last two weeks in September. She takes a long-handled grubbin' hoe and digs a trench across the width of the bed and drops the seeds in by hand four to six inches apart, dependin' on how they bounce. Lots of times she'll drop two or three seed instead of one. They'll grow, so we don't worry about it. She covers 'em with about an inch of dirt, then she digs another row eight inches below the first, and so on. About 12 pounds of seed'll plant one-third of an acre. I figure between 75 percent and 90 percent of our seed germinates.

Oscar digs a shallow trench and drops in the seeds individually in a manner similar to the Hankins method. Photo courtesy of Dennis McClintic.

Care During the Growing Season

All I do during the growin' season is try to keep the weeds down and the rodents out.

I've never had any trouble with insects, and I don't use chemical sprays or anythin' else for disease. I've never had to water, either.

I weed by hand once a year—any time before the weeds flower and go to seed, which usually is mid-August or September. I only worry about pullin' the weeds big enough to flower. To keep the rodents out, I have to trap the moles. The moles won't eat the ginseng. They just like the grubs and worms, which I've got plenty of. But once they get a run started through a bed, it's like a highway through there. The rats use the mole run to get to the roots, and it's hard to get the critters out. So, as soon as I see mole sign in the spring, I set my traps to catch 'em. I once caught 45 moles out of one run! [Author's note: When Oscar speaks of "rats," he is referring to rats, mice, and voles, collectively.]

Oscar buries a stratification bag fashioned from aluminum screen filled with a seed/sand mixture. Photo courtesy of Dennis McClintic.

Anyway, I lose some roots to rodents each year, and I also usually have problems with disease infectin' the tops of the plants. The disease starts in August or September and keeps spreadin' right on up 'til frost, when the tops die off. It doesn't kill the older plants…it just makes the tops die, which stops root growth that year. If it hits early, sometimes I lose a lot of seed too. But the next spring a new top comes up. I don't do anything with my beds during the winter. There's nothing that needs doing. [Author's note: Although Oscar's diseased plants were never tested, they were probably infected with Alternaria blight and/or downy mildew, *Peronospora parasitica.* Oscar had only limited problems with disease, almost certainly because he spaced his plants well apart and his excellent site seemed to provide all the conditions necessary to produce healthy specimens with not only substantial root systems but also with strong immune systems able to resist the various evil fungi that can attack ginseng.]

Harvesting and Storing the Seeds

Sometimes the berries start turnin' red in late July, but usually not 'til August or September. As soon as they ripen, my wife starts picking 'em…about once a week all fall. I figure we get four to twelve seeds off a three-year-old plant, on up to about 50 seeds off a good, healthy, old one.

She puts the berries in a sack made out of window screen and keeps it in a cool spot under a tree. When it's full, I close up the sack—to keep the rats out—and bury it in a deep, dark, shady place so the seeds won't dry out. I cover the screen bag with about four inches of dirt. The side of a bank is the best storage site, 'cause heavy rain will drain right off.

The seeds won't germinate and sprout 'till the second spring after you pick the berries, so I just leave 'em in the ground… never fool with 'em 'til I'm ready to plant the next fall. Course, lots of times I'll dig a bag up in between 'cause people want to see some seeds. One season I couldn't get all my beds planted in the fall and had to wait 'til March to sow. Those beds came up the first of May just like the others we'd set out in the fall.

I always test my seeds in a tub of water just before I plant 'em. The berry pulp's long gone by then and I can tell if they're good seed. Only the ones that sink to the bottom of the tub are worth anythin'. The floaters I scoop off and throw out in the woods. The sinkers I lay in the shade to drain before I sow 'em.

Digging and Drying the Roots

To get anythin' for 'em, you have to wait at least four years before you dig your roots. Five years is probably best. Growin' much longer is wastin' time, I think, unless you need to keep harvestin' seed off 'em. [Author's note: Although Oscar was fortunate in that even his tilled ground seemed to naturally produce roots with a good deal of wild character, the market has changed, and today he would probably need to wait at least six years before harvesting in order to maximize the wild look of his roots and get a high percentage of the wild price.]

I dig the roots in the fall, after the berries are picked. It's simple to do. I find the dead top, then I just stick a spade fork under it, loosen the ground up, grab hold of the old stem, and pull it right out of there. That way, you don't cut up the root. Buyers look for that and knock down the price.

I bring the roots to the house and drop 'em in a big ol' tub. When it's just about full of ginseng, I fill it the rest of the way with water and wash each root by hand. I pick out a handful at a time and get the dirt off with my fingers. Then I rinse 'em in a small bucket of clean water. I make sure there're no leaves or other trash clingin' on the roots. I'm gentle—I don't scrub or scrape too hard—'cause I don't want to damage the skin.

As soon as they're rinsed, I lay 'em out on a screen-bottomed tray in the shade to drain. I don't want the big parts of any of the roots to touch each other until they're almost dried, or else they'll rot where they're touchin'. So I lay 'em out only one layer thick, and the screen lets air get around 'em.

When I have a tray full, I slide it onto one of the bottom racks in my drying room. I always slip a new tray in at the bottom so any water still on the roots will drip on the floor. Then I go back and finish washin' and layin' out the rest of the roots in the tub.

Oscar inspecting roots during the drying process. Photo by Kim Fadiman.

If I leave 'em in the water too long, they'll start to rot.

The room I do the dryin' in, I built onto my house long before I began growin' ginseng. I used to use it, and still do, for cannin' fruits and vegetables and sausage, for dryin' apples and beans, and for anythin' else that would mess up the main house. When I started raisin' 'sang, I built a dryin' rack along one wall with scrap lumber and $20 worth of screen for my tray bottoms. The trays are about 2' × 4' and two inches deep. The room's only 12' × 14' but, if I want to, I can dry 80 trays (over 300 pounds) at one time. I've never had it plumb full, though.

I keep the temperature in the dryin' room at 90° for about the first 12 hours after the roots are washed. You got to get rid of most of that wetness right away, or the roots'll ruin on you. I use a wood stove, and I have a thermometer hangin' in the room where I can keep an eye on it. Some folks'll put their roots right next to, even on, a stove... but that's a mistake. That'll burn the roots or dry 'em too fast, and you won't get as good a price.

After the first night, I try to keep the temperature about 70°... but in cold damp weather—'specially early on dewy mornings or whenever it's rainin'—I fuel up the wood stove to heat the room back up to 90°. If I don't, the roots seem to take back moisture from the air and mold quick as anything. Once the temperature's up to 90°, I open the door—if I have to—to see the room stays under 95°.

It needs about two weeks of dryin' before most of the moisture is out of the roots. After that, I can stack all the trays together on my top rack—or dump two or three trays into one—to make room for a fresh tray on the bottom rack. It may take another three or four weeks before the roots are completely dry. You can pretty well tell they're dry if all the sponginess is gone, and they break with a snap, and the break is a little waxy lookin'.

Makin' It Pay

After the roots are completely dry, I pack 'em in a clean cardboard box 'til I'm ready to sell. I don't seal it completely tight, so they can still get air, and I make sure the rodents can't get to it. Dried ginseng keeps fine at room temperature, but not in a basement or cellar or even by a window... they'll suck water right back into themselves, and mold.

The price you get is always fluctuatin', so you got to be a cautious seller. The folks I sold mine to this year gave me a better price 'cause I didn't use chemicals. I'd advise anyone against usin' chemical fertilizers. The roots grow too fast, the plants are more susceptible to disease, and the Chinese won't like 'em so much. You're just not helpin' yourself to use chemical fertilizers. [Author's note: See Resources for organic buyers.]

I think the future of ginseng is as solid as anythin' right now, but I'd still tell anyone wantin' to grow it to be cautious and not to get too much into it at first. I figure in ten years about all I've spent is $1,200. The seven-strand barbed-wire fence that I put around our six acres to keep out pests—animals and humans—cost $600... and during the first five or six years, I reckon I paid another $600 all told for seeds.

'Course, the biggest expense is going to be your own labor. I'm lucky: My wife does a whole lot of it and my kids helped some in the beginnin'. They're even talkin' 'bout goin' into it with me now. But it's hard work! Ever since that morning when I dreamed my mountain was covered with green ginseng, I've been workin' toward gettin' the whole hillside planted. I'm now half done... still got another three acres to go.

9

Ginseng Resources

Provided in this chapter are sources of products, information, and services that a ginseng farmer might find useful. I have listed only suppliers of products and services with whom I'm familiar. The inclusion of a firm is not a guarantee of reliability, and absence certainly does not imply disapproval. Thus, the following lists are not complete, but only intended to be helpful. Full contact information for each listing can be found in the Comprehensive Resources Directory.

Ginseng Root Buyers (by mail)

An alphabetical listing of established buyers follows. See Resources Directory for full addresses. Include a self-addressed stamped envelope (SASE) with all inquiries. Follow their individual shipping instructions carefully to get their best possible price.

- American Botanicals
- GAIA Herbs, Inc.
- Herbalist & Alchemist, Inc.
- Hershey's International, Inc.
- Hsu's Ginseng Enterprises, Inc.
- Tony Ison
- Lowe Fur and Herb, Inc.
- Ohio River Ginseng & Fur, Inc.
- Quality Northern Ginseng Co., Inc.
- RRAWR, Inc. (wild roots only)
- Ridge Runner Trading Company, Inc.
- Springland Trading, Inc.
- Strategic Sourcing, Inc.
- Yat Chau (U.S.A.), Inc.

Organic Buyers (usually only small quantities)

- Eclectic Institute
- GAIA Herbs, Inc.
- Herb Pharm
- Nature's Cathedral
- Wise Women Herbals
- Red Moon Herbs

Organic Consultants and Suppliers

- ARBICO
- Peaceful Valley Farm Supply (over 3,000 organic products and advice on how to use them)
- Roger Sego (organic ginseng growing and product recommendations)

Supplier/manufacturer of Ginseng Farming Machinery

- Buetsch Implement Co., Inc. (carries a full line of equipment, some of it available used, including ginseng washers, diggers, straw shredders, bed rakes, sprayers, hillers, seeders, and tractors)

Suppliers of Artificial Shade

The following are sources for polypropylene shade cloth. Send a large SASE and request instructions for installation when you inquire. Most suppliers also carry the paraphernalia needed for installation.

- Gintec Shade Technologies (US and Canada)
- Marathon Feed Inc.
- Shelter Shade

Suppliers of Seeds and/or Seedling Roots

A listing of established seed and/or seedling dealers follows. A legal-sized SASE is always appreciated when a price list is requested.

- Bob's Goldenseal & Ginseng (Robert Tipp, Owner)
- Catoctin Mountain Botanicals (Steve Galloway)
- Companion Plants (Peter Borchard, Owner; seedlings only)
- Emerald Castle Farms
- Green Gold Enterprises, Inc. (W. Scott Persons, President)
- Harding's Ginseng Farm (Larry Harding, Farm Manager)
- Horizon Herbs, LLC (Richo Cech, President)
- Hsu's Ginseng Enterprises, Inc. (Paul Hsu, President)
- Johnny's Selected Seeds (Beth Jensen)
- Mike Dammen
- Mountain Gardens (Joe Hollis, seeds, small quantities)
- North Carolina Ginseng and Goldenseal Co. (Robert Eidus)
- Ozark Mountain Ginseng (Dennis Lindberg, Grower)
- RRAWR, Inc. (Bruce Phetteplace, wild New York genetic stock only)
- Richters Herbs
- Wild Grown

Ginseng On-site Consultants

For expenses plus a fee, the two individuals listed below will travel to your site and advise you on nearly all aspects of ginseng farming. Their time and knowledge is expensive, however, and you will probably get the most for your money if you utilize their expertise well before sowing your first seed. Learning that you have a poor site for growing green gold can save you a great deal of time, effort, and money. In the few locations where 'sang growing is fairly common, the local county extension and/or soil conservation agents are sometimes knowledgeable—and their recommendations are free. In addition, suppliers of seeds and seedling roots will often provide some free advice to their customers.

- Robert Beyfuss
- W. Scott Persons (Green Gold Enterprises, Inc.)

Soil and Tissue Analysis

Most states have laboratories that will analyze your soil or your plant tissue, diagnose problems, and recommend nutrient amendments or disease treatments. They perform these services rather cheaply or even free of charge to residents, and their knowledge of ginseng is improving steadily. Inquire before sending samples the first time, then follow the laboratory's sampling, packing, and shipping instructions explicitly in order to obtain the best information. The following private firm is efficient enough to compete and offers some more sophisticated services:

- A & L Labs (multiple locations)

Ginseng Growers Associations

- The Association of Ginseng Growers of British Columbia (TAGG)
- The Ginseng and Herb Cooperative (WI)
- Ontario Ginseng Growers Association
- Vermont Ginseng Association

Other Ginseng-related Organizations (Nonprofit)

- American Herbal Products Association (Industry perspective on regulatory issues and sustainable use of wild medicinals)
- Organic Materials Review Institute (OMRI)
- Rural Action (Their planting stock program provides planting stock, educational resources, and assistance to growers)
- Smoky Mountain Native Plant Association (Supports growers of native Appalachian herbs)
- West Virginia Trappers Association (Holds regular auctions of ginseng and other woodland botanicals)

10

Ginseng References

Little was written about American ginseng farming prior to about 1985, and much of that was never made available to the general public or, if it was, is now long out of print. For example, research presented at the first six National Ginseng conferences was only given to the participants and a few other interested parties immediately following the conferences. Although some of the early literature may now be difficult to locate, I have listed a few such references, because I did rely on them as I prepared this book, and I wish to give the authors recognition.

Most of the references provided below are more recent and, while often not easily accessible, they are obtainable. Titles are listed because I drew from them and/or because I think the reader may find them interesting and useful resources. There is much to be gained from these writings, and I urge you to make the extra effort (perhaps with the help of your local librarian) to obtain whatever additional information interests you.

Ginseng Farming

Beyfuss, R. (n.d.). The Practical Guide to Growing Ginseng. Acra, NY: Cornell Cooperative Extension of Greene County. [Author's note: A 63-page primer on growing ginseng in backyards or woodlots. Available from Cornell Cooperative Extension, 6055 Route 23, Acra, NY 12405, for $7.00, including postage.]

Carol, C., and D. Apsley. (n.d.) Growing American ginseng in Ohio: An Introduction. Ohio State University Extension Fact Sheet [Author's note: A four-page primer available online at ohioline.osu.edu]

Davis, J. M. 1997. Ginseng: A Production Guide for North Carolina. Publication AG-323. Raleigh, NC: North Carolina Cooperative Extension Service, North Carolina State University.

Dickman, M. 1983. *How to Raise Two Cash Crops for Profit: Ginseng and Golden Seal.* Willow Springs, MO: Dickman.

Hankins, A. 2000. Producing and Marketing Wild Simulated Ginseng in Forest and Agroforestry Systems. Publication 354-312. Petersburg, VA: Virginia Cooperative Extension, Virginia State University.

Harding, A. R. 1972. *Ginseng and Other Medicinal Plants*, rev. ed. Columbus, Ohio: A. R. Harding.

Hosemans, Fred, and Charlene Hosemans. 1993. *Ginseng Growing in Australia*. Gembrook, Victoria, Australia: Gembrook Organic Ginseng Publications.

Jacobson, M., and E. P. Burkhart. 2007. Opportunities from Ginseng Husbandry in Pennsylvania. Forest Finance #5. University Park, PA: College of Agricultural Sciences, Research, and Cooperative Extension, Pennsylvania State University.

Oliver, A. L. 1998. *Ginseng Production Guide for Commercial Growers*. Victoria, BC: The Associated Ginseng Growers of British Columbia and British Columbia Ministry of Agriculture.

OMAFRA staff. 2003. Production Recommendations for Ginseng. Publication 610. Guelph, ON: Ontario Ministry of Agriculture, Food and Rural Affairs. ServiceOntario .ca/publications

Persons, W. S. 1994. *American Ginseng: Green Gold*. Asheville, NC: Bright Mountain Books.

Smallfield, B. M., and J. M. Follett. 2004. Ginseng, a Grower's Guide for Commercial Production. Christchurch, NZ: Institute for Crop and Food Research.

Ginseng Reports and Conference Proceedings

Bailey, W. G., C. Whitehead, J. T. A. Proctor, and J. T. Kyle. 1995. "The challenges of the 21st century." Proceedings of the International Ginseng Conference, Vancouver, 1994, Simon Fraser University, Burnaby, BC.

Beyfuss, Robert L., ed. 2000. "Ginseng production in the 21st century." Conference proceedings, Cornell Cooperative Extension of Greene County, Cairo, NY.

Follett, J. M., and A. M. Templeton. 2003. "The globalization of ginseng." Proceedings of the Third International Ginseng Conference, Melbourne, Australia.

"Ginseng market opportunities in the Asia Pacific Region for the British Columbia ginseng industry." 1991. Report of the British Columbia Ministry of Economic Development, Small Business and Trade.

Hausbeck, M. K. 2013. "Control of diseases, pests, and weeds in cultivated ginseng." Available online at reggies.msu.edu (go to bottom of page).

Josiah, Scott J. 1998. "Farming the agroforest for specialty products." Proceedings of the North American Conference on Enterprise Development through Agroforestry, Center for Integrated Natural Resources and Agricultural Management, University of Minnesota, St. Paul, MN.

McMahan, Linda. 1981. "The trade, biology, and management of American ginseng, *Panax quinquefolius*." Staff report of the International Convention Advisory Commission, Washington, DC: TRAFFIC North America.

Proctor, J. T. A., and W. G. Bailey. 1987. "Ginseng: industry, botany, and culture." *Horticultural Reviews* 9:187–236.

Robbins, Christopher S. 1998. "American ginseng: The root of North America's medicinal herb trade." TRAFFIC North America.

Weber, H. Chr., D. Zeuske, and S. Imhof, eds. 1998. "Ginseng in Europe." Proceedings of the First European Ginseng Congress, Philipps-Universitat, Marburg, Germany.

Ginseng Research (not included in any of the above)

Burkhart. Eric P. 2013 "American ginseng (*Panax quinquefolius* L.) floristic associations in Pennsylvania: Guidance for identifying calcium-rich forest farming sites." *Agroforestry Systems* 87 (5): 1157–1172.

Burkhart, Eric P. 2011 "Conservation through cultivation: Economic, socio-political and ecological considerations regarding the adoption of ginseng forest farming in Pennsylvania." Ph.D. Dissertation, Pennsylvania State University, University Park, PA.

Grubbs, H. J., and M. A. Case. 2004. "Allozyme variation in American ginseng (*Panax quinquefolius* L.): Variation, breeding, system, and implications for current conservation practice." *Conservation Genetics* 5:13–23.

Konsler, T. R. 1984. "Responses of American ginseng (*Panax quinquefolium* L.) to kind of bed mulch and plant spacing through six growing seasons." Presented at the Sixth National Ginseng Conference, Guelph, ON.

Konsler, T. R., T. J. Monaco, T. J. Sheets, and R. B. Leidy. 1988. "Response of American ginseng to foliar applications of 2,4-D." *Journal of the American Society for Horticultural Science* 113(3) (May): 360–62.

Park, H., S. M. Park, and S. H. Jeon. 2002. "Production and quality of mountain ginseng." Proceedings of the Eighth International Symposium on Ginseng—Advances in Ginseng Research, Seoul, Korea.

Proctor, J. T. A., D. Louttit, and J. M. Follett. 2001. "Controlled temperature, aboveground stratification of North American ginseng seed." *HortTechnology* 11:100–103.

Shoemaker, P. B., and T. R. Konsler. 1982. "Fungicide evaluation for ginseng blight (*Alternaria panax*)." Presented at the Fourth National Ginseng Conference, Lexington, KY.

Stoltz, L. P. 1980. "Embryo development of ginseng seed at various stratification temperatures." Presented at the Second National Ginseng Conference, Jefferson City, MO.

Stoltz, L. P. 1982. "Mineral nutrition of ginseng." Presented at the Fourth National Ginseng Conference, Lexington, KY.

Ginseng's Medicinal Properties

Attele, A. S., Y. P. Zhou, J. T. Xie, J. A. Wu, L. Zhang, L. Dey, W. Pugh, P. A. Rue, K. S. Polonsky, and C. S. Yuan. 2002. "Antidiabetic effects of *Panax ginseng* berry extract and the identification of an effective component." *Diabetes* 51:1852–58.

Benishin, C. G., R. Lee, L. Wang, and H. J. Liu. 1991. "Effects of ginsenoside Rb1 on central cholinergic metabolism." *Pharmacology* 42:223–29.

Dharmananda, S. 2002. "The nature of ginseng: Traditional use, modern research, and the question of dosage." *HerbalGram* 54:34–51.

Hikino, H. 1991. "Traditional remedies and modern assessments: The case for ginseng," in *The Medicinal Plant Industry*, ed. R. O. B. Wijesekera, 149–66. Boca Raton, FL: CRC Press.

Hobbs, C. 1996. *Ginseng: The Energy Herb*. Loveland, CO: Botanica Press.

Lee, F. C. 1992. *Facts About Ginseng: The Elixir of Life*. Elizabeth, NJ: Hollym International.

Lewis, W. H. 1986. "Ginseng: A medical enigma," in *Plants in Indigenous Medicine and Diet: Biobehavioral Approaches*, ed. N. L. Etkin, 290–305. Bedford Hills, NY: Redgrave.

Murphy, L. L., R. S. Cadena, D. Chávez, and J. S. Ferraro. 1998. "Effect of American ginseng (*Panax quinquefolium*) on male copulatory behavior in the rat." *Physiology & Behavior* 64:445–50.

Murphy, L. L., and T. J. F. Lee, 2002. "Ginseng, sex behavior and nitric oxide." *Annals New York Academy of Sciences* 962:372–77.

Murphy, L. L., J. A. Rice, and M. T. Compardo. 2003. "Water-based American ginseng extract (*Panax quinquefolius*) inhibits human breast cancer cell proliferation in vitro and tumor growth in vivo." Submitted to Cancer Letters.

Upton, R., ed. 2012. *American Herbal Pharmacopoeia: American Ginseng Root, Standards of Analysis, Quality Control, and Therapeutics*. Scotts Valley, CA: American Herbal Pharmacopoeia.

Other Related Literature

Agroforestry Notes. 1999. "American ginseng production in your woodlot." and "Economics and marketing of ginseng." [Author's note: Single copies are available free from the USDA National Agroforestry Center.]

Bilger, B. 2002. "Wild sang." *The New Yorker*, July, 38–45.

Bourne, J. 2000. "On the trail of the 'sang poachers—special report." *Audubon*, March-April, 84–90.

Burkhart, E. P. 2004. Nontimber Forest Products from Pennsylvania 1: American Ginseng. The Pennsylvania State University, University Park, PA.

But, P., and T. But. 1984. "Hong Kong: The buyer of American ginseng." Address to the Sixth North American Ginseng Conference, Guelph, ON.

Corbin, J. 1997. "A study of American ginseng habitat in western North Carolina and east Tennessee." Report to the Plant Protection Division of the North Carolina Department of Agriculture (December).

Eliot, Doug. 1979. "They call it green gold." *Wildlife in North Carolina*, May, 7–13.

Evans, B. L. 1985. "Ginseng: Root of Chinese-Canadian relations." *Canadian Historical Review* 66:1–26.

Gronewold, S. E. 1984. "Yankee Doodle went to Canton." *Natural History* 2:62–74.

Hankins, A. 1997. "The Chinese ginseng industry." *The Business of Herbs* 25.

Hardacre, V. 1974. *Woodland Nuggets of Gold: The Story of American Ginseng Cultivation.* Northville, MI: Holland House.

Hu, S. 1977. "Knowledge of ginseng from Chinese records." Reprinted from *The Journal of the Chinese University of Hong Kong* 4.

Johannsen, K. 2006. *Ginseng Dreams: The Secret World of America's Most Valuable Plant.* Lexington, KY: The University Press of Kentucky.

Lass, W. E. 1969. "Ginseng rush in Minnesota." *Minnesota History* 41 (60): 249-266.

Manget, L. 2012. "Sangin' in the mountains: The ginseng economy of the Southern Appalachians, 1865-1900." *Appalachian Journal* 40:28-56.

Nash, G. V. 1895. "American ginseng." USDA Division of Botany Bulletin (16), Washington, DC.

Nash, G. V. 1898. "American ginseng, its commercial history, protection, and cultivation." USDA Division of Botany Bulletin, rev. ed. (16) Washington, DC.

Oliver, A. L. 2013. *The "Green" Gold Rush: A History of Ginseng in British Columbia.* Kamloops, BC: Overland Press.

Taylor, D. A. 2005. *Ginseng, the Divine Root.* Chapel Hill, NC: Algonquin Books of Chapel Hill.

Yepsen, R. B., Jr., ed. 1984. *The Encyclopedia of Natural Insect and Disease Control.* Emmaus, PA: Rodale.

Websites of Interest

"American ginseng production in your woodlot." (Agroforestry Notes). unl.edu/nac

British Columbia Ministry of Agriculture. "Ginseng Production Guide for Commercial Growers," agf.gov.bc.ca/speccrop/ginseng/ginseng

CITES (Convention on International Trade in Endangered Species) information, cites.org

"Ginseng in Pennsylvania." Pennsylvania State University, dcnr.state.pa.us/forestry/wildplant/ginseng.html

"Ginseng Wisdom." Trade environment database (TED) webpage, american.edu/projects/mandala/TED/ginseng.html

Hankins, A. "Producing and marketing wild simulated ginseng in forest and agroforestry systems." Virginia Cooperative Extension. Publication 354-312, ext.vt.edu/pubs/forestry/354-312/354-312.html

Let future generations of ginseng growers continue to provide quality roots for this profitable market while reducing the pressure on the dwindling wild populations in our forests. Photo courtesy of Mark Haskett.

New York State Department of Environmental Conservation webpage on American ginseng, dec.state.ny.us/website/dlf/privland/forprot/ginseng

North Carolina State University ginseng production guide, ces.ncsu.edu/depts/hort/hil/pdf/ag-323.pdf

Ontario Ministry of Agriculture, Food and Rural Affairs (OMAFRA) website for ginseng, omafra.gov.on.ca/english/crops/ginseng

Pennsylvania Department of Environmental Protection website for compost tea, dep.state.pa.us/dep/deputate/airwaste/wm/recycle/tea/tea1.htm

Pennsylvania State University website for ginseng, dcnr.state.pa.us/forestry/wildplant/vulnerable_plants.aspx

PART THREE

Other Species of Green Gold: Goldenseal and Ramps

Twenty-six years ago, I was fortunate to obtain a very special position as a faculty member with North Carolina State University. I call it special for two reasons. First, I am located at a research and extension center in the Blue Ridge Mountains of western North Carolina. This is beautiful country, abundant in plant life, and rich in history and culture. Second, my predecessor, Dr. Tom Konsler, whose innovative research on ginseng is referred to in Part Two, made my research into other native medicinals more readily accepted by the university. At that time only a handful of universities had any kind of medicinal herb research program. However, with the support of Dr. Konsler, my technician George Cox, and the herb growers in the region, I was able to pursue a career studying how to cultivate native medicinal herbs while educating others about how to grow them profitably.

I chose a hot topic to study! At that time, the use of medicinal herbs was restricted to the few people who shopped in health food stores, practiced yoga, and read *Prevention Magazine*. Today, most Americans use herbal supplements that are readily available in discount stores and even through some major health insurance programs. What a difference 26 years has made!

I am so glad my coauthor, Scott Persons, gave me the opportunity to share with you what I have learned about growing these plants—other species of "green gold," to use his term. When I write in my professional role as an extension specialist, it is expected that the information is based on research. Unfortunately, there is still not much research on forest herbs, and it takes a long time to complete a study on most of these plants. Over the years, however, I have learned a tremendous amount from growing and observing these plants, both in my studies and in my home gardens. I have gained generations worth of knowledge by listening to people share their experiences while sitting on back porches or the tailgates of trucks. The research-based information is here, but in addition you'll get all the fascinating information I've picked up from these wonderful mountain people and a few northern flatlanders, as well as my own opinions on the best way to grow forest herbs.

I am a strong supporter of organic agriculture and urge you to consider growing all forest herbs organically. In addition to the many environmental and health reasons for doing so, there is a growing market for organic herbs, and usually a premium is paid for them. There are many good books and websites on organic agriculture and certification. Some are given in the Resource Directory, including my own website on organics: ncorganic.org.

Many of the most popular medicinal herbs are forest plants, and most of the forest herbs used to make commercial products are collected from the wild. This sudden rise in demand for forest herbs has had a dramatic effect on wild plant populations and the ecosystems in which they grow. Development of forested land has also greatly reduced natural habitats. As Scott explained in chapter 2, ginseng was the first North American herb to be harvested so severely that it nearly disappeared in some areas. This encouraged some enterprising people to begin cultivating it, and over time it became a big industry in the northern United States and Canada.

Goldenseal was the next native medicinal to suffer from being overharvested. In the early 1900s, goldenseal was also cultivated, but unlike ginseng, it did not develop into a lasting industry. Now there is a resurgence of interest in growing goldenseal, and it is a commercial crop in a few select areas. None of the other plants covered in this section can really be classified as "crops," so you will be one of the pioneers if you try to commercially grow any of these useful and interesting plants.

I would like to take this opportunity to thank the two people who together helped me find the first funding to get my medicinal herb research program off and running. They are Steven Foster, well-known author and photographer of medicinal herbs, and Mark Blumenthal, founder of the American Botanical Council. I am also grateful to Nature's Way in Utah for having the courage to be the first to provide me with funding. It was through their financial support that the research on the effects of lime, nitrogen, and phosphorus on goldenseal was done. Finally, I thank my family for their incredible patience, love, and understanding as I wrote this book and the two revisions. It always took much longer than any of us anticipated, and they left me to my computer and papers for many, many nights and weekends as I struggled to get the words "just right." My husband, Glen, was always encouraging and cooked many a meal so I could write, and my children, Shannon and Sean, were wonderful about "keeping it down" so Mom could work. And, of course, I want to thank Scott for giving me this marvelous opportunity. I know I would not have taken on this task without his invitation and encouragement.

11

Goldenseal: Its History, Range, Description, Uses, and Government Regulation

History and Range

Goldenseal (*Hydrastis canadensis* L.) is a highly valued medicinal herb that has been collected in the forests of North America for hundreds of years. Many Native American tribes, including the Crow, Cherokee, Iroquois, Meskwaki, Seminole, and Blackfoot, extensively used goldenseal rhizomes and fibrous roots. (From here on, in discussing goldenseal, "root" will refer to both rhizomes and roots together.) Goldenseal root preparations were used to treat a variety of ailments and for several non-medicinal uses.

It did not take long for European settlers in North America to discover the benefits of goldenseal. The Thomsonians (followers of Samuel Thomson, 1769–1843) are reported to be the first of the settlers to use goldenseal on a regular basis. The first comprehensive review of the medicinal uses of goldenseal is believed to have been published in 1852, and recognition of goldenseal as an official drug occurred with its listing in the 1860 *United States Pharmacopeia*. Commercial goldenseal products became available about the same time. To satisfy the rising demand for all the goldenseal roots needed to make these products, the first large-scale commercial goldenseal cultivation began in Washington State at the Skagit Valley Golden Seal Farm in 1905.

Over the years, a large number of colorful and descriptive names have been given to goldenseal. Among those still in common use are yellowroot, ground raspberry, yellow puccoon, wild circuma, eye-balm, yellow paint, wild turmeric, and yellow-eye.

The historical range for goldenseal in the United States is very broad, extending from as far north as Vermont and Wisconsin, south to Alabama and Georgia, and from the East Coast, west to Kansas. Goldenseal is also native to Ontario, Canada. Although harvesting and development have eliminated many wild populations of goldenseal, patches still grow in the moist, rich, mesic cove hardwood forests of this broad

Wild-simulated planting of goldenseal at the National Center for the Preservation of Medicinal Herbs in Rutland, Ohio. Photo by Jeanine Davis.

area. The core of the range where goldenseal grows naturally now appears to be Illinois, Ohio, Indiana, and eastern Kentucky.

By the mid-1880s, there were reports in the literature about damage to wild goldenseal populations caused by overharvesting of the plants and cutting of the forests where they grew. In my own state of North Carolina, the decline continued and for many years goldenseal was classified as an endangered species, making harvest from public lands illegal. The state removed goldenseal from the protected plant species list in 2010, but collection from public lands is still not permitted. Protected plant lists change frequently, but as of this writing in 2013, goldenseal was listed as endangered in Connecticut, Georgia, Massachusetts, Minnesota, New Jersey, and Vermont; threatened in Maryland, Michigan, and New York; of special concern in Tennessee; and vulnerable in Pennsylvania. Goldenseal is also listed as threatened on a federal level in Canada and as a species at risk in the province of Ontario.

The status of goldenseal is of concern on an international level, and in 1997 goldenseal was listed in Appendix II of the *Convention on International Trade in Endangered Species of Wild Fauna and Flora* (CITES), an international treaty monitoring trade in threatened and endangered species. This listing imposes strict international controls on goldenseal trade that are designed to protect the species and encourage sustainable use. Unfortunately, this has not stopped people from collecting goldenseal, and wild populations of the plant continue to decrease.

As a result of the serious threat to native populations, the nonprofit organization United Plant Savers chose goldenseal as their "poster child." This organization has done an incredible job of educating consumers and the industry about the plight of native medicinal plants that are at risk of extinction due to overharvesting and loss of habitat. They strongly encourage the use of cultivated goldenseal over wild-harvested material. They teach people how to grow goldenseal, and they give out free goldenseal as an incentive to start new plantings. United Plant Savers developed two lists of plants: an "at risk" list of plants that they feel are truly at risk of being endangered and should be cultivated immediately, and a "to watch" list of plants that are still abundant in some areas, but have been harvested to the point of being rare in others. Goldenseal is included in the "at risk" list.

Freshly dug cultivated goldenseal root.
Photo by Jeanine Davis.

Government Regulation

Because of the CITES status of goldenseal, there are regulations surrounding the growing, collecting, and selling of goldenseal worldwide. Before you can export (out of the country) cultivated or wild-collected goldenseal roots or rhizomes, whole, parts, or powdered, you must obtain a CITES permit or certificate. Finished products, such as capsules and extracts are not regulated. In the United States, to secure the required documentation, you must submit an application and a $100 processing fee to the United States Fish and Wildlife Service's (USFWS) Office of International Affairs, Division of Management Authority. From your application, it will be determined if the goldenseal was legally acquired and if export of the goldenseal will be detrimental to the survival of the species. For wild-collected material, you must prove that it was collected legally by showing copies of state permits and/or the landowner's permission. If you purchased wild-collected material that you want to sell, you must have an invoice with a purchase date and the name, address, and phone number of the person who sold you the material. For cultivated material, you must be able to prove that the goldenseal roots or seeds came from legally and non-detrimentally (non-detrimental to the species) acquired parental stock and that the plants were cultivated in a controlled environment for four years or more without any augmentation from the wild. Export of wild goldenseal from Canada is not permitted. To export cultivated goldenseal from Canada, you must obtain an export form from the CITES section of the Environment Canada website and follow the detailed information for each province.

If you plan to grow goldenseal, whether you expect to export it or not, I highly recommend that you get an invoice for your

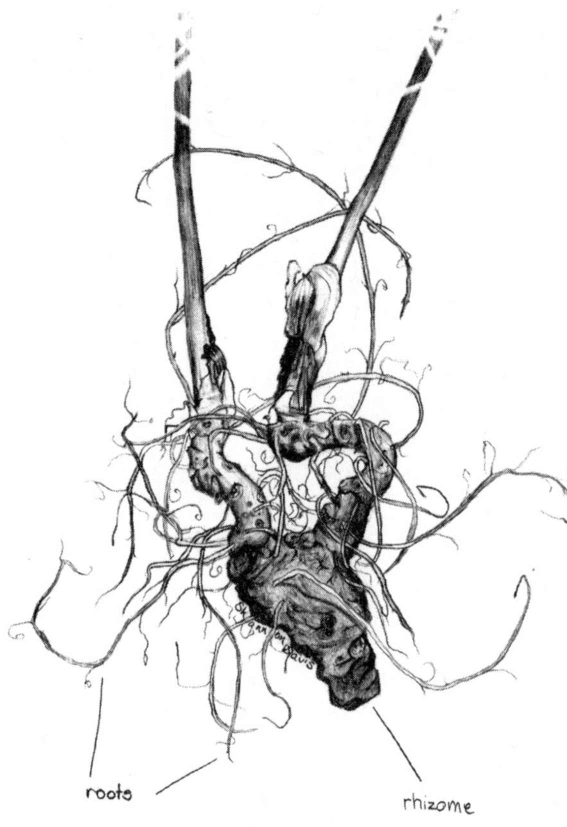

The goldenseal "root" consists of a large fleshy rhizome and a mass of small fibrous roots. Pencil drawing by Shannon J. Davis.

purchase of seeds or plants and store it in a safe place. I also suggest that you take photographs of the plants being planted, growing, and during harvest. It is also a good idea to let your county agent or some other state, provincial, or federal employee see your plantings, so if there is any question, you have a witness. Even if you never have any plans to export the roots, the person you sell them to might and having the necessary documentation for them to export could make or break your sale. For permit applications, or to get more information on the CITES regulations, contact the Division of Management Authority for the United States or Canada as listed in the Comprehensive Resource Directory in the back of this book. Leave plenty of time to get your permits, too. Growers and dealers report that the process can take 60 days or more. In addition, individual states and provinces may have additional rules and regulations concerning goldenseal.

Plant Description and Life Cycle

The botanical name for goldenseal is *Hydrastis canadensis* L. It is a member of the buttercup family, Ranunculaceae, in the subfamily Hydrastiodoideae. It is an herbaceous perennial, which means that the portion of the plant above ground dies back each autumn, but the underground rhizome and roots live on for many years, and the plant sends up a new top each spring. In most areas, the plant emerges in mid-March to early May from buds on the perennial rootstock.

The underground part of the plant is composed of a brown horizontal rhizome, ½ to ¾ inches thick, which is marked by little cup-like depressions where the annual stems have fallen away. The interior of the rhizome is bright yellow, and its exterior is covered with a mass of yellow fibrous roots that can also produce buds. The rhizome usually lies just below the soil surface, and the fibrous roots can reach a foot or more deep in the soil and spread horizontally five feet or more from the main rhizome.

Goldenseal flower. Photo by Jeanine Davis.

Mature plants (at least three years old) are six to fourteen inches tall and have two or more erect hairy stems, usually ending in a fork with two leaves (metric conversions are in Interesting and Helpful Calculations in the back of the book). The leaves are three to twelve inches wide and three to eight inches long. They usually have five to seven lobes, are palmate in shape, and double-toothed on the edges. As soon as the plant emerges in early spring, a flower bud develops at the base of one of the two leaves. A small, inconspicuous, white flower opens as the leaves unfold. Plants started from seed usually flower when three to four years old, whereas plants propagated from cuttings may flower the first year. The flowers can self-pollinate or be pollinated by a variety of bees and flies. Each plant can produce a single, green, raspberry-like fruit that turns red and ripens in July. Goldenseal spreads into the surrounding area through growth of the rhizomes and fibrous roots. Two buds for the next season's top growth usually form on the rhizome near the base of the stem. If it has been a stressful growing season, the plant may die down soon after the fruit matures. In a good growing season, the plant may stay green and healthy looking until frost.

Ripe goldenseal fruit. Photo by Jeanine Davis.

Uses and Medicinal Properties

Traditional Uses

Cherokee Indians used the bitter-tasting root of goldenseal for many purposes, e.g., as an antiseptic, a general health tonic, and to treat snakebite. Iroquois Indians reportedly took it for whooping cough, pneumonia, and digestive disorders. Many Native Americans used the yellow juice of the root as a face paint and clothing dye. Early American pioneers valued goldenseal as an eyewash and as a treatment for sore throats, mouth sores, and digestive disorders. It has also been used to treat hemorrhoids, acne, and painful menstruation. Goldenseal is now a top-selling herb in North America and can be found in many formulations in stores across the country. Current uses include treatment of nasal congestion, mouth sores, digestive disorders, eye and ear infections, and as a topical antiseptic.

Medicinal Properties and Current Uses

Goldenseal roots contain a number of alkaloids, but the medicinal properties are usually attributed to the two major isoquinoline alkaloids—hydrastine and berberine. These are usually present in dried roots in concentrations of two to ten percent. Berberine is the compound that gives goldenseal its distinctive yellow color and bitter taste. Berberine has been shown to have a wide range of effects on the human body. For example, it can stimulate digestion and bile secretion, lower blood pressure, reduce muscle spasms, and inhibit growth of bacteria. Berberine has demonstrated antibiotic activity and produced results comparable to those of standard antibiotics in clinical trials testing the effectiveness of various treatments for common gastrointestinal infections. In several of these studies, berberine was the best treatment, and unlike with many synthetic antibiotics, berberine does not contribute to the development of yeast infections, which can often be as difficult to treat as the original bacterial infection. This is probably the explanation for the widespread use of goldenseal products to treat travelers' diarrhea, food poisoning, giardia, and cholera.

A 2003 study demonstrated that goldenseal root extract inhibited the growth of *Helicobacter pylori*, a difficult to eradicate bacterium that causes stomach ulcers and may lead to the development of stomach cancer. Other studies clearly demonstrate the antibacterial properties of goldenseal against dangerous microorganisms such as Staphylococcus, Streptococcus, *Escherichia coli*, and Pseudomonas. Berberine has immunostimulatory, anti-inflammatory, anticonvulsant, and sedative activities. What is perhaps most exciting is that it has exhibited anticancer activity directly, by killing tumor cells, and indirectly, by stimulating white blood cells.

Hydrastine, the other major alkaloid in goldenseal, has its own distinct properties. For example, it has been shown to reduce blood pressure. However, there are also reports that it can constrict peripheral blood vessels and stimulate involuntary muscles.

Goldenseal has long been used as a topical antiseptic and to treat fungal conditions such as ringworm. Research supporting this use shows that berberine and hydrastine are "contact disinfectants." This means that, unlike systemic antibiotics, these antimicrobial alkaloids are not absorbed into the bloodstream. They inhibit only the bacteria, fungi, and parasites with which they come into direct contact. Such contact usually occurs only on the surface of the mucous membranes.

Many herbalists and goldenseal users believe goldenseal helps boost the immune system and increases the efficacy of other medicinal herbs. Research supports the claim that goldenseal enhances immune function and, at least in the laboratory,

Jeanine Davis propagating botanicals in 2001 at the North Carolina State University Mountain Horticultural Crops Research Station in Mills River, North Carolina. Photo by Jackie Greenfield.

inhibits the growth of some strains of influenza. I supplied the goldenseal root for a study conducted by Chad Cecil for his Ph.D. research at North Carolina State University which demonstrated that goldenseal extracts and berberine inhibited growth of the H1N1 influenza virus.

Some people ingest goldenseal to mask the presence of illegal drugs in urine tests. Experiments have confirmed the drug-masking capability of goldenseal, but laboratories can also now screen the urine for the presence of the herb or its constituents. Historically, the Cherokee used goldenseal to improve appetite. More recently, anorexia has been treated with goldenseal. It was also an ingredient in an insect repellant. Goldenseal's reputation as a general tonic or panacea earned the plant one of its names, "poor man's ginseng."

Goldenseal leaves also contain significant amounts of berberine and hydrastine. There are some weight loss products and teas that utilize goldenseal leaves. As extraction technologies improve, more manufacturers may use goldenseal leaves, a more sustainable product to produce than goldenseal roots since the plant does not need to be destroyed to provide the raw material.

12

Goldenseal Growing Instructions: Methods, Care, Protection, Harvesting, and Marketing

Growing Methods

Goldenseal has been cultivated since the early 1900s, and many of the current recommendations for growing goldenseal are surprisingly consistent with those found in a 1914 United States Department of Agriculture (USDA) Farmers' Bulletin. Goldenseal production will probably be most successful in areas where goldenseal is native. Success in other areas will depend on how well those conditions can be duplicated. My years of research and growers' experiences form the basis of the information in this section. We found that goldenseal is not hard to grow. Producing it for a profit is the challenge. Goldenseal is in demand, however, and since it is listed on the CITES Appendix II list and on the United Plant Savers "at risk" list, more growers should be able to produce this plant profitably.

Choosing a Production System

As Scott described in his section of this book on ginseng, there are three production systems to choose from: intensively cultivated under artificial shade, woods-cultivated, and wild-simulated. Please refer to his section for details on those systems. Since most goldenseal is grown from rhizome cuttings instead of from seed, establishing a wild-simulated planting of goldenseal is more difficult than for ginseng. As a result, most commercial growers of goldenseal use the woods-cultivated or artificial-shade method.

Site Selection

Site selection is the most important factor for producing healthy goldenseal. It must be grown in an area that has sufficient winter cold to meet the dormancy requirements of the rhizomes, so it probably will not thrive any further south than northern Georgia. Notice that I said "thrive." It will grow further south, but it will be difficult to do so commercially because extended periods of high heat stress the plants and if the winters do not provide adequate chilling

A woods-cultivated goldenseal planting site in Kentucky. Photo by Jeanine Davis.

for the rhizome, it might not emerge in the spring. Referring to the 2012 USDA Cold Hardiness Zone Map, it should grow most reliably in zones 4 to 7. In Canada, goldenseal should be hardy up to the southern reaches of zone 3. Goldenseal likes moisture and rich soil, so it can definitely be grown in much of the Northeast, mid-Atlantic, upper Midwest, and Pacific Northwest in the United States. It is not recommended for the Deep South or in desert regions of the country.

Goldenseal grows best in a rich, loamy soil with good airflow and water drainage. It likes moist soil, but not soggy soil. Therefore, in areas with questionable drainage, plant goldenseal on a slight slope to improve water drainage, particularly surface run-off. Never plant it in a heavy, poorly drained soil. Likewise, try to avoid soils with high clay content because they often don't drain well. The best soil is a dark, loamy forest soil—the kind found in old deciduous forests—soil you can dig up with your bare hands, that is cool, moist, friable, full of pieces of organic matter, and smells earthy and alive.

If you want to grow goldenseal in the forest, look for a site where other woodland herbs are growing such as mayapple, trillium, bloodroot, or black cohosh. Do not select a site where there is no undergrowth because it is probably too dark for goldenseal. Conversely, try to avoid sites where the undergrowth is particularly thick, such as in a rhododendron thicket, because the soil will not be the best and the effort required to remove those plants and their roots will be too costly. A site with deeply rooted, mixed hardwoods is preferred to a solid stand of conifers or other shallow-rooted trees that can compete with the goldenseal for moisture and nutrients. In my experience, goldenseal plantings established under oak, poplar, walnut, and basswood have usually been successful.

If you choose to grow goldenseal under some kind of artificial shade constructed in an open field, it is important to either choose a site with few weeds or gain control of the weeds before planting the goldenseal. Grasses can be a serious problem if you plant into a pasture without adequately turning the soil to kill existing weeds and weed seeds. As with a forest site, look for well-drained, moist, rich soil.

Try to find out the history of the land you want to grow on. Have other crops been produced there before? If so, were there any disease or insect problems with that crop? What kind of agricultural chemicals were used? In particular, were any persistent herbicides applied to that field? You do not want to plant goldenseal in an area known to be infested with soil-borne diseases, especially Rhizoctonia. Also, to reduce the risk of disease, do not replant goldenseal immediately after growing a crop of goldenseal. I have successfully grown goldenseal after ginseng, black cohosh, and bloodroot.

Providing Shade

Goldenseal needs to be grown in the shade. Natural forest canopy or some kind of artificial-shade structure can provide this. I prefer the natural forest but have grown goldenseal successfully under a polypropylene shade-cloth structure for over 20 years. In a natural forest, you want a mixed, mostly deciduous forest with a high canopy. The high canopy will allow for good airflow, which is important for foliage health. Look for a site with mostly mature trees. Their canopies will be high, and their shade is usually adequate to have prevented bushy undergrowth from thriving. Deep-rooted trees will not get in the way of working up the soil or compete with the goldenseal plants for moisture and nutrients the way shallow-rooted trees can. The shade level should be approximately 75 percent. This can be difficult to determine without a light meter. If you do not have access to a light meter, a very crude estimate of whether there is the minimal amount of light needed to grow goldenseal is being able to read this book at your site at midday in the middle of the summer. A slightly more sophisticated method is this one described by Dave Apsley and Chip Carroll in a 2004 Ohio State University Extension Factsheet on growing ginseng: "A quick method to approximate the amount of shade that is on a given site is to place 10 or more white paper plates at even distances on the ground at approximately noon on a sunny summer day. Count the number of plates that are at least half shaded. Next divide the number of shaded plates by the total number of plates placed on the ground. Multiply this number by 100. If this number is 70 or greater the site is probably shady enough to grow ginseng."

A field-cultivated goldenseal planting site in Ontario, Canada. Photo by Jeanine Davis.

Artificial shade can be provided by a wood lath structure, a polypropylene shade structure, or vining plants trained over a support structure. Scott provides detailed plans in his ginseng section for building a shade structure. When you design the shade structure, be sure to provide for adequate air circulation. Make the structure seven feet tall or higher. If you add side curtains, leave two opposite ends open to the prevailing breeze.

I conducted a study at the Mountain Horticultural Crops Research Station in western North Carolina in which goldenseal was grown under polypropylene shade cloth at four levels of shade—30 percent, 47 percent, 63 percent, and 80 percent. Plant stand counts (the number of plants that came up compared to the number of roots that were planted) and annual survivability were highest under 47 percent and 63 percent shade. On the other hand, the more light the plants were exposed to, the more problems we had with foliar diseases and weeds. These were especially bad in the 30 percent and 47 percent shade treatments. The highest root yield after four years of growth was obtained with 63 percent shade, but the highest root alkaloid content was obtained with 30 percent shade. The effect on alkaloid content is important because some buyers pay a premium for high hydrastine or total alkaloids. When root yield was multiplied by root alkaloid content, it was demonstrated that the highest alkaloid yield per unit land area was obtained with 63 percent shade. If you are being paid a premium for high hydrastine or total alka-

Little polypropylene shade structures where we grew goldenseal and other woodland medicinals under different levels of shade. Photo by Jeanine Davis.

loid concentrations in your roots, I would recommend trying to provide about 63 percent shade for goldenseal. However, I must point out that, for over 20 years, some of our prettiest and healthiest goldenseal which produced high root yields was produced under a 78 percent lath-weave polypropylene shade-cloth structure. Taking all this into account, I recommend using 63 to 80 percent shade for growing goldenseal.

Site Preparation

Taking the time and making the effort to prepare your site properly in the beginning will make your work much easier as the years go by. First, begin your preparations early. If at all possible, start preparing your site weeks or months before you intend to plant. In a forest site, remove small undesirable trees, tree roots, weeds, and other undergrowth. In an open field site, turn the soil several times, allowing weed seeds to germinate between turnings. Collect soil samples for soil tests. If time allows, grow a cover crop to add organic matter and nutrients and to smother out weeds. I like to grow a combination of plants like crimson clover, vetch, and rye. Be sure to turn these into the soil or mow them before they go to seed.

If you find it necessary to control difficult weeds with herbicides, be careful to choose ones that do not have a long carryover time in the soil. I try to grow goldenseal organically whenever possible, but sometimes a planting must be established quickly in a weedy site, and the use of an herbicide may be considered the only practical solution. Since you must grow the goldenseal for three or more years before harvest, you can still have that goldenseal certified organic if you start the process after the herbicide application (I know that sounds like cheating, but it does meet the regulations for organic certification for a perennial crop). In those cases where using an herbicide is considered necessary, the herbicide Roundup may be the best choice because it is systemic, kills almost all weeds, and does not persist as an herbicide in the soil. Use Roundup only before the soil is worked up or any plants are put in the ground. Keep in mind that Roundup is a powerful chemical and will kill almost any plant if it touches the foliage, so be very careful when using it! Do not get it on any plants that you do not want to destroy. Always apply herbicides on a calm day and strongly consider buying or making a shield to control where the spray goes. If you are currently certified organic, use of Roundup will void your organic certification for that particular field, forcing you to wait three years before the field and the crop growing there can be certified organic.

Once the land is cleared of weeds, underbrush, and debris, if you plan to grow in a woods-cultivated or artificial-shade system, till the soil and add any soil amendments (fertilizer and lime), following the recommendations from the soil tests you conducted earlier (see next section). If you are going to add aged manure or compost, this is the time to do it. To promote good water drainage and to warm the soil early in the spring, build raised beds or dig trenches between your beds. Scott gives an excellent description of how to do the trench method in his chapter on woods-cultivated ginseng. If the soil is deep enough and free of many roots, you can build raised beds that are 2–6" tall and 3–4' across (metric conversions are in Interesting and Helpful Calculations). In a field site, raised beds can be built with a bed-shaper that is pulled behind a tractor. In a forest setting, if you cannot get a tractor onto the site, raised beds can be constructed by pulling up tilled soil with a hoe or shovel. Be sure to slightly crown the center so water will run off. Whichever method you use, leave sufficient space between beds to allow easy walking, pushing a wheelbarrow, and kneeling for weeding and picking fruit. Sometimes it is just not practical to build raised beds. In western North Carolina, there are many mountain sites that have only a thin layer of soil over rock. The soil is too shallow there to build raised beds or dig trenches. In these situations, try to work up the soil as best as you can and define your planting areas with long straight branches, little survey flags, rocks, or landscape timbers.

If you plan to use the wild-simulated method, pulling back the leaf litter and incorporating any soil amendments will take place at time of planting.

Fertilization

Several months before planting, collect soil samples for fertility testing and nematode assays. Many states have their own testing laboratories that will analyze your soil samples for a nominal fee. Call your county extension office, land-grant university, department of agriculture, or ministry of agriculture for this information. In addition, a large number of commercial soil testing laboratories are located all over both countries. If you choose to use a commercial lab, use one close to your area. Labs use extraction methods appropriate for the soils in their region. The method used by a lab outside your region may not be appropriate for your soil and could result in an inaccurate reading. Your county extension agent or the Soil Science Department at your state land-grant institution can provide more information on this topic. Along with your soil sample, you will need to submit a form that will ask what crop you want to grow. I would be very much surprised if you found "goldenseal" on that form. For most soil testing laboratories, it will be necessary to identify the crop as "native ornamental" and write in "goldenseal." Better yet, call the laboratory and ask for their advice. If your soil test results reveal low organic matter at the planned site, you can increase it by growing a cover crop or through the addition of compost. Leaves from the future growing area can be collected, shredded, and composted. Any good quality compost, however, should work just fine.

Soil pH and Calcium

My studies on western North Carolina forest soils show that soil pH has a dramatic effect on goldenseal growth, root yield, and root alkaloid content. In these trials, goldenseal produced the highest root yield when grown in our acidic mountain soil amended with 2.7 to 5.2 tons of lime per acre, which resulted in a soil pH of 5.5 to 6.0. Goldenseal will grow in a wide range of soil pH. At a low pH (4.5), goldenseal grows very slowly, but the root alkaloid levels will probably be high. At a high pH (>6.5), goldenseal grows very quickly, but root alkaloid levels will probably be low and the plants may be more susceptible to disease. In this same study, the response of goldenseal to different rates of nitrogen and phosphorus, supplied as calcium nitrate and superphosphate were also examined. Plant survival and root yields decreased as the rate of nitrogen increased. Additional phosphorus reduced plant survival rates, but had no effect on root weight of the surviving plants. Based on these results and grower experiences, I urge caution when using synthetic fertilizers on goldenseal. Using dolomitic limestone to raise a low soil pH to a pH of 5.5 to 6.0 will help your plants grow a little faster. However, if your soil tests recommend the addition of phosphorus, try using a slow-release natural product, such as rock phosphate, instead of superphosphate. To add some nitrogen, rely on organic sources such as compost, soybean meal, and poultry litter products.

Because ginseng often responds favorably to additional soil calcium applied as gypsum, some goldenseal growers are also adding gypsum to their soils. Results from my studies and observations of many goldenseal farms do not support this practice. In my study, gypsum was applied as

Goldenseal plants emerging in the spring in the soil pH, nitrogen, and phosphorus study. Photo by Jeanine Davis.

Goldenseal roots after three years of growth in soil amended with lime. Values shown were the target pH values. A pH value over 7.0 was not achieved. Photo by Jeanine Davis.

recommended for ginseng, at a rate of 2,000 to 5,000 pounds of calcium per acre. The addition of calcium resulted in reduced leaf number and size, increased disease incidence and earlier dieback, and lower root weights than plants grown in soil without additional calcium. I have observed that the plants in growers' fields where large amounts of gypsum were added responded similarly. Plant nutrition is complex and influenced by the amounts and ratios of minerals in the soil, the soil type, and soil pH. Adding high amounts of calcium to the soil without benefit of a soil test can affect the availability and uptake of other nutrients by the plant in unexpected and possibly undesirable ways. I do not recommend adding additional calcium to the soil for goldenseal unless the soil pH needs to be raised or a soil test indicates it is needed.

Fertilization after Planting

Once goldenseal has been planted, I have rarely found it necessary to add more fertilizer to the soil. However, goldenseal often responds favorably to a light foliar feeding in the spring with an organic fish or seaweed fertilizer. These are not big sources of nitrogen, phosphorus, and potassium; a common analysis is only 2-3-1, but these products are rich in other nutrients including oils, amino acids, vitamins, hormones, and enzymes. Mix with water, and using a pump-up backpack sprayer, apply it to the foliage to the point of runoff. If more nutrition is desired, a non-organic option is to dissolve two tablespoons of a 20–20–20 fertilizer in a gallon of water and spray as above. Be sure to follow label recommendations for foliar feeding. If you apply a solution that is too concentrated, you can burn the foliage.

Propagation

Goldenseal can be propagated by several different methods. The easiest and most common way is to cut the thick underground rhizome of a mature plant into pieces. Another method which growers have mixed success with involves planting pieces of the long slender fibrous roots that contain tiny buds. Small dormant plants, which are usually two-year-old rhizomes with roots, can be purchased for transplanting. You can also grow goldenseal from seed. To produce roots large enough to harvest for market, it usually takes from three to five years from rhizome pieces, three to six years from transplants, four to six years from fibrous root cuttings, and five to seven years from seed.

Obtaining Plant Material

When you purchase or dig goldenseal to use as planting stock, keep in mind that goldenseal is a species at risk. Help protect threatened populations of goldenseal by finding out where your planting material originated from. If you intend to buy from a nursery (the preferred method), ask them where they obtained their plants. The most desirable situation is that the nursery grows all its own goldenseal. If you are only going to grow a few goldenseal plants in your yard, you should have no trouble finding enough nursery-grown potted plants. If, however, you want to put in a tenth of an acre or more of goldenseal, it may be a challenge to find a nursery with an adequate supply of plants it has grown itself. You might also find the cost of large volumes of nursery-raised material to be prohibitive. In this case, you might want to purchase your material from a raw botanicals buyer (dealer) who buys and sells bulk herbs and roots. These buyers usually handle dried material, but they can often get fresh material on request. Try to find a buyer with a good reputation for obtaining plant material legally, in compliance with state and federal laws, and who practices and supports sustainable harvesting. This may not be an easy task. There are also a number of companies selling goldenseal plants and rhizomes on the Internet. Check them out carefully before sending them your money. Any legitimate company will clearly post their name, physical address, phone number, and "About us" information. If they don't, and you are still tempted to buy from them, send them an email requesting that information as well as their federal tax identification number. It has been my experience that companies that are not upfront on their websites about who they are and where they are located do not respond to those emails.

Since many ginseng dealers also buy and sell other roots and herbs, they can be a source of large quantities of goldenseal. If you live in a state in which wild ginseng is harvested and sold, there will be a state office in charge of the federally mandated ginseng protection program. In North Carolina, for example, this is the Plant Industry Division of the North Carolina Department of Agriculture and Consumer Services. As part of the ginseng program, the state maintains a list of all the ginseng dealers in the state. In many states, this list is available to the public online. Once you make contact with one of these dealers, you should insist that the material you purchase is obtained legally and harvested sustainably. To meet the CITES regulations for trade in goldenseal, you need to have a receipt for your planting stock. The receipt should include information about the state and county from which the goldenseal was harvested. If the goldenseal is to be harvested from public land, ask for a copy of the permit. If it will be harvested from private land, get a copy of the landowner's letter of permission.

The least expensive way to obtain your goldenseal planting stock is to dig your own material. In many states and throughout Canada, you cannot dig goldenseal from any public lands. In states where goldenseal is more abundant and it is legal to dig from public lands, you will probably need to obtain a permit from the state or federal forest service. If you want to dig goldenseal on private land, be sure to get the landowner's permission in writing. Conducting or assisting with plant rescues on land that is going to be developed is another opportunity for obtaining goldenseal. Call the Nature Conservancy, your state plant conservation office, or other environmental organizations in your area to find out who organizes plant rescues and how to obtain a list announcing when they will be held and how the material will be handled. For some organized plant rescues, all the plants collected must be used for conservation purposes, and using them to grow and harvest commercially would not be acceptable.

If you purchase your rootstock, buy only dormant plants. This means that plants should be dug in late fall, during the winter, or in very early spring. The buds should be well developed, but they should not have sprouted. If you are digging your own roots, it is still best to dig the plants when they are dormant. If you are doing a plant rescue during the late spring or the summer when the plants are actively growing, handle them with great care. Replant them as quickly as possible and keep the soil moist. If it is late summer or fall, you should see well-developed buds on the rhizome. Cut the top of the plant off and hold the root in a root cellar or refrigerator until cool weather arrives. If it is late fall or winter when you dig, either replant immediately or put the dormant roots into cold, moist storage until planting time.

Inspecting and Storing Propagation Material

If you dig your own planting stock, gently clean off the soil and inspect the rhizome and roots. If sticky clay or mud covers the roots, it will make your task easier if you wash them first by holding them under running water or swishing them around in a tub of water. If you purchased the roots, smell them when you first open the bag or box in which they were shipped. They should have a very earthy and distinctive goldenseal smell to them. I cannot describe it, but you will quickly learn what goldenseal smells like. If you smell anything like mold, mildew, or that sick-sweet smell of decaying plant material, inspect every plant carefully. If more than just a few of the plants are affected, you should reject the shipment.

Whether you dig the roots yourself or purchase them, check the roots for any signs of insects or disease. If the roots have small, black lesions, do not use any part of that plant for planting stock. It may have a root disease that you do not want to introduce into your new planting site. Also look for little, round nodules on the roots. This could indicate the presence of root knot nematodes. In fact, do not use any material that looks weak, unhealthy, or strange in anyway. It is just not worth the risk. All the rhizomes should be thick and hard. If the rhizomes or fibrous roots are squishy at all, they are starting to decay. If you break a root or scratch it with your fingernail, it should be bright yellow without any brown streaking inside.

If you need to store the roots (rhizomes with roots, root pieces, or transplants) either for the winter or for just a short time, pack them in burlap sacks or cardboard boxes. Do not try to store the roots in sealed plastic bags. If you will not be able to check the roots regularly, line a cardboard box with a plastic trash bag, and loosely fill it with roots. Moisten the roots and loosely close the top of the bag so there is some exchange of air between the inside and outside of the bag. Do not use a twist tie or put a knot in the top of the bag. If you do, the roots will start to decay very quickly. If it will be for more than a few days, pack them in moist sphagnum moss to help retain moisture. The moss also provides some protection against fungal growth. Keep the roots damp but not too wet. Once every week or two open the boxes or bags, smell the roots, stir them up with your hands, and moisten them if they are starting to get dry. The roots must be kept cool in storage. I have access to walk-in coolers that are kept at 40°F and are ideal for storing roots. If you only have a small number of roots, you can store them in your refrigerator, but buy a thermometer and make sure the temperature where the roots are kept stays between 35°and 40°F. You do not want to freeze the roots, and many refrigerators are close to freezing—especially in the back of the bottom shelf where it is likely you will put your roots. If you have a large quantity of roots and no walk-in cooler, you can store them during the winter in a root cellar or in a cool corner of an earthen cellar, if the temperature stays between 35° and 40°F. Most basements are too warm, and an unheated outbuilding is probably going to fluctuate too much in temperature.

Propagation by Transplants, Rhizome Pieces, or Root Cuttings

By far the easiest and most reliable methods for propagating goldenseal are using transplants or planting pieces of the rhizome. Transplants come ready to plant. Whole rhizomes need to be cut into ½" or larger pieces. Each piece should have healthy fibrous roots and, ideally, a bud. In general, the larger the piece you plant, the faster you will have a harvestable-size root. However, if you are trying to grow commercially you want to make your rootstock go as far as possible. My research on propagating goldenseal has shown that pieces as small as one gram will produce a plant, but the smaller the piece, the longer the plant must grow before the root reaches a harvestable size. That study also showed that the presence of fibrous roots on the rhizome piece proved to be more important for plant survival than the presence of a highly visible bud. There are often tiny buds present that are not readily apparent on the rhizome. They will usually sprout if there is no large bud present, or if the large bud is broken off, damaged, or dies. Some rhizome pieces that do not have large buds will not send up a top the first growing season but will do so the following season. So, for a commercial operation, cut the rhizomes into pieces ½" to 1" in length, being sure to include some roots, and try to have at least one obvious bud per piece.

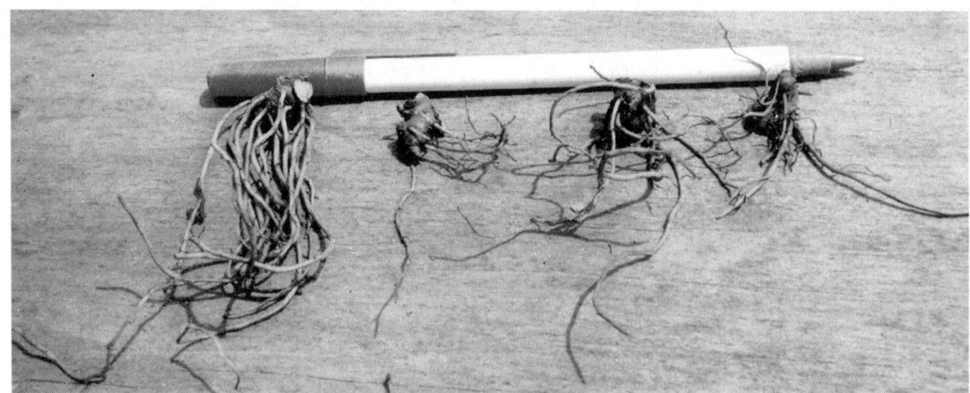

Goldenseal propagation by cutting rhizomes. The piece on the far left is a good example of a healthy rhizome piece for propagation. The piece second from the left has too few fibrous roots. The other two pieces are acceptable. Photo by Jeanine Davis.

If you are growing under an artificial-shade structure or in a woods-cultivated system, make narrow trenches 2–3" deep across the raised beds and plant the transplants or rhizome pieces with the bud facing up. Bury them with soil so they are at least 2" below the soil surface. Do not leave the bud exposed; cover with several inches of organic mulch, as described later. If you are planting in a wild-simulated system, dig small holes 2–3" deep, set the rhizome piece or transplant in with the bud facing up, and cover with soil.

Another method of vegetative propagation involves using root cuttings and a process called layering. To use this method, look for strong, fibrous roots with buds or large swellings that look like they might turn into buds. Cut the fibrous roots into 2- to 4"-sections, with the bud or swelling in the middle, and layer them under 1" or 2" of soil in a nursery bed in late fall or early spring. Cover with several inches of mulch. The fibrous root pieces that you plant should form small tops or large buds during the summer. These little plants can then be set into production beds the following autumn. Although several growers have reported success with this method, my results have been very mixed. However, if you have lots of fibrous roots and some time, it may be worth trying.

Seed Propagation

Propagation of goldenseal from seed can be difficult and unpredictable. However, Richo Cech of Horizon Herbs and several grower friends of mine are convinced that the highest root weight increase is obtained by using seed propagation. So, if you have access to seed, it is worth trying to grow it. Storing goldenseal seed in a cooler or a stratification box buried in the soil can result in poor germination and in a delay in germination by a year compared to fresh seed that is sown immediately after it is collected from the plant. In my research, germination rates of purchased seed that had been stored in a cooler or buried in the soil ranged from zero to ninety percent the first spring after seed harvest. Generally, most of the seed that has been stored will germinate the second spring after seed harvest. In contrast, seed that is sown soon after harvest from the plant will often have a 75 percent or higher germination rate the first spring after planting. Interestingly, Henri Hus gave this same advice in a 1907 Missouri Botanical Garden report. Sowing fresh seed, however, is usually practical only if you already have a bed of goldenseal and can collect your own seeds.

Many growers and researchers have experimented with goldenseal seed and have obtained similar results. Like ginseng, goldenseal seed has very specific temperature requirements for germination. The consensus is that goldenseal seed requires a warm period followed by a cold period before it will germinate. We have conducted seed germination studies on goldenseal at the Mountain Horticultural Crops Research and Extension Center in western North Carolina for many years. A wide range of seed extraction methods, seed disinfection treatments, storage temperatures, and

These goldenseal plants all grew from seeds from one berry, sown at the same time. Seeds emerged from nine to twenty-one months after sowing. Photo by Jeanine Davis.

planting times were tested. Storage temperature and time of planting had the most dramatic effect on seed germination. In all of these studies, the highest germination rates the first spring after seed harvest were obtained from seeds that were sown outdoors immediately after harvesting from the plant (in late summer). The next two best treatments were very similar in their resulting germination rates. Those treatments were (1) to hold the seed at 70°F for 30 days and then move them to 40°F until planting in the spring or (2) hold the seed at 40°F until planting in the spring. In these latter two treatments, most of the seeds did not germinate until the second spring after seed harvest. Final germination rates by the second season for these methods ranged from 25 to 80 percent.

In New Zealand, Follett, Douglas, and Littler did a study on goldenseal seed germination using two soaking solutions and several cold stratification regimes. They found that soaking fresh goldenseal seed for two hours in gibberellic acid (GA3) (0.5 grams of the actual compound in a liter of water) accelerated germination. (Gibberellic acid is a naturally occurring compound that is readily available; ProGibb T&O is an example of a common commercial product.) They also found that stratifying seed at 40°F for eight weeks or longer delayed and reduced germination.

Richo Cech of Horizon Herbs in Oregon has an innovative way to handle the seeds. He calls it the "mesh bag at the back of the sink method." After he cleans the seed, he hangs them in a mesh bag by the sink and rinses them twice a day in cold water for three weeks before storing them at cool temperatures or sowing them in a nursery bed.

More information on seeds is covered in the section on seed collection, storage, and planting later in this chapter.

If you purchase goldenseal seed, call the company and ask if they produce their own seed and, if not, where they purchase it from. Also ask how it has been handled and stored. My experience has been that germination rates are higher when I buy directly from the person who produced the seeds. I suspect that seeds from companies that have to purchase goldenseal seeds from other growers have often been allowed to get too dry or too warm during transportation and handling, reducing viability and germination. Goldenseal seeds tend to vary in weight from year to year and location to location. There is an average of 27,000 seeds in a pound, but based on what growers, wild seed collectors, and researchers have reported and my own experience, this can range from 22,000 to 34,000 seeds per pound.

Planting and Spacing

Time of Planting

Depending on where you live and what method of propagation you choose to use, goldenseal can be planted in the fall or in the spring. Fall planting of transplants, rhizome pieces, and root cuttings has been successful in all growing areas and is always recommended. Early spring planting of transplants and rhizome pieces has been quite successful in the Southeast. Spring planting of root cuttings should also work. I have even had success with planting in midsummer in North Carolina. Many years ago, we were putting in a large number of test plots. We had about 25 pounds of goldenseal rhizomes and one-year-old seedlings left over. These had been dug in late October, put in moist sphagnum moss, and held at 40°F. We finally got around to planting the excess material in early July. The plants grew incredibly fast, and by the next spring they could not be distinguished from the plants that had been planted several months earlier at the "proper time." When to plant seeds is a little more complicated and will be explained in the section on seed collection, storage, and planting.

Planting Methods

Goldenseal rhizomes, transplants, or root cuttings are usually planted by hand. If you have raised beds, to prevent damaging the beds as you plant, kneel either in the aisles or on a 6" wide board spread across the width of the bed. Make narrow trenches 2–3" deep across the raised bed using a hand trowel. To space the plants evenly, mark the spacing you want to use on the edge of the board with a permanent marker, e.g., make a mark every six inches. When our beds are narrow enough to reach across from the aisles instead of kneeling on a board spread across the bed, we use a wooden stake that is as long as our beds are wide (we use tobacco or tomato stakes) and make the markings for the in-row spacing on it. We also mark a four-foot-long stake with the

Planting goldenseal by hand using a marked stake for proper spacing. Photo by Jeanine Davis.

between-row spacing, lay it beside the bed, and drag it along as we move. If the in-row spacing is different from the between-row spacing, we use a different color marker for each one to avoid confusion.

If you are planting a large area under artificial shade, you can use a vegetable or tobacco transplanter that is pulled and powered by a tractor. Unfortunately, I am unaware of any small-scale people-powered transplanters for use in a woods-cultivated setting.

Goldenseal seed is usually sown into a nursery bed as described in more detail below. You can sow it by hand, but if you have large quantities of seed, I strongly recommend using a small hand-push seeder as Scott describes in chapter 5 for ginseng seed. Very large volumes of seed can be planted in beds under an artificial-shade structure with a tractor-mounted precision seeder.

Plant Spacing

It is common practice to plant goldenseal on a 6" × 6" spacing, and our research supports this as being the spacing that provides the greatest increase in root weight under the local conditions in western North Carolina. Spacing any closer tends to crowd the plants, making them more susceptible to disease and decreasing root weight. Wider spacing is an inefficient use of tilled area and makes digging more difficult. If you are not trying to obtain maximum yields in the shortest period of time, or you want to grow the plants for longer than five years from rhizome pieces or six years from transplants or root cuttings, you can space the plants farther apart, up to 12" × 12".

Mulching

Goldenseal should be mulched to hold in soil moisture, reduce weed growth, moderate temperatures, and provide winter pro-

Susan Bryson collecting data from a study on goldenseal grown with different mulches. Photo by Jeanine Davis.

tection. The mulch layer should be several inches deep at planting time. Mulch will decompose and compact at different rates depending on the type of mulch used and may need to be replenished every year or two. In areas where the soil tends to freeze and thaw, several inches of extra mulch should be provided to protect the roots during the winter. Depending on the location and amount of exposure, up to 6" of mulch may be piled on the beds for the winter. If extra mulch is added, it should be raked back to a depth of about 2" before the plants emerge in the spring.

Goldenseal is commonly mulched with whole or shredded leaves, hardwood bark chips, a hardwood bark and sawdust mixture, or straw. Although straw is used successfully in many northern states and Canada, it has been my experience in North Carolina, and that of many growers in the Southeast, that straw performs poorly under our conditions. Straw tends to hold moisture near the crown of the plant, which often results in rot of the lower stem and crown. In my research studies, slugs have also caused significantly more damage in straw-mulched plots than in any other mulch treatment. Of all the mulches we have tried, hardwood bark and pine bark mulches have performed the best under artificial-shade structures. Leaf mulch is recommended for forest plantings. Because sawdust is readily available in my area (usually just for the price of hauling it away), I tried growing goldenseal with a fresh sawdust mulch. The first year, the plants were pale yellow and obviously nitrogen deficient; the decomposing mulch competed with the young plants for nitrogen. There was, however, almost no slug damage in the sawdust-mulched plots compared to the six other mulch treatments included in the study. After the first year, the plants grew so well that after four years of growth, root yields were equivalent to those obtained from the hardwood bark plots. Using aged sawdust instead of fresh sawdust would probably prevent the yellowing problem the first year.

Irrigation

When grown under a forest canopy in the moist climates of the eastern United States and Canada, goldenseal rarely, if ever, requires irrigation. In the drier regions of the West, or under drought conditions, if the plants are not irrigated, they will drop their foliage and go dormant earlier than usual. This usually does not harm the plant but will reduce root growth and can delay harvest for a year or two. If that kind of delay is not acceptable, site selection should include consideration for how irrigation will be supplied when necessary. There are some very innovative forest irrigation systems being used for ginseng, goldenseal, and shiitake mushrooms. In the mountains of western North Carolina, gravity-fed systems are often built using water from nearby streams. These systems rely on the pressure developed by water flowing down a slope to force water through micro sprinkler heads. Some growers who do not have a water source close by put a tank near their beds and fill it with water from carboys that they haul in on four-wheelers. Then they use soaker hoses to carefully place water right next to the plants. This is only practical for very small plantings. In large artificial-shade gardens, drip-irrigation or standard overhead irrigation systems with risers and oscillating sprinkler heads are usually used.

Slug damage on goldenseal leaves. Photo by Jeanine Davis.

Pest Control

Under natural conditions and when grown in small isolated plots in the woods, goldenseal suffers few attacks from diseases or insects. The major problem in many plantings is slugs. Slugs can quickly devour an entire bed of seedlings, kill the crowns of small plants, make Swiss cheese of the foliage of large plants, and eat all the fruit. Slug control can be difficult, and successful methods are often site-specific. Control methods that have been successful at some sites include putting out beer traps, or spreading diatomaceous earth (must be replaced after every rain), a mixture of lime and wood ashes, or commercial slug and snail bait around the plants. If the populations of slugs are intolerable, it may be necessary to remove the mulch from around the plants. Years ago, I was involved in a research project with Dr. Fumiomi Takeda, Research Horticulturist at the USDA Appalachian Fruit Research Station in Kearneysville, West Virginia. We looked at the use of natural clay products for slug control. Preliminary results with foliar sprays were promising. I am not aware if any of those products have ever been made available for purchase, but the kaolin clay product Surround sprayed on the foliage has worked effectively for some growers. Be aware that this product will coat the leaves with what looks like a slurry of mud.

Moles and voles may also damage beds and should be controlled with traps or by bordering the beds with wire mesh set 8–12" deep in the soil. See chapter 5 for more information on slug and rodent control. If you feel you must use baits that can be toxic to pets, livestock, and wildlife, apply them in such a manner that non-target animals cannot gain access to them.

Slug trap that can be filled with beer or other bait. Photo by George Cox.

Weeds

If the planting site was properly prepared and the beds adequately mulched, weeds will probably not be a serious problem in the woods, and hand weeding several times during the season will usually be adequate. If, however, the forest canopy is thinned and/or the soil greatly disturbed during site preparation, weed growth could be stimulated, particularly grasses. Weeds can be a much more serious problem under a new artificial-shade structure since these structures are usually erected on land that was recently covered in grasses and weeds. Follow the recommendations provided by Scott in chapter 5 for controlling weeds prior to planting a crop. A good thick layer of mulch is still the best control mechanism.

Nematodes

Root knot nematodes can severely reduce growth and root yields of goldenseal. Always have a nematode assay done along with your soil test when evaluating a site for goldenseal, even in a forest. The only reliable method for nematode control at this time is soil fumigation, and this is usually not practical or desirable for a goldenseal site, especially in the woods. Soil fumigation is not an organic practice and kills all the beneficial soil micro-organisms along with the pathogenic ones. Therefore, if pathogenic nematodes are present at a damaging level, choose another site. Sometimes areas of established plantings exhibit nematode damage. Clearly mark those sites, remove and destroy the infested plants, and be careful not to move infested soil on shoes or tools to uninfested areas. It may be worth your time to apply some organic soil amendments, natural nematicides, or biological controls to try to clean up those areas. Do not consider replanting any crop there until you have conducted a nematode test and it comes back clean.

Diseases

When I first started growing goldenseal, the only disease commonly reported on it was Botrytis leaf spot. This was not considered a serious problem and usually only affected plants late in the season, shortly before the tops would die off naturally anyway. If it did become a concern, removal of the affected foliage and the mulch in heavily infected areas provided reasonable control. As more goldenseal was cultivated, however, reports of more threatening diseases increased. Most of these diseases occur in large plantings under artificial-shade structures, and are rarely seen in forest plantings. Diseases now commonly reported on goldenseal include Alternaria (see photo in color section), Rhizoctonia, and Fusarium. Under artificial-shade structures, damping-off of seedlings can be a problem, especially in spots where there is excessive drip from off the structure, as under joints and seams. There have also been several unconfirmed reports of Phytophthora root rot. In North Carolina, however, goldenseal has been grown successfully in old ginseng beds known to be infected with *Phytophthora cactorum.*

In 2002, Dr. Rick D. Reeleder, formerly of Agriculture and Agri-Food Canada in London, Ontario, presented a poster on research he had conducted to see if goldenseal was susceptible to several major ginseng diseases. He reported that neither *Cylindrocarpon destructans* nor *Phytophthora cactorum* infected goldenseal (both of these are described in chapter 5). Goldenseal seedlings were susceptible to damping-off from *Rhizoctonia solani* (isolated from ginseng), but the mature plants were not. Most interesting was the 2004 report of a new pathogen of goldenseal, *Cylindrocladium colhounii.* Dr. Reeleder reported that it causes a significant rot of goldenseal roots in commercial plantings in Canada. It causes the foliage to wilt and die but does not usually kill the plant. Thus, new foliage is produced each year, but it dies back early, and the roots do not grow properly. I have not had experience with this disease, but watch for it in all my plantings.

Disease Prevention and Management

As Scott so clearly explained in his discussion of ginseng diseases, the best approach to disease control is prevention. Site selection is the most important factor, followed by use of clean plant material and healthy soil. My experience has been that if goldenseal plants are pushed to produce high yields quickly, e.g., they are planted densely, fertilized heavily, and grown under artificial shade, they are more susceptible to disease than plants that are allowed to grow in a more natural manner. A more natural manner means plants that are growing in the woods and are spaced far apart in soil that is well drained and rich in organic matter, and where the air movement is good and a wide variety of other kinds of plants are growing nearby. This is not to say that you cannot grow goldenseal intensively. You can, but as Scott describes for ginseng, you should then be prepared to apply fungicides on a regular basis. This is a problem because there are even fewer pesticides registered for use on goldenseal than on ginseng. In fact, during a recent search of the *Greenbook*, which contains the complete labels for all the pesticides registered with the United States Environmental Protection Agency (EPA), I could not find any pesticides labeled for goldenseal. That does not mean that there are not special exemptions or rules, such as a 24-C label, for goldenseal in certain states. Check with your state land-grant university or department of agriculture to find out what the situation is in your state or province.

Organic growers use a variety of natural products to help control diseases on goldenseal during very wet periods. These include compost teas, hydrogen peroxide, copper, mild soap solutions, and horsetail tea. I cannot vouch for any of these treatments and they don't all meet the letter of the law for pesticides, but some growers swear they work, and I have seen some beautiful organic and biodynamic goldenseal plantings.

Disease Diagnosis

No matter which approach you take to disease management, if you encounter a problem, it is important to know how to identify it. If your plants become sick, dig a plant in the early stages of the disease, being sure to include some soil from around the roots, and submit the specimen to a plant pathology laboratory for diagnosis. Most land-grant universities offer a plant disease clinic of some sort, usually with free or very inexpensive services. Call your county extension agent or other agricultural consultant to find out what is available in your state or province. If such a service does not exist in your area, find the closest one that does offer that service and contact them for information on how you can make use of their service. Often there will be a fee of maybe $25 to $75 for nonresidents to use a public service, but it will be worth the cost.

Because you are growing an uncommon crop, it is likely that the plant pathologist trying to diagnose your disease problem will have little or no experience with your plant. You can help the pathologist by providing him or her with the list of known goldenseal diseases included in Joe-Ann McCoy's Disease List at the end of this book. Having this list available may improve your chances of getting a diagnosis, because the pathologist should quickly be able to narrow your problem down to just a few of the known possibilities. Of course, there is always the chance that you will have the first known incident of a disease on that plant, but let's just hope that's not the case.

You cannot work to control or prevent a problem you do not understand. Once you have a diagnosis, spend some time learning all you can about the disease—what conditions it needs to thrive, and what conditions are unfavorable to its development. To my knowledge, a book has not yet been published on diseases of native North American medicinal herbs, so you will have to make use of other references. I have found that college-level vegetable and nursery crop disease books usually include all the diseases of concern with these plants, at least to the genus level. As with plants, diseases are identified by genus and species. One of the ginseng root rots is caused by *Phytophthora cactorum. Phytophthora* is the genus name and *cactorum* is the species name. Apparently, goldenseal does not get *Phytophthora cactorum*, but there is some evidence that it might get some other species of Phytophthora. All Phytophthoras have a similar life cycle and environmental requirements. Information is readily available about controlling *Phytophthora cactorum* on ginseng or *Phytophthora capsici* on peppers. Reading about them will teach you about those diseases and how they are controlled on other crops. With that information you can probably devise some methods for preventing or controlling Phytophthora on goldenseal. Be certain to keep good records on what you do, so you can repeat them if necessary.

Seed Collection, Storage, and Planting

Seed Collection and Cleaning

Once your plants are mature enough to produce fruit, you should consider collecting seed for your own use or to sell. To collect seed from goldenseal, harvest fruit when fully ripe (dark red). Put the fruit in a bucket, and add water to about twice the depth of the fruit. Mash the fruit by kneading with your hands or with a tool such as a potato masher, being careful not to crush the seed. The goal is to start the process of releasing the seed from the fruit pulp. Cover lightly with a damp cloth or a piece of plastic wrap, and let the pulp, seed, and water mixture sit and ferment at room temperature until the flesh can be easily removed from the seed. This usually takes several days, but do not let it sit any longer than necessary, so check daily. Then pour off the dirty water, and rinse with fresh water until the seeds are clean. Fill the container one last time with water, and remove all debris and any seeds that float. The most viable seeds will be the "sinkers." (Although, a local grower recently told me that he got about 40 percent germination from the floaters. So you might want to consider sowing those out separately.) An alternative method is to mash the fruit, and then spread the pulp and seeds out on a screen (a window screen will work) and spray with a high-pressure stream of water. This often requires rubbing some of the pulp through the screens with your hands and definitely works best with very ripe fruit. For the large-scale producer, there are seed cleaners available that will do all this in one step. If you use one of these latter two methods, I still recommend testing the seed for viability by making sure they sink in water.

Seed Storage

Mature goldenseal seeds are small, round, black, and hard (see photo on the next

Goldenseal seed are shiny, black, round seeds that should never be allowed to dry out. Photo by Tom Konsler.

page). Like ginseng seed, they should never be allowed to dry out. If you sow fresh goldenseal seed, they will not sprout right away. Goldenseal seeds need a warm period followed by a cold period, and time (probably for inhibitors to be leached from the seed) to germinate (see earlier section on seed propagation). It is best to sow fresh seed immediately into a nursery bed (explained in the next section). If this is not practical and you need to store the seed for a period of time, make simple envelope-style pouches out of window screen, securing the sides with staples (a desk stapler works just fine). Mix the seed with some moist, clean playground sand, pour into the pouches, staple the tops, and bury the pouches about 2" deep in a shaded protected area. Cover with mulch and make sure the area stays moist. In both cases, the site should be similar to where you would want to grow goldenseal, i.e., shaded with well-drained, moist soil. The seeds in the pouches can be sown that fall through early winter or very early the next spring. The least desirable method is to store the seed in the refrigerator. This will most likely reduce germination and delay it by a year.

In the two earlier versions of this book, I mentioned the use of a stratification box, as is used for ginseng seed. I do not recommend that method any longer because the risk of loss to decay is high, germination is often delayed until the second spring after sowing, and germination rates are usually very low.

Planting Seeds

Goldenseal seeds are best sown in a nursery bed and grown out for a year or two before moving to the production beds. Due to the unpredictable nature of goldenseal seed, and the fact that some seeds may come up a year after the others do, direct-seeded production beds often have uneven stands and plants of different ages. In a well-prepared, shaded nursery bed, sow 10 to 12 seeds per foot, between ¼" to ½" deep, in rows 3–4" apart. If you are sowing a large volume of seed, a mechanical seeding device is highly recommended. Cover the seed with from ½" to 1" of soil and tamp down lightly with a rake or hoe. Cover the area with several inches of mulch to keep the soil moist and prevent the seeds from experiencing excessive freezing and thawing. If you sow your seeds in the spring, keep the mulch layer thin so the seedlings can push through it. For fall-planted seeds, gently rake back some of the mulch in the spring so the seedlings can emerge.

Rodents may eat your seeds and slugs may eat your seedlings. If this is a problem, a method that has worked well for me is to cover the nursery bed loosely with metal window screen anchored with landscape pins. Cover with a few more inches of mulch and keep the bed moist. If you planted in the fall and used the screen, in the spring remove the mulch you placed over it so you can observe the emerging seedlings. Make sure the screen is loose enough for the seedlings to emerge and grow. Consider propping the screen up as the seedlings grow to protect them during the first growing season. If it appears to be restricting growth, remove it, but then watch closely for slugs. The plants can be moved to the production beds when they are one to two years old.

A few people I know grow their seedlings in flats and transplant them into the field when they are about one year old. The only downside to this method is that you must monitor the moisture content of the medium in the flats during the entire time seeds—and then plants—are in the flats. This may be two years or more. The couple of times that I used this method at the research station, fresh seeds were sown into flats with one- to two-inch cells filled with a peat-based soilless medium. The flats were kept moist and held in a cooler at about 40°F for several months. The flats were then put in a shaded area in the greenhouse and kept moist. Germination the first spring was about 75 percent, and we set the dormant little plants into their final production beds after only one season of growth. Other growers put their flats in the woods and cover them with leaf mulch. These flats are sometimes stored for three years before the plants germinate and grow large enough for transplanting. An industry colleague told me there is a group of growers in Wisconsin who has figured out a way to efficiently and profitably grow goldenseal from seed sown in the field. I do not know what their method is, but hearing this from a reliable source encourages me to keep working at a method of my own. If a reliable average germination rate of 75 percent could be achieved, it would definitely be worthwhile and cost effective to sow seed. To reduce the problem of the medium drying out, use cell-less flats or flats with large cells.

Harvest and Post-harvest

Harvest

When goldenseal plants have fully occupied the beds in which they were planted (usually after three to five years) either harvest the roots or divide the plants. If left undisturbed, the plants will start to crowd themselves out, and the oldest roots will eventually die. The best time to dig the roots is in the fall through early winter after the tops have died down. If you have a market for the leaves and stems, harvest them in late summer or early autumn while the foliage is still green. You need to dig the roots carefully, trying to keep the many fibrous roots intact. Small plots can be dug by hand with a fork, but large fields will require some kind of mechanical digger. Modified potato, horseradish, and bulb diggers can be used to bring the golden-

Staff at the Mountain Horticultural Crops Research Station harvesting goldenseal roots by hand. Photo by Jeanine Davis.

seal to the soil surface where it can be easily gathered by hand into picking buckets. Whichever method you use, strive to keep the fibrous roots intact, gently shaking off excess soil before gathering the roots into picking buckets or sacks. Whether you plan to dry your roots or sell them fresh, it is important to keep them moist and turgid at this point. If you plan to keep some of your harvest for dividing and replanting, select large healthy plants, and have a container available to keep them moist and cool, or have beds prepared to replant into immediately.

The time of year that you choose to dig goldenseal will affect the ratio of fresh weight to dry weight. For example, in my area, if you dig goldenseal in the spring, from April through early June, it may take five to seven pounds of green root to make one pound of dried root. In the fall, from October through December, it may only take 2½ to 2¾ pounds of green root to make one pound of dried root.

Making Yield Estimates

For your records, do not forget to get a good estimate of your yields per unit of land area. This can easily be accomplished by measuring off an area of bed, harvesting the roots from that area, and weighing them. Be sure to keep them separate during the drying process so you have an estimate of your fresh weight–dry weight ratio.

If you know exactly how much actual bed space your total harvest will include,

Dried goldenseal roots. Photo by Jeanine Davis.

then you can easily estimate your final yield. As an example, if you harvested 4 pounds of roots from 20 square feet of bed space, and the total bed space to be harvested is 100 square feet, then multiply the pounds of root from the measured area by the total amount of bed space you are going to harvest. Then divide by the square feet of the measured test area. So, for my example, 4 × 100 = 400; 400 ÷ 20 = 20. I would expect to harvest 20 pounds of fresh root from the 100 square feet of actual bed space.

A way to get a rough estimate of yields per acre is to estimate the total bed space in your production area. For example, if you have four-foot-wide beds, on five-foot centers, and mark off a five-foot long section of bed, you will be harvesting roots from 25 square feet of your production area. That includes the one-foot walkway between your beds. To estimate your yield per acre, multiply the pounds of root from the measured section by 43,560 (the number of square feet in an acre). Then divide that number by the square footage of the measured area. So, if you harvested 4 pounds of fresh root from the 25 square feet, 4 × 43,560 = 174,240; 174,240 ÷ 25 = 6,970 pounds of fresh root per acre. Figuring on 30 percent dry weight, 6,970 × .30 = 2,091 pounds dried root per acre.

Washing and Drying Roots

Because your goldenseal roots are destined to be consumed by people, you must devise a way to efficiently and thoroughly clean your harvest. This is not an easy task. Goldenseal's fibrous mass of roots holds soil and pieces of mulch much more tenaciously than a ginseng root. If you have a very small

After the roots have gone through the root washer, it is important to make sure every root is free of dirt before it goes into the dryer. We use an additional hand washing step at the research station. (Echinacea roots in photo.) Photo by Kelly Gaskill.

harvest, you can carefully wash the roots by spreading them over a screen outdoors and spraying with water from a food-grade hose. Work the roots between your hands, being careful to get out the pieces of mud that tend to get caught right up against the rhizome. Remove all dirt, breaking larger roots if necessary, but do not use a brush. If you have a large volume of roots to clean, you will need a root washer. There are commercial ginseng, carrot, and other root washers available that consist of a drum that turns and tumbles the roots as water is sprayed over them. Many newer ones filter and recycle the water. Simple root washers can be easily constructed. A new cement mixer with a plastic drum is another option; tumble the roots while spraying with water for three to five minutes before dumping them out on a screen. This will loosen much of the dirt from the roots and make spraying them clean much easier. There are many do-it-yourself (DIY) root washer plans and videos on the Internet.

There are plans available on how to build herb dryers from scratch or by modifying existing structures. Photo by Jeanine Davis.

Spread the clean, wet roots on clean, food-grade screens and dry in a well-ventilated area in the shade or in a forced-air dryer. Simple dryers can be constructed

Tobacco barn converted into a herb dryer. Photo by Jeanine Davis.

This simple dryer is constructed of wood and recirculates hot air from the bottom of the unit, through the herbs on the screen shelves, and out the top. Photo by Jeanine Davis.

from small sheds or rooms in barns. There are plans to build two dryers, one about the size of a large chest freezer and one constructed from a utility building available in a leaflet on my work website (see References and Resources). Bulk tobacco drying barns and peanut drying wagons can also be modified to successfully dry goldenseal roots. Because of new food safety regulations being instituted or discussed, design all washers and dryers of materials that can be thoroughly cleaned and sanitized. The

Cardboard barrel for international shipments.
Photo by Jeanine Davis.

dryer plans in the leaflet described above do not meet that criteria as of September 2013.

The key points in drying are to keep temperatures low, usually between 85° and 100°F, and to provide good air flow around the roots. Very humid areas may require higher temperatures up to 130°F to dry the roots, but try to keep the temperature as low as possible. If the roots are dried too hot and fast, the outside of the root dries first, leaving the inside of the root moist. It is then very difficult to extract moisture from the interior of the root. The quality of the roots may be destroyed, and the roots will bring a substantially lower price than properly dried roots. Roots will lose about 70 percent of their weight during drying. Test for dryness by breaking a large root; it should snap cleanly.

Yields and Packaging

Yield estimates for goldenseal are highly variable due to the tremendous differences among growers in their cultivation methods, including the amount of planting stock used, plant spacing, bed and row spacing, and years to harvest. Commonly reported yields for production under artificial-shade structures average around 2,000 pounds of dried root per acre after three to four years of growth from rhizome pieces, but yields as low as 800 pounds per acre and as high as 4,300 pounds per acre have been reported. In a three-year growing period, the rootstock planted will increase in weight anywhere from three- to ten-fold; remember that this is on a fresh weight basis. In the forest, growth is not as rapid, and a three- to five-fold increase in fresh root weight can be expected over a period of three to four years. Yields reported for woods-cultivated goldenseal range from 500 to 1,600 pounds of dried root per acre. I do not have reliable estimates for wild-simulated goldenseal production.

If you are selling fresh roots, they must be kept cool and delivered to the buyer very quickly. The few growers I know who sell fresh root to be made into herbal products transport the roots in big plastic bins loosely covered with damp cloths. If the roots will be used for planting stock, they are often packed in moist peat moss or sphagnum moss. Dried roots should be packed loosely into cardboard cartons or barrels or polysacks. Burlap sacks are no longer recommended. Polysacks are made of woven polypropylene and come in a wide range of sizes including large bulk sizes that have the footprint of a standard pallet. These are often referred to as FIBCs, Flexible Intermediate Bulk Containers. Polysacks are very strong and are suitable for food storage. Always store the containers of dried roots in a cool, dry area secure from insects and rodents.

Marketing

This discussion of marketing goldenseal might well have been the beginning of this chapter instead of the end because if you cannot sell your goldenseal, why grow it? Fortunately, finding a buyer for goldenseal has not been a problem for most producers. Part of being a successful herb grower is keeping an eye on the market in order to gauge when to plant more goldenseal and when to hold back. This means you have to educate yourself on the medicinal herb market, who the major buyers are, what the trends are in the industry, and what the supply and demand picture is, not just now, but four and five years into the future. It really helps if you can develop a close relationship with a buyer that you trust.

Before you plant a seed, transplant, root cutting, or rhizome piece, you need to have a good idea of what market you are seeking and who your buyer will be. If you intend to sell your crop to a botanical industry supplier for incorporation into food supplements, extracts, and other products for human consumption, you should find out what good agricultural practices (GAPs) need to be followed (see Appendix 4) and if your crop has to be certified organic. Does the buyer want fresh root or dried root? Is there a minimum alkaloid level? How should the roots be cleaned, dried, and packaged? When and how is delivery expected? If you are selling your roots as planting stock, are you targeting homeowners, nurseries, or large-scale growers and what nursery licenses or inspections are required? If you are going to export your products, do you meet all the CITES requirements? All of these factors are very important and influence your costs of production, risks, profit potential, market size, and marketability.

There are many different markets for medicinal herbs. The traditional buyer of roots and herbs is a raw botanicals supplier, sometimes referred to as a dealer, consolidator, or bulk herb buyer. This buyer may work independently or be the employee of a larger corporation. The raw botanicals supplier buys unprocessed botanicals directly from the farmer and wild-harvester and usually sorts, cleans, grades, tests, and packages them to sell to manufacturers of herbal products. They often do some minimal processing such as cutting and sifting or grinding. The supplier will ensure the identity of the botanical and usually have it tested for biochemical constituents, heavy metals, pesticides, and microbial contaminants. The raw botanical supplier may specialize in one or two herbs, such as ginseng and goldenseal, or may serve as a "one-stop shop" and offer a hundred or more raw botanical products. Some herbal products manufacturers also buy directly from growers and wild-harvesters. Whereas the raw botanicals supplier will often buy small quantities of botanicals from a large number of growers and harvesters and consolidate them at the warehouse, many manufacturers prefer to buy from growers who can supply large quantities of a single product. There are many markets for wholesale quantities of herbs and roots, ranging from several pounds to hundreds of pounds in size. These include small manufacturers of natural products, herbal schools and colleges, alternative health care practitioners including herbalists, and natural food stores.

The Internet has made it possible for everyone to have a market presence on the wholesale or retail level. Many growers venture into the market with a website offering a variety of herbs in one ounce, eight ounce, and one pound packages. They can offer very personalized service to their customers who in turn provide feedback to help the grower improve their products and service. Sales of medicinal herbs and simple value-added products such as extracts, salves, and lotions have become popular at many farmers' markets and tailgate markets.

The nursery market is another outlet for medicinal herbs on a retail and wholesale level. There is a shortage of nurseries that can provide large quantities of seeds or planting stock for commercial growers. As the demand for more domestically produced herbs increases, the demand for high quality planting stock and seed will surely increase as well. In the meantime, there is a ready market among home gardeners and herb enthusiasts for herb plants and seeds. I know many nurseries that make a large percentage of their annual sales at spring and fall herb festivals. The most successful of these businesses have a wide selection, good signage, and a knowledgeable sales staff to answer questions. More information on markets is provided in Appendix 1, Forest Botanicals Bought and Sold.

Economics of Production

Prices

Prices paid for goldenseal in 2013 were highly variable depending on the product, market, and whether it was certified organic or not. I was pleased to see there was little, if any, difference in cultivated versus wild goldenseal prices; that means we have done a good job convincing consumers that cultivated goldenseal is as good as wild. Growers selling dried goldenseal root to raw material suppliers (i.e., bulk buyers or dealers) who buy direct from the grower received an average price of $22.60 per pound (prices ranged from $16 to $35 per pound). They received considerably more for certified organic roots, with an average price of $68.50 per pound—three times as much. Prices paid for dried leaves to those same buyers were much more consistent, with growers receiving an average of $4 per pound.

Growers who sold goldenseal wholesale direct to the manufacturer or a company that packaged and resold herbs received an average of $85 per pound for dried, cut roots. There was only a 13 percent premium for certified organic, with an average paid of $96 per dried pound.

Those growers who sold retail, often at farmers' markets or over the Internet received an average price of $109 per pound for dried root, whole or cut. If it was certified organic, the average price was $115 per pound; providing a 5.5 percent premium.

Nurseries selling wholesale received an average price of $76 for 100 plants. Retail prices averaged $3.58 per plant. Not many growers sell goldenseal seeds, but the few that did were asking anywhere from $60 per pound on a wholesale level to $2,700 per pound retail.

Prices really haven't changed much over the past ten years. In 2004, dealers were paying from $22 to $30 per pound for dried goldenseal root. That dropped to $11 to $25 per pound in 2007, compared to $16 to $35 per pound in 2013.

The Joy of Growing Goldenseal

Maybe you have no intention of ever selling any of your goldenseal. Maybe you find pleasure just in growing this beautiful plant knowing that you are helping conserve a threatened plant. When I lived in a house nestled in the woods, I really enjoyed nurturing a small patch of goldenseal out behind the house. I encourage anyone who has some woods, even in a subdivision, to try to grow a little goldenseal. It's a marvelous little plant, not difficult to grow, and you will feel good knowing that you are helping to protect a threatened plant. Growing goldenseal just for the joy of it is covered in more detail in the Home Garden part of this book.

Budgets

If you are serious about trying to make a profit growing goldenseal, you need to have a business plan and an enterprise budget. Here are a few sample budgets designed to give you a picture of the costs required to grow goldenseal, including labor, equipment, and services. These budgets are only hypothetical examples. Your expenses, yields, and profits are going to be different. There are many important expense lines omitted from these budgets that must be taken into consideration, such as taxes, equipment maintenance, advertising, and utilities. The labor lines in particular vary greatly from farm to farm. Many family operations do not pay for labor and don't account for their time. This would substantially reduce the costs of production on all of these budgets. I reduced the hourly wage from the previous edition of this book in response to comments from growers that the wage was too high because they usually recruit unskilled labor, e.g., high school kids, to do the kinds of tasks included in the budgets. The price for the planting stock is based on the price a raw botanicals supplier would pay for fresh, certified organic root.

In the past few years, there have been several academic articles published on the economics of growing woodland medicinals. Every one that I have read so far shows that the only woodland herb that can be grown profitably in the woods is ginseng.

Table 12. Four-year Hypothetical Farm Enterprise Budget for One-tenth Acre of Certified Organic, Woods-cultivated Goldenseal Using the Lowest Yields Reported by Growers or Obtained from Research and Demonstration Plots

One-tenth Acre	Year 1	Year 2	Year 3	Year 4	Total
Labor					
Site preparation & planting: 60 hrs × $7.25/hr	$435				$435
Maintenance, mulching, weeding: 100 hrs × $7.25/hr	$225	$166	$167	$167	$725
Harvesting, washing, drying: 30 hrs × $7.25/hr			$29	$189	$218
Total Labor	$660	$166	$196	$356	$1,378
Operating Expenses					
Planting stock (certified organic roots): 60 lbs × $17/lb	$1,020				$1,020
Fertilizer and lime	$68				$68
Mulch	$124		$124		$248
Purchase of used or self-constructed equipment for washing & drying:				$744	$744
Organic certification (assume this represents a percentage of the total cost for the farm)	$124	$124	$124	$124	$496
Packaging				$124	$124
Total Operating Expenses	$1,336	$124	$248	$992	$2,700
Total Expenses (Labor + Operating)	$1,996	$290	$444	$1,348	$4,078
Gross Income					
Dried root sales: 54 lbs × $70/lb				$3,780	$3,780
Dried herb sales: 15 lbs × $6.25/lb			$94	$94	$188
Fresh seed sales: 0.2 lbs × $427/lb			$85	$85	$170
Total Gross Income with Low Yields	$0	$0	$179	$3,959	$4,138
Net Profit (Gross Income – Total Expenses)	- $1,996	- $290	- $265	$2,611	$60

This budget does not include all fixed costs (e.g., insurance, taxes, and equipment maintenance) or marketing costs. U.S. federal minimum wage (2013) was used for the labor rate.

These are good papers and I am not criticizing the work that's been done. Those papers illustrate an important point; expect a wide range of variability when you grow woodland medicinals. This is not like growing corn or soybeans where you buy patented seed best suited to your environment, plant the seed in uniform rows, are prepared to control most insects and diseases, and (except when Mother Nature throws a drought or flood at you) produce fairly predictable yields. With woodland medicinals, the growing environments are vastly different: seeds and planting stock are from wild populations or not far removed; everyone's plant populations are different; damage from insects, diseases, mammals, or birds is impossible to predict; and the number of

Table 13. Four-year Hypothetical Farm Enterprise Budget for One-tenth Acre of Certified Organic, Woods-cultivated Goldenseal Using the Highest Yields Reported by Growers or Obtained from Research and Demonstration Plots

One-tenth Acre	Year 1	Year 2	Year 3	Year 4	Total
Labor					
Site preparation & planting: 60 hrs × $7.25/hr	$435				$435
Maintenance, mulching, weeding: 130 hrs × $7.25/hr	$292	$217	$217	$217	$943
Harvesting, washing, drying: 130 hrs × $7.25/hr			$22	$921	$943
Total Labor	$727	$217	$239	$1,138	$2,321
Operating Expenses					
Planting stock (certified organic roots): 60 lbs × $17/lb	$1,020				$1,020
Fertilizer and lime	$68				$68
Mulch	$124		$124		$248
Purchase of used or self-constructed equipment for washing & drying				$744	$744
Organic certification (assume this represents a percentage of the total cost for the farm)	$124	$124	$124	$124	$496
Packaging				$248	$248
Total Operating Expenses	$1,336	$124	$248	$1,116	$2,824
Total Expenses (Labor + Operating):	$2,063	$341	$487	$2,254	$5,145
Gross Income					
Dried root sales: 160 lbs × $70/lb				$11,200	$11,200
Dried herb sales: 50 lbs × $6.25/lb			$313	$313	$626
Fresh seed sales: 1.8 lbs × $427/lb			$769	$769	$1,538
Total Gross Income with High Yields	$0	$0	$1,082	$12,282	$13,364
Net Profit (Gross Income – Total Expenses)	–$2,063	–$341	$595	$10,028	$8,219

This budget does not include all fixed costs (e.g., insurance, taxes, and equipment maintenance) or marketing costs. U.S. federal minimum wage (2013) was used for the labor rate.

years from planting to harvest varies. Compound that with all the different products, markets, and prices the grower will likely have to deal with and you can see why there is no right or wrong answer here.

I did want to address this issue, however, and it is important not to paint an unrealistically rosy picture. So the woods-cultivated budgets presented here are based on yields growers have shared with me and those I have obtained over 25 years of growing goldenseal in research and demonstration plots. One budget is based on the lowest yields and one on the highest. The prices are the high end of what growers actually received in 2012 to 2013.

The third budget is for certified organic goldenseal grown under artificial shade. It

Table 14. Four-year Hypothetical Farm Enterprise Budget for One Acre Certified Organic Goldenseal Grown under Artificial Shade

One Acre	Year 1	Year 2	Year 3	Year 4	Total
Labor					
Labor to build shade structure, prepare soil, plant, and weed: 875 hrs x $7.25/hr	$3,110	$1,078	$1,078	$1,078	$6,344
Labor for harvesting, washing, drying, and packaging: 125 hrs x $7.25/hr				$906	$906
Total Labor	$3,110	$1,078	$1,078	$1,984	$7,250
Operating Expenses					
Planting stock (roots): 600 lbs x $17/lb	$10,200				$10,200
Fertilizer, lime, and foliar nutrient sprays	$3,110	$1,364	$1,364	$1,364	$7,202
Plant protection: including organic fungicides, deer repellant	$341	$341	$341	$341	$1,364
Fuel and oil for tractors and equipment	$136	$105	$105	$136	$482
Machine hire/rental	$434			$713	$1,147
Mulch	$2,480	$620	$620	$372	$4,092
Repairs and maintenance	$520	$390	$390	$651	$1,951
Shade structure materials	$12,400	$248	$248	$248	$13,144
Utilities				$992	$992
Organic certification: percentage of whole farm certification cost	$124	$124	$124	$124	$496
Total Operating Expenses	$29,745	$3,192	$3,192	$4,941	$41,070
Total Expenses (Labor + Operating)	$32,855	$4,270	$4,270	$6,925	$48,320
Gross Income					
Dried root sales: 1,800 lbs x $70/lb				$126,000	$126,000
Total Gross Income	$0	$0	$0	$126,000	$126,000
Net Profit (Gross Income – Total Expenses)	–$32,855	–$4,270	–$4,270	$119,075	$77,680

This budget does not include all fixed costs (e.g., insurance, taxes, and irrigation), packaging, depreciation, or marketing costs. U.S. federal minimum wage (2013) was used for the labor rate.

is an updated version of a budget that Jackie Greenfield, former research associate in my program, created in 2003 based on extensive grower interviews, existing budgets, yields, and operating budgets from our research and on-farm studies. I changed the hourly wage as explained above, and updated costs for materials and services. Since so few growers are selling fresh root right now, I also changed the income to reflect dried root sales only. Prices for the roots were the high end of what growers actually received in 2013.

13

Goldenseal Growers' Stories

Vestal Shipman: Small-scale Forest Goldenseal Farmer

I wrote the following story about Vestal Shipman for the 2005 edition of this book. Since that time Vestal and his wife moved to Georgia to be closer to family. Reading over the original story, I wondered if Vestal had retired completely, since, according to my calculations, he would be 85 years old in 2013. Not having a current phone number for him, and not expecting to find anything, I did a Google search for his name and was surprised to see a new newspaper classified ad pop up with his name on it. It was the opening of ginseng season and his ad said "Paying Top Price! High Mountain Ginseng." So I gave him a call and we had a wonderful conversation. His original interview is included here because it is as relevant as ever, followed by an inspiring update.

Vestal's Story, As Told to Jeanine Davis in September 2003

I can't remember when I first met Vestal Shipman, but it seems like we have known each other for a very long time. Vestal grows goldenseal in the mountains of eastern Tennessee. Over the years, we've met at growers' meetings, talked on the phone, and corresponded by email. Our favorite topics are goldenseal and ginseng. We have even been speakers at a goldenseal grower educational program together. I discussed my research on growing goldenseal, and Vestal explained how he effectively and efficiently grows it on his own farm.

I knew Vestal would provide an interesting viewpoint on how to grow and market goldenseal. He is a former schoolteacher, who produces goldenseal to supplement his retirement income. I met with Vestal and his wife at their place on a warm

September morning. We sat on the front porch of their home as the morning mist burned away and talked as we watched the hummingbirds frantically sip at the feeders before flying south for the winter. Here is Vestal's story.

Vestal inspects his goldenseal beds. Photo by Jeanine Davis.

Vestal's Herb Business

Vestal is involved in several aspects of the native botanicals industry. He cultivates goldenseal and wild-harvests ginseng and other herbs. He also buys green roots from local wild-harvesters, carefully dries them, and sells them to local dealers. Vestal always makes a special effort to educate local wild-harvesters on how to harvest ginseng roots and replant seeds so there will be ginseng there in the future for their grandchildren to enjoy and harvest.

Vestal and his friends wild-harvest herbs from the mountains of eastern Tennessee and western North Carolina. In that area, wild populations of goldenseal are very scarce compared to ginseng. Vestal only knows of three wild patches of goldenseal in the whole region, and he does not harvest from them. In his opinion, wild-harvesting of goldenseal should be illegal. He would like to see cultivated goldenseal completely replace wild goldenseal in the marketplace, so he encourages people to grow it, as he does.

How Vestal Grows Goldenseal

Vestal's first planting of goldenseal came from a healthy-looking patch he found on what appeared to be an old homesite. It was quite common for the early mountain settlers to maintain a small bed of goldenseal for their own use, and many patches of goldenseal now growing in the Appalachian Mountains were originally in family medicinal herb gardens. He collected a few roots and seeds from that patch and planted them in the forest on his own property. They grew so well that he decided to plant many beds of goldenseal. Since he also sows all the seeds produced by his plants, he now has goldenseal growing all over his property. Not surprisingly, he has found that it grows much better on northern slopes than on western slopes.

Vestal has become quite adept at handling goldenseal seed, and many of his experiences support what I found in my research. He reports that the best way to achieve good germination is to get the seeds from the berries on the plant into the soil as quickly as possible. He always tries to sow the seed the day after he harvests the berries. Because he plants the seeds in July, when it is usually hot and dry, he sows the seeds in an area where he is certain the soil will naturally remain moist until fall. This is often a northern slope. He pulls back the natural leaf mulch and lightly rakes the surface of the soil. Then he scatters the seeds thinly over the soil. Vestal is not in a hurry to harvest roots, so he gives each plant lots of space to grow—up to one square foot per plant. Then he covers the site with 3–4" of leaf mulch. In early spring, as the seeds begin to germinate, he carefully removes most of the leaf mulch with a leaf rake, leaving only an inch or two behind. This makes it easier for the young seedlings to make it to the surface. Vestal has tried other methods of handling seeds, but has not had much success. For example, once he refrigerated the seed for several months before planting, but after he sowed them they either rotted or just never germinated.

Most of Vestal's land has the kind of good, rich forest soil that goldenseal likes, so he does not have to fertilize it much to get good plant and root growth. He has found that a little potash added to the soil before planting seems to improve growth. He plants his roots and seeds in seven-foot-wide sections, which include a narrow walkway on each side. He does all the work by hand, with a rake and a digging tool. He relies on the rain to keep the soil moist and depends on the natural leaf mulch the trees provide to keep his beds covered.

Vestal does very little to his goldenseal once he has it planted except to admire it once in a while. If he needs to let a bed of goldenseal grow longer than four years, he thins it during that fourth year to keep the plants from choking themselves out. He either replants those he removes in their entirety or first cuts some of the roots off the rhizomes and washes, dries, and sells the roots.

A keen observer, Vestal keeps good notes on various methods he has tried to grow goldenseal. For example, he had heard somewhere that goldenseal, with its natural antifungal properties, might provide ginseng some protection from diseases, so he planted some ginseng and goldenseal together in the same bed. He left a good amount of space between the plants, because he thought the spreading habit of the goldenseal might take over the ginseng. So far, the goldenseal is not choking out the ginseng, and best of all, there is no disease in the ginseng. He admits this is not a scientific study and only a tiny planting, but it adds to the body of evidence that goldenseal has some disease-fighting properties.

Vestal does not have any pest or disease problems with his goldenseal. He thinks this is probably because he does not try to push his plants to grow quickly. The goldenseal I saw that September morning looked exceptionally healthy, especially for so late in the season. Vestal reports that even on the dry west side of his mountain,

Goldenseal early in the growing season with green developing fruit. Photo by Jeanine Davis.

if he can get it established, goldenseal grows well. Weeds are not a big problem for him either. A little bit of hand weeding the first year after he has planted rhizome pieces, or the first few years after sowing seed, is all that is necessary.

Vestal grows goldenseal under a natural canopy because he already has forested land. It does not require a large capital investment to put in a planting, and the plants thrive under those conditions. He did want to point out, however, that he has a friend who grew goldenseal under a shade-cloth structure, and it grew twice as fast as Vestal's did in the woods.

When the plants are four years old or older, Vestal digs his goldenseal roots by hand. He says that digging roots when the ground is dry leaves lots of little feeder roots behind. This is not a bad thing because the next spring he has eight to ten little plants coming up where there used to be one. Thus, his beds are mostly self-perpetuating. He just has to fill in the gaps if he wants to have a nice full bed of goldenseal in the same place in another year or two. Vestal has never actually calculated his goldenseal root yields because he has not harvested any large patches of it yet. He estimates from what he has harvested so far that he is probably getting 600 to 700 pounds of dried root per acre.

How Vestal Markets His Crop

Finding a market for his goldenseal is not a problem, and Vestal says that just about any herb dealer in his area will buy both the root and herb. He did say that the demand for goldenseal is not as great as it was a few years ago, but he still gets a good price for it. In 2003, he often got $30 per pound of dried root. He sells all his goldenseal to local buyers who have a market for his high-quality Tennessee-grown goldenseal. He does not try to sell directly to big companies because he does not have large amounts of roots to sell at any one time and does not want to get into competition on that market with the large-scale artificial-shade goldenseal growers in the north. Unlike ginseng, he has found there is no difference in price between goldenseal grown under artificial shade and that grown under wild-simulated conditions, as long as the quality is good.

Vestal also has a market for goldenseal herb—that is, the foliage of the plant. He harvests the herb by grabbing it by the handful and pulling with a quick, sideways motion. This takes a lot of handwork, but he says it goes quickly, and he finds that it pays well. In 2003, he got $7 to $8 per pound of dried herb. Vestal harvests about two pounds of dried goldenseal leaves per 0.01 acres (about 400 square feet), which translates into about 200 pounds per acre. Thus, he figures he can make $1,400 to $1,600 per acre from the foliage each year.

Vestal really enjoys growing goldenseal because, in addition to being profitable, it is a beautiful plant and root. He gets great pleasure from watching it grow and from handling it. Goldenseal is not a big money crop for him at this time because he is not harvesting large quantities of roots yet, and he is still expanding his acreage. His plan is to have a couple of acres of goldenseal in production at all times and to replant new areas as he harvests mature patches so he can have a continuous supply of roots for sale. He is also thinking of getting a website for selling organic goldenseal roots and seeds. He has no doubt that goldenseal will be a reliable profit maker for him in the near future.

Update from Vestal

I spoke on the phone with Vestal in September 2013. It was ginseng season and he was in Tennessee, where he spends about two and a half months each year buying and selling ginseng. He enjoys doing it, and it provides a good supplement to his retirement income. He shared the actual figure with me, but I'll just say it is well worth his time. The rest of the year, Vestal and his wife live in Atlanta, relaxing and enjoying life. A young man takes good care of his Tennessee property while he is away.

Vestal's Ginseng Business

Ginseng has always been good to Vestal, so he tries to be good to it in return. He has earned his place in the ginseng industry, spending many years learning the business inside and out. Vestal handles ginseng a little differently than many other dealers. He buys it green and dirty and washes, grades, and dries it himself. He has a dedicated customer base and knows what each buyer wants. By doing all the post-harvest handling himself, he is able to provide them with only the roots they want to buy. As a result, he gets higher prices and his customers are satisfied.

Vestal said there was a bigger market for green ginseng this year than usual. Other buyers have told me the same thing. He sold his own home-grown, wild-simulated, green ginseng before the ginseng season opened for $250 per pound. Since Tennessee only has two categories for ginseng, wild and woods-grown, he had to sell it as woods-grown in order to sell it legally out of season. Woods-grown prices are usually much lower than wild or wild-simulated prices, but the price he received for green woods-grown was comparable to what was being paid for green, wild ginseng once the season opened.

Vestal has 100 to 150 people who dig ginseng for him. He teaches them about sustainable harvesting and replanting for future generations. Vestal encourages ginseng diggers to mimic nature and plant the whole berry, not individual seeds. He tells them to gather the berries, dig a ½"-deep trench, spread the whole berries out in the trench, cover it lightly with soil, and top it off with a good thick layer of leaves. He says the flesh of the berry keeps the seed moist, giving them a better chance to survive than if the seeds were planted individually.

When diggers bring in roots that are too small for him to buy, he gives them a lesson that he hopes will teach them not to dig any more small roots. He sets one of the little roots on the scale and points out how it is so light it doesn't even register on the scale. Then he sets a legal-sized root on the scale and shows how it is heavy enough to be worth $20 or so. He tells them to leave the little valueless roots in the ground until they grow into bigger $20 roots. Then he tells them to take those little roots back and replant them; when the roots are bigger, *then* he'll buy them.

Vestal reminds all his ginseng diggers that someone probably planted all the ginseng that they are digging now, so they should plant it back for others, too. Ginseng digging has been going on in Cocke County, Tennessee for over 200 years. Vestal figures that 90 percent or more of the ginseng growing in the woods there was planted by someone.

Vestal's Goldenseal Patch

Vestal has not been spending as much time with his goldenseal as he did in years past, but he is still in the business of growing and selling it. Since 2003, he more than quadrupled the size of his planted area, mostly by spreading the seed his plants produced. He is quite pleased with how healthy the patch looks and loves showing it off to people. He does have a market for goldenseal right now and is considering paying a young man to dig it for him. He could get $25 per dried pound for roots which would make them both some money. The nice thing about goldenseal is, he can have it all dug out this fall and next spring there will be goldenseal coming up throughout the area. It is a perpetual goldenseal patch.

Vestal's Woodland Gardens

Vestal grows a wide variety of woodland botanicals. Ginseng is his major crop; he has always grown more of it than anything else. Goldenseal is his second largest crop, but he also grows ramps, wild ginger, and false unicorn (which he calls star root).

Table 15. Vestal's Estimated Five-year Budget for One-half Acre of Forest-grown, Wild-simulated Goldenseal from Planting to Harvest in 2003 and 2013*

	2003			2013		
For One-half Acre	**Price/unit**	**Expense**	**Income**	**Price/unit**	**Expense**	**Income**
Planting stock: 50 pounds	$20/lb	$1,000		$18/lb	$900	
Labor to plant and maintain beds: 40 hrs	$10/hr	$400		$11/hr	$440	
Labor to harvest, clean, and dry: 200 hrs	$10/hr	$2,000		$11/hr	$2,200	
Total Costs		$3,400			$3,540	
Expected Yield						
100 pounds herb per year starting at the third year x 3 years	$10/lb		$3,000	$15/lb		$4,500
300 lbs of dried fiber roots and rhizome	$30/lb		$9,000	$25/lb		$7,500
Gross Return			$12,000			$12,000
Net Profit at the End of Five Years			$8,600			$8,460

*No equipment or land costs are included. After harvest, the fiber root left in the soil regrows and another crop will be ready to harvest in four to five years.

There is a good market for false unicorn root right now, and we discussed how to grow it from seed. He had bought 48 pounds of root the week before we talked for $60 per dried pound.

Marketing

Vestal believes that a big part of his success is the fact that he has always stayed in close communication with his buyers. He tries to build a trusting relationship with all of them and regularly talks to his buyers who are scattered across the country and world, e.g., California, Pennsylvania, Taiwan, and China.

Advice to New Growers

Vestal encourages people who want to grow woodland botanicals to start with ginseng. He recommends planting seeds because planting young roots is too risky, in his opinion. He is concerned that young roots could be infected with disease that could then become established in the ginseng patch. He pointed out that seed is pretty inexpensive. He recommends getting a small patch of ginseng growing, and then expanding it with the seed those plants produce. Of course, you want to plant your ginseng where you can keep a close eye on it.

If Vestal was a young man again, he said that he would grow more medicinal herbs, and a wider variety of them than he did. He moved to Cocke County, Tennessee in 1990 with $5,000, and was able to turn that into a substantial amount of money as a grower, but if he knew then what he knows now, it could have been much more. He knows money can be made from growing medicinal herbs, but you have to educate yourself, get help from the right people, produce a

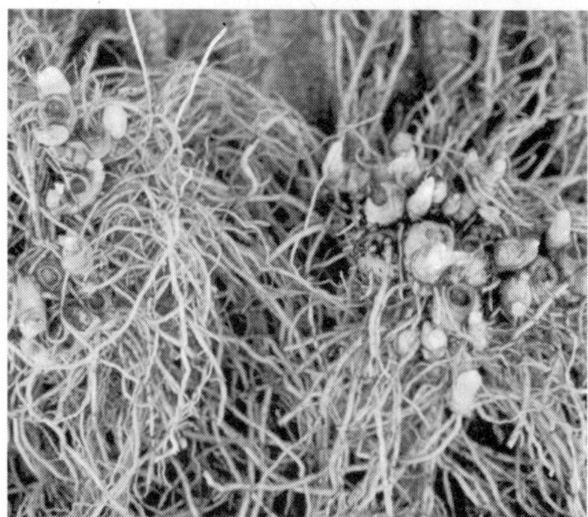

Steve Saint-Onge (*top*) is in touch with his goldenseal, from newly dug roots (*middle*) to early green fruits (*bottom*). Photo courtesy of Steve Saint-Onge.

quality product, and build good relationships in the market.

Steve Saint-Onge: Organic Goldenseal Grower

I interviewed Steve over the course of several conversations in 2002 and 2003. That interview is included here. I also spoke with him again in the summer of 2013 for an update. That is included at the end of the original interview.

Steve's Story, As Told to Jeanine Davis in 2002 and 2003

When I met Steve Saint-Onge, he was a fit-looking, outdoorsy man in his forties. He was dressed like a hiker, in shorts, an environmentally themed T-shirt, and hiking shoes. Steve first contacted me through email after finding my website. That website has introduced me to some of the most interesting people, and Steve is one of them. Steve is an organic goldenseal grower who was awarded a grant to study the effects of soil pH on goldenseal. In the spring of 2002, he drove from Vermont to western North Carolina to pick up several sacks of olivine that he needed for his study. Olivine is a mined rock that is an excellent natural source of magnesium, but it is only mined in two places in North America. One of those places is in western North Carolina. Since I live only about 45 minutes from where Steve was going to pick up the olivine, I requested that he come by for a personal interview. I asked him to tell the story of how his life led him to be an organic goldenseal grower. Steve is a good example of how, if you choose to, you can change your life growing herbs.

Steve grew up in northern Vermont, near the Canadian border. Even as a child, he had a green thumb and put it to use by working in the gardens of various people in town. He enjoyed growing useful and beautiful plants, but already at that young age, he worried about the chemicals used to grow plants and their effects on the environment. While attending Catholic high school, he read the book *Silent Spring*, by Rachel Carson. The book had a profound effect on him, and, as a result, he began studying and practicing organic agriculture. As a sophomore in high school, he decided to go to college to study horticulture. He started his college career in the Department of Horticulture at the University of Vermont, but quickly found that he hated it! He thought that the beginning courses he was forced to take were irrelevant to anything he wanted to do and, worst of all, he wasn't growing anything. Also, the prevailing attitude there was for conventional chemical-based agriculture. He found no support for, or interest in, organic agriculture.

Steve left college after only one year, and that is when his life took an abrupt turn. He soon found himself doing social work in New York City with Mother Teresa's Sisters. There, in the South Bronx, he did outreach with the elderly, and in his spare time, he gardened.

In the mid-1980s, Steve went back to the University of Vermont and designed his own program in Transpersonal Psychology. His goal was to work with people and plants in a way that would be beneficial to both. After completing his bachelor's degree, he moved to Boston, Massachusetts, to take a position in human services. While there, he did graduate work in neuropsychology and schizophrenia at Harvard. Steve enjoyed his work, but he did not like living in the city, so he bought some land in northern Vermont. For several years, he farmed that land, improving the soil and planting blueberries, raspberries, and strawberries, while still living and working in Boston. In 1994, he moved permanently to his farm, but he still worked in Boston. It was a long commute!

How Steve Got Into the Herb Business

For as long as he could remember, Steve had an interest in medicinal herbs. He grew up hearing the story about when his father contracted dysentery as a young child and was cured by an aunt using a traditional Native American herbal preparation. In 1988, Steve met the well-known herbalist Rosemary Gladstar at a conference. She had recently moved to Vermont from California. Steve was very impressed with Rosemary's knowledge of herbs and began studying with her. Soon they were doing psychological research and herbal studies together. He earned an herbal certificate from Rosemary and realized that he wanted to make herbs a major part of his life. He started by planting a one-quarter acre herb garden. It was about this time that Steve learned that the great aunt who had saved his father's life was, in fact, a Native American, and thus, that he was of Native American heritage. He has been fascinated by the Native American use of herbs ever since.

In the late 1990s, Steve attended an organic growers conference in Vermont. He was looking for a marketable organic crop to grow, but wanted something that required less-intensive hand labor than vegetables. He heard Bob Beyfuss from Cornell Cooperative Extension give a talk on ginseng and goldenseal. Steve came home from that conference with armloads of handouts and an interest in trying to grow both of these wonderful plants. Using Bob's information, Steve did a site evaluation and found he had forested areas that fit the profiles for both ginseng and goldenseal. He ordered his first planting stock from Scott Persons at Tuckasegee Valley Ginseng and decided to grow everything organically. His first plantings were put in during a very dry year, and he did not irrigate them. As a result, all the ginseng died, but the goldenseal survived. The next year, Steve chose a moister site and put in more ginseng. He also doubled the amount of goldenseal on his farm. The plantings were in the woods, but near his house for security.

Steve did everything Scott Persons and Bob Beyfuss told him to do, and the ginseng grew nicely this time, but the goldenseal grew exceptionally well. He kept doubling the size of his goldenseal plantings each year, and in 2001, he put in 4,000 new goldenseal plants. The plants and the market all looked good.

How Steve Grows Goldenseal

Partly due to personal preference, and partly due to the practicalities of working in the forest and growing organically, Steve does most of his cultivating, transplanting, weeding, and harvesting by hand. He has become quite efficient at it and thinks in the long run that it is a practical method for him. He also enjoys working closely with the plants and the soil. He described a kind of intuitive farming that comes from the close association he has with his plants as he digs in the rich, fragrant humus of the forest with his bare hands. This has also provided a market advantage for him. Many of his buyers purchase his herbs specifically because of the way he grows them.

When Steve first started farming goldenseal, he chose a new planting site based on appearance. He looked for a well-drained site with rich-appearing soil. He prepared the land by hand, using a pick-axe to remove roots and rocks. Initially, he did not do any kind of soil testing or add any soil amendments. Now, following expert advice, he tests his soil and usually adds a little wood ash, gypsum, and compost before planting, incorporating them into a nice planting bed with a rototiller. He does not think it is necessary to build raised beds for the goldenseal because he always plants in a well-drained area. He sets the transplants by hand and then covers them with a thick layer of composted leaves. He cannot provide any advice on disease or insect control, because he has not had any problems so far. This is not uncommon for growers who try to duplicate nature when they grow their woodland herbs, i.e., choosing ideal growing locations, not spacing the plants too closely, and being sparing with the fertilizers. He grows his plants for four years, weeding the beds and replenishing the mulch as needed. He collects seed from his plants and has started a nursery of stock plants.

Steve harvested his first commercial root crop in the fall of 2002 and sold it fresh to two buyers. Currently, he dries only a small number of roots in a home drying unit for his own use, but he has plans to build a dryer that will handle much larger quantities. The goldenseal business is going well for Steve right now, and the prices he gets for his roots are excellent. The retail price for organic goldenseal root in the spring of 2003 was $200 per dried pound. He got $95 to $100 per dried pound of root wholesale, and already had his fall 2003 crop sold when we spoke. His marketing strategy is to search out local herbalists and grow specifically for them. Steve also decided that he wanted to conduct scientifically valid research on his farm. He read the literature and came up with several ideas to improve goldenseal root yields by incorporating some organic soil amendments and adjusting the soil pH. Friends and colleagues encouraged him to submit a grant proposal to the federal Sustainable Agriculture Research and Education (SARE) program, so he did, and it was funded. He hopes his research will help him and others to grow profitable yields of high-quality goldenseal in a manner that is healthy for the plant and the land on which it is growing.

Advice to New Growers

Steve recommends that other farmers consider growing organic goldenseal, but he urges that they proceed cautiously. Growing goldenseal, especially organically, is not always easy. It takes a willingness to learn how to do it properly and time to observe the plants. Several years ago, some farmers in his area tried growing goldenseal under artificial shade and had serious problems with weeds and disease. Their experiences discouraged others from growing commercial goldenseal in his region. This has given him a market edge, since he is one of only a few goldenseal growers in the area. He also learned from their mistakes and does not plan to duplicate them.

Steve loves what he is doing now. He is an herbalist and an organic herb grower. He spends his time in the woods, doing what he likes to do, being where he wants to be, and helping people, all at the same time. He is gradually working up to making his whole living from the farm. He is taking it one year at a time with the final goal of developing an eco-farm where people can learn to live ecologically and practically. He reminds people to be diversified in their production and to have a good backup plan to pay the bills if something does not work out just right the first time. His advice for beginners is to start small, have fun, and only grow what you like to grow.

Update: Interview with Steve in August 2013

I caught up with Steve by phone, and we spent about an hour talking about a wide variety of topics. I was saddened to learn that he has been struggling for many years with the tick-borne illness, Lyme disease. He wanted me to mention the Lyme disease in his story because it is a risk for anyone who spends time in the woods. He urges everyone to check themselves daily for ticks, to learn the symptoms of Lyme disease, and to see a doctor immediately if they have any symptoms.

Steve is still a small-scale certified organic herb grower. Presently he has more black cohosh than anything else, and it is very healthy. The plants readily self-seed, expanding the patch naturally.

Even though it has been an exceptionally wet season, his goldenseal is holding up well, and at the time of our conversation it was all covered in bright, red berries. He collects his own seed and is experimenting with sowing it in trays of sand. The goldenseal research beds that he established in 2002 still exist, although some are clearly doing better than others. Jewelweed has become his biggest weed, and it is taking over everything in his woods, particularly the goldenseal. Fortunately, the deer love jewelweed and Steve has always had an abundance of deer. For once, they are useful because they actually eat the jewelweed growing up around the goldenseal without damaging the goldenseal.

Because of Steve's health issues, he now relies heavily on the assistance he gets from interns who volunteer to work on his farm and learn how to grow organic herbs from someone who has been doing it for a long time.

John Kershaw: Large-scale Artificial-shade Goldenseal Grower

I interviewed John by telephone for the first edition of this book in 2003. Goldenseal was his primary herb crop at that time, although he was experimenting with other woodland medicinals. I had not spoken to John for a few years, so I was excited to have a good excuse to call him up and chat for a while. I learned that he has expanded his plantings of other medicinals and had more information to share about them. Since this section of the book is about goldenseal, below are the 2003 and 2013 interviews minus the discussions about the other woodland medicinals. I put that information from both interviews in chapter 28, Other Forest Botanicals Grower's Stories.

2003 Interview with John About His Farm and Growing Goldenseal

John Kershaw and his wife, Michele, own Goldcap Farm in Burford, Ontario, Canada. I have communicated with John by telephone for many years. I update him on my research on goldenseal, and he keeps me current on what he and other large-scale Ontario herb growers are doing. John's operation is about as different as it can be from that of Steve Saint-Onge and Vestal Shipman, but I wanted to include it to show that growing goldenseal can be a large-scale, commercial farming business.

Goldenseal grown under artificial shade in Ontario, Canada. Photo by John Kershaw.

John lives in the major ginseng production area in Ontario, Canada, but unlike most of the goldenseal growers in Ontario, John was not previously a ginseng grower. While he was recovering from an injury and on leave from his job for five months, he did a lot of thinking about possible new income opportunities. He read extensively on a wide range of topics and crops. Goldenseal looked particularly interesting, and his financial analysis suggested it could be quite profitable; so in 1997, he planted his first crop of goldenseal.

John's Basic Growing Methods

The first goldenseal John planted was under a wood post and polypropylene shade-cloth structure. He also tried a small planting in the woods, but he did not take long to decide that it was not the best method for him to use. In his situation, he thought he could make a better profit from artificial-shade-grown goldenseal then he could from woods-grown goldenseal. His experience in the woods was that it was too difficult to prepare the soil, took many hours of hand labor, the mosquitoes were awful, and production was very low compared to the yields he obtained under artificial shade.

John grows certified organic goldenseal under standard polypropylene shade-cloth structures supported by 8'-long posts that are sunk 1' to 1½' into the ground. He installs side curtains on the west sides of his structures to maintain the best airflow while preventing damage from the late afternoon sun. He is a firm believer in the benefits of raised beds, and he uses them

Artificial-shade structure similar to the kind John uses. Photo by Jeanine Davis.

for every crop he grows. He puts very little fertilizer on his goldenseal, working some organic 4-4-8 fertilizer, a little iron, and some dolomitic lime into the bed when he plants, and that is it for the life of the crop. In his experience, goldenseal is not a heavy feeder and only removes a little potash from the soil. He has been studying soil microbiology lately, as described by Dr. Elaine Ingham of the Soil Foodweb. In the future, he plans to make use of products and practices that encourage growth of beneficial soil bacteria, fungi, and mycorrhizae.

John determined that a 6" × 8" plant spacing works best for him. This means he plants about 650 pounds of goldenseal rhizome pieces to the acre. John buys dormant, mature goldenseal rhizomes, usually three years old or older, to use as planting stock. In 2003, at $20 Canadian (US$15.35) per pound, planting stock is a major expense in establishing a new planting. These rhizomes are a variety of ages, sizes, and shapes. He cuts the rhizomes into 1" to 1½" pieces, trying to include a large bud on each piece, although it is not necessary to have an obvious bud for the piece to grow. He trims some of the long fiber roots from the rhizome pieces, so they are easier to handle during planting, and uses a mechanical transplanter and lots of hand labor to plant the rhizomes. (It takes eight people approximately nine hours to plant one acre of goldenseal.) Finally, he applies a thick layer of oat or barley straw as mulch. He uses the small square bales of straw and applies it all mechanically. Rye will also work if he cuts it green.

A few buyers are interested in purchasing goldenseal with high levels of the alkaloids, berberine and hydrastine, and will pay a small premium for it. For example, one buyer wants a minimum of 3 percent hydrastine. Others look for total alkaloids to be at least 7 percent. John has been trying to improve the alkaloid content of the goldenseal he produces by making small adjustments in his cultural practices, such as the composition and amount of fertilizer he applies. The berberine levels in his fiber roots and rhizomes are already high, so he is concentrating on the hydrastine. On his farm, the rhizomes naturally have higher hydrastine levels (4.2 to 4.3 percent) than the fiber roots (3.2 to 3.8 percent), so he is trying to increase the hydrastine in the fiber roots. This is in contrast to my experience and that of some other growers I've talked to who find no difference in alkaloid content between the fiber roots and rhizomes. John is not alone, however; in his book, *Growing At-Risk Medicinals*, Richo Cech cites an old research paper that states goldenseal rhizomes contain much higher levels of alkaloids than the fibrous roots. John has also noticed that in heavy soils, more rhizome is produced than fiber. I have made this same observation. Most buyers purchase goldenseal based on taste, smell, and appearance of the dried roots, but the few buyers looking for goldenseal with particularly high alkaloid levels will often pay a premium for it. In 2003, that translated into $22 per pound Canadian (US$16.88), and that is the market for which John is producing.

John grows some goldenseal from the seed he produces, and his experiences are similar to others reported in this book. He gets a small percentage of seed to germinate the first spring after sowing, with most emerging the next year. To improve plant stands and uniformity, he plans to use a precision seeder to sow his goldenseal seed in the future. He has had good success with seed and thinks it is because he keeps his seedbeds irrigated. In his experience, both the seed and seedlings need to be kept well watered to survive. Growing from seed, however, will probably never replace growing from rhizome pieces because production from seed is still not completely reliable and takes several more years for the plants to reach harvestable size. That said, however, John is intending to replant previously harvested areas with seeds because it is less expensive to produce from seed. He already has a succession of plantings that will be ready to harvest each year for the next few years, so he can afford to wait the four to five years it takes from sowing seed to harvest.

John mentioned another reason for growing from seed that I had never heard before. Some growers in Ontario who have been producing goldenseal from their own rootstock year after year are starting to see a decrease in yields and vigor. This is probably due to how fast they turn a crop in that area, three years from planting to harvest. In that short time, the rhizomes probably do not have adequate time to build up high levels of food reserves before they are dug, cut, and replanted again. Also, many growers tend to sell their biggest, most vigorous rhizomes instead of saving them for replanting. Growing goldenseal for an extra year or two before digging, selecting the largest, most vigorous rhizomes for replanting, maintaining a planting stock nursery for propagating new plants, or growing from seed should prevent this problem.

Dealing with Pests and Diseases

John has not experienced a serious pest problem on his goldenseal. Turkey and deer do some damage on occasion, and he does have some slug problems, but nothing like what we experience in the South. He has found that dolomitic lime, applied around the perimeter of his beds, provides the best slug control.

The main diseases he has confronted are Rhizoctonia and Botrytis. He reports that he has obtained good control of Rhizoctonia by diluting the organic insecticide Sharpshooter with water and applying it to the soil as a fungicide. I must point out that this is not a recommended use of the insecticide, and I have not been able to find any published research in support of this practice. However, John told me that he knows the developer of Sharpshooter who told him how to dilute the product and use it for this purpose. Sharpshooter is an approved organic product made from citric acid, acetic acid, and sodium lauryl sulfide (basically, vinegar, lemon juice, and soap). John also makes a compost tea to combat Botrytis and says it works really well for him. There has been a lot written about the use of compost tea for disease control in recent years, and some research supports it. John also removes as many of the diseased leaves from his plants as he practically can and has found that the fibrous roots and rhizomes of the goldenseal plant continue to grow even if almost all the leaves have been removed. I would like to remind the reader that I am not recommending that you do what John does, because I do not know how situation specific his results are; however, you might want to experiment with some of his methods. Also, to remain within the letter of the law, you must follow the label instructions on any agricultural chemical, whether it is organic or not.

Harvesting and Drying

John digs most of his goldenseal after the third year of growth from time of planting the rhizome cuttings because, under his conditions, the plants start to choke themselves out by the fourth growing season. He gets exceptionally high root yields per acre, higher than anyone else I know. He estimates that the roots he plants will increase in weight by four- to ten-fold in three years. On John's farm, one acre will produce 6,000 to 7,000 pounds of green root. Since his green roots produce about 30 percent dried root, he counts on yielding at least 2,100 pounds of dried root per acre. At a price of $40 Canadian (US$30.69) per pound in 2003, this translated into a gross return of $84,000 Canadian (US$64,449) per acre. John guesses that his maximum potential yield is 2,700 pounds of dried root per acre, and he is trying hard to reach it. Usually, the dried goldenseal that John sells consists mostly of fiber roots because he replants most of the rhizomes he produces. To do this, he cuts the excess fiber roots off the rhizomes shortly after he harvests them. The rhizomes are cut and replanted, whereas the fiber roots are washed, dried, and sold. He does this because he is still expanding his plantings each year, but he points out that the market prefers rhizomes. According to John, many buyers believe the rhizomes are better than the fiber roots, but interestingly, they won't pay more for them. Indeed, John's rhizomes contain about 25 percent more hydrastine than the fiber roots, but, as I mentioned earlier, he is trying to raise the level of hydrastine in the fiber roots. Anyway, once he has met his goal of the total number of acres of goldenseal he wants to have in production, he will increase the amount of rhizome in his finished product. John cleans his green rhizomes and fiber roots with root washers he built, modeled after standard ginseng root washers. He dries his goldenseal in tobacco kilns at 90° to 95°F.

Marketing

John has a good market for certified organic goldenseal, so all his crops are certified by the Organic Crop Producers and Processors, a major organic certifier in Canada. Choosing to be an organic herb grower has turned out to be a good market decision for John, because he has always been able to sell his goldenseal. He also reports that he gets higher prices for his organic goldenseal than other growers are getting for their conventionally grown goldenseal. In early 2003, the average price for goldenseal in his area was $40 Canadian (US$36.69) per pound. He earned an extra $10 Canadian (US$7.67) per pound for certified organic goldenseal.

John feels that building lasting relationships with buyers is very important because when the market is tight, the buyers tend to buy from the growers they know and trust. He finds his markets by reading industry publications, surfing the Internet, and talking to other growers. He knows that good marketing takes time and effort, and he is willing to make the investment. Successful marketing often requires doing some detective work and learning where the buyers are. John also goes to great lengths to ensure that he always delivers a quality product, which often results in repeat sales to very satisfied customers.

In today's market, almost all buyers require that herbs meet certain quality standards and be tested for chemical constituents, contaminants, dirt, and moisture. Many times, the company buying the herbs will offer to run the laboratory analyses, but some growers do not trust the buyer to provide an accurate analysis and prefer to pay an independent lab to test their crops. John says that he has faith in the lab analyses provided by his goldenseal buyers, but for many of his other crops, he also pays for an independent test. This is not because he does not trust the buyers' labs, but because there still is so much variability between labs in testing of some of the other herbs he grows that he wants a second opinion.

John says he was very fortunate to have gotten into goldenseal production at just the right time. He believes that the major growth stage for the cultivated goldenseal market is over, and that the industry has matured. He does not think a new grower can expect to make a lot of money from large-scale artificial-shade-grown goldenseal production right now, at least not growing for the regular goldenseal market. He feels he will continue to make money at it because he has already made the major investments in shade structures, equipment, and planting stock. He also has his markets and gets a premium for a certified organic product. In fact, when we first talked, he was planning to plant three more acres.

Advice to New Growers

John's advice to new growers is to have a market before you start and keep in contact with your buyers throughout the year. Don't let them forget you. Also, watch how much you invest in a crop. Producing goldenseal under shade cloth is not cheap. For example, producing one-third of an acre of goldenseal—from planting through drying the root—cost $13,000 to $15,000 Canadian (US$9,975 to $11,510) in 2003. He admits it is really scary to invest that much money into a crop, but he is willing to take the risk because he knows he has a market for it. He advises growers to keep good records and make projections on costs and returns. Also, remember when you harvest your crops, you don't want to sell everything you produce because you need to keep some of it to replant for your next crops. You have to keep replanting to keep your business going, and planting just from seed takes too long. Last, but not least, he advises that you and your spouse or partner have other steady jobs!

John's plans for the future include working toward having a total of 20 acres of medicinal herbs in production at any one time. He is looking for new market opportunities, but has decided against making value-added products because, in his opinion, there are just too many regulations on manufacturing value-added food products in Canada.

I must add that John is not an average person. He seems to have endless energy and is completely committed to what he is doing. He stays current on the industry and reads three hours a day. He highly recommends that everyone in the business of growing plants for our changing society read the book, *Boom, Bust, and Echo.* In addition to his herb business, he also works a regular job, 45 to 50 hours a week. And, as if that was not enough, he milks 140 goats each day. Oh, and he has four children. Personally, I get tired just talking to John!

Update on Growing Goldenseal from Interview with John Kershaw in 2013

I called John one evening in September, and we had so much to catch up on that we talked for two and a half hours! That was great fun. Part of the reason we had to talk for so long was that he has diversified into more herbs. I put that part of the discussion in the Chapter 28, Other Forest Botanicals Growers' Stories. What follows here is about his farm in general and growing goldenseal.

Changes on the Farm

John is still a certified organic farmer in Burford, Ontario, but his business has

Table 16. John's Budget for One Acre of Certified Organic Goldenseal Grown under Artificial Shade for Three Years in 2003 and 2013*

	2003			2013		
For One Acre	**Price/Unit**	**Expense**	**Income**	**Price/Unit**	**Expense**	**Income**
Planting stock: 600 lbs	$19.18/lb	$11,508		$18/lb	$10,800	
Labor: 780 hrs	$6.91/hr	$5,390		$10.50/hr	$8,190	
Materials and equipment		$8,440			$10,500	
Total costs		$25,338			$29,490	
Expected yield: 2,000 lbs of dried fiber roots and rhizome	$32.23/lb		$64,460	$33/lb		$66,000
Gross return			$64,460			$66,000
Net profit at the end of 3 years			$39,122			$36,510

*John's budget for one acre of certified organic goldenseal grown under artificial shade is based on three years from planting of rhizome pieces until harvest. The cost of the shade structure, dryer, fertilizer, and custom work are included in the labor, materials, and equipment costs. It is in U.S. dollars.

changed some over the past ten years. The "buy local" and "buy domestic" (from North America) movements have presented more opportunities for him. There have also been some problems and challenges. The medicinal herb industry has been up and down, so he constantly has to re-evaluate what he is producing.

I was sorry to learn that the honeybee problems we have here in the United States have reached Canada, too. John had 12 hives last winter and lost them all. He is building those back up again, but it is expensive and time consuming. His livestock operation is expanding, too. He still has a herd of goats, but has added cattle and Berkshire pigs. Next, he wants to get some lowline Angus. I had never heard of them before. He explained that they are a small breed of Angus cattle; only about three and a half feet tall. They produce a high ratio of meat to bone, their meat is tender, and they are small and gentle. That was good information for my own farm. I recently got two miniature donkeys (my standard-sized donkey had passed away). Now I want to look into getting lowline steers once our full-sized Black Angus leave the farm, for you-know-where, this fall. I think it would be especially fitting to have little steers since the name of our farm is "Our Tiny Farm." But I digress.

Growing Goldenseal

After 16 years, goldenseal is still an important crop for John. He has learned that goldenseal can be an easy crop to grow if you don't fuss over it too much. When he has babied it, he has more problems with disease. So he tries not to over-manage it. One big change he has made is that he now uses rye straw instead of oat straw as the mulch. The rye seeds do sprout and grow, making the beds look weedy, but he has fewer problems with Botrytis and Alternaria, so he thinks it is a good trade-off. To prevent even more rye seed from being dispersed, next spring he is going to try to clip off the tops of the rye plants with a flail mower before the goldenseal is up too high

In 2003 John was trying to improve his goldenseal seed germination. He said he has not made great progress on that, although some years are definitely better than others. So perhaps there is something about the conditions the year the seed is produced that affect germination in the following years. He has discovered that the most cost-effective method to grow goldenseal from seed is to just scatter fresh seed across the surface of a well-prepared seedbed and to cover with straw.

He continues to take an integrated, organic approach to pest control on goldenseal and all his herbs. He has observed that the farm guinea hens seem to be playing an important role in the ecosystem of Goldcap Farm. They have free access to the farm (and neighbors' properties). As they move through his goldenseal beds, they eat bugs and slugs without causing any harm to the goldenseal. In fact, there has been a large increase in Japanese beetles in the area, but his farm has had little problem with them.

In 2013 John reported yields of around 6,000 pounds of fresh goldenseal root to the acre, which translates into about 2,000 pounds dried. Yields haven't changed much over the past ten years.

Some growers and experts working with the industry say that wildlife, mostly deer, turkeys, and groundhogs, are causing extensive damage to cultivated herbs and are major contributors to the loss of natural populations of ginseng, goldenseal, and other native botanical herbs. It has been John's experience that turkeys cause more damage than deer, particularly in cultivated plantings. Rye and wheat straw is used as a mulch in ginseng and goldenseal gardens, and the turkeys like to scratch in it looking for seeds and worms to eat. Their scratching damages the herb roots.

It was good talking to John again, and I have made myself a promise to visit him before the next revision of this book is written. That way, we'll get some pictures, too.

14

Ramps: History, Description, and Uses

Range and Tradition

Ramps, *Allium tricoccum* or *Allium tricoccum* var. *burdickii*, also known as wild leeks and wood leeks, are native to the forests of eastern North America. They grow in patches within rich, moist, deciduous woodlands and bottomlands, usually in the mountains, from as far north as Canada, west to Missouri and Minnesota, and south to North Carolina and Tennessee. You have probably seen ramps many times without knowing it. Ramps have smooth, broad lily-of-the-valley-like leaves that emerge from the forest floor in very early spring. The dark green leaves stand about ten inches tall. They are attached to an underground bulb, which looks like a young onion with roots. At that time of year, the bulbs taste like sweet spring onions with a strong garlic-like aroma. Tasty as they are, they have a notorious reputation for imparting a distinct garlic-like odor to the consumer that lasts for several days. Many adults tell stories about how they would eat ramps when they were kids just so the teacher would kick them out of school because they smelled so bad, allowing them to spend a beautiful spring day outside. As one of the first plants to emerge in the spring, ramps have traditionally been consumed as the season's first "greens." Many traditions have evolved around the annual gathering and preparation of ramps. Throughout the mountains of eastern North America, from Canada south to Georgia, annual spring ramp festivals are held. These festivals are major tourist attractions, and the communities in which they are held actively promote them. Many of these festivals serve as fund-raisers for fire departments, emergency medical services, and civic groups.

Ecological Concerns and Opportunities for Growers

Ecological Concerns

A tremendous quantity of ramps is consumed at ramp festivals, which often serve tens of thousands of people at a time. What

Mature ramps in Richwood, West Virginia. Photo by Jackie Greenfield.

most people are not aware of is that almost all of the ramps eaten at these festivals are gathered from wild populations in local forests. As the popularity of these festivals has grown, the intensive annual harvesting has begun to seriously damage the wild populations of ramps. If these trends continue, many ecologists are concerned about the long-term survivability of wild ramps. Studies in Canada and Ohio demonstrate that excessive harvesting can seriously damage individual populations of ramps. Years ago, gatherers would only take a small number of bulbs from a population for use by their immediate family. Now the demand for ramps at these festivals is so great that entire populations are often dug at one time. In addition, gourmet cooks have discovered ramps. Ramps have been a featured item on the Food Network channel, the "Emeril Live" cooking show, and in Martha Stewart *Living* magazine. Many white-tablecloth restaurants serve ramps as a novelty dish in the spring. The lowly little ramp is now airfreighted all over the continent and served on fine china to well-dressed patrons. And most recently, the national "Wild Foods" movement has resulted in ramps being sold at farmers' markets and specialty stores. Magazine articles, blogs, and cooking shows encourage people to get outdoors and harvest their own ramps. Ramps have come a long way from being a staple for pioneer families or a tasty addition to trout and potatoes cooked by a fisherman over an evening campfire. Putting the spotlight on ramps has increased the demand for large, consistent supplies of this wild forest plant.

Protected Plant Status

Ramps are included on many "Species at Risk" lists in North America. Where they are listed, there is usually some legislation pertaining to their collection from public and private lands. Collection from public lands might be limited or require a permit. Collection from private land might require permission from the landowner. Usually these rules do not impact cultivation of ramps except for the requirement that planting stock should be obtained legally and preferably from commercial sources.

Legislation concerning ramps varies by state and province, and country. This information can be very difficult to find for your state or province depending on which one you live in. I suggest doing an Internet search on "species at risk," "endangered species," and "plant conservation" with the name of your state or province. The federal protected plant lists for the United States and Canada are fairly easy to find. For the United States search for "USFWS endangered species list" and for Canada search for "Canadian Species at Risk Public Registry."

Examples of some of the designations I found were that New Hampshire, Maine, Tennessee, and Rhode Island categorize ramps as plants of Special Concern; New York lists *Allium tricoccum* var. *burdickii* as endangered; and Quebec lists ramps as vulnerable. In Quebec you are allowed to collect 50 bulbs per person per year from wild populations. In the National Forests in North Carolina you can collect five pounds each year for personal use. For commercial collection, you need to obtain a permit, pay $0.50 per pound, and limit collection to 500 pounds.

Opportunities for Growers

In an effort to meet the rising demand for ramps and conserve native populations, I strongly encourage forest owners to consider cultivating ramps. I don't think there will ever be as large a market for ramps as there is for ginseng or goldenseal, but I believe a grower can make some money from ramps without too much effort. Cultivated ramps can be sold to festival organizers, restaurants, caterers, roadside stands, and specialty stores. They can also be sold over the Internet to individual consumers and future growers. A number of value-added products can be made from ramps to extend the market season and provide even more income. If farmers provided a readily available, affordable, reliable supply of high quality ramps, native populations would have the opportunity to regenerate and multiply as cultivated ramps gained recognition and popularity among consumers.

Plant Description and Life Cycle

Ramps are long-lived perennials that begin growing in very early spring. They have an underground bulb that survives from year to year. In early spring, usually March, the bulb sends up new leaves. Ramps are easy to spot when they first emerge because they are often the only fresh green plants in the forest at the time. They usually grow in cool, shady areas with damp soil and an abundance of decomposed leaf litter and other organic matter. Each plant has one to three long, narrow leaves that come together at the neck of the bulb, much like a

Flowering ramps with seeds developing on left flower.
Photo by Jackie Greenfield.

lily. The leaves die back as the days lengthen and the temperatures rise.

The most commonly consumed portion of the ramp is the bulb, which grows during the short five-to-eight-week period that the leaves are alive and producing sugars to send to the bulb for storage. In the Southern Appalachian mountains, the bulbs are collected for consumption during April and early May, when they are soft, sweet, and flavorful. Usually toward the end of May, the leaves start to turn yellow and die back and a flower-stalk with a single bud, known as a scape, emerges on some plants. Occasionally, the scape develops while the leaves are still present. This bud develops into an inflorescence (cluster) of small cream-colored flowers, usually in late June. The flowers can self-pollinate or be pollinated by a variety of small insects. By July, some of the flowers produce three-lobed capsules, with each lobe containing a single seed. By late August, the capsule becomes thin and papery and splits open, exposing the seeds. Unless you know where to look, it can be difficult to find the round, shiny black seeds sitting atop a leafless stalk. Eventually, the seeds fall to the ground where some will eventually germinate near the mother plant. Most ramps do not bloom every year, and in some areas the plants seem to be synchronized to bloom only every other year. In addition, not all flowers produce seeds. And sadly, most of the seeds never make contact with the soil, so they do not germinate and grow. Those that do germinate often take two seasons to do so and look just like tiny blades of grass when they emerge.

By October, most of the seeds have fallen and the bulbs have hardened. The only thing visible at this time may be the old seed stalk. Over the winter the bulb feeds on the sugars that it stored as it prepares to send up new leaves in the spring. If you dig a bulb in February, you might find it divided into two, and sometimes three, bulbs. Plants that flower have a much higher tendency to produce more bulbs through division than plants that do not flower. In the wild, native populations of ramps usually form colonies that can cover vast areas in the forest. If a site has been left undisturbed, the bulbs can become so dense that other plants cannot grow among them.

Uses and Special Properties

Years ago, mountain residents consumed ramps as a spring tonic because they provided necessary vitamins and minerals that the body needed after long winter months without any fresh vegetables. People living in the mountains harvested the bulbs shortly after the leaves appeared and served them in a variety of dishes. Ramps are often fried up with eggs or potatoes and can be incorporated into pancakes, soups, and stews. Some people still consume them as a healthful tonic, but most eat them because they are a tasty novelty that signifies the advent of spring.

The health benefits of ramps might turn out to be more than just folklore. A study published by Dr. P. Whanger and colleagues in 2000 showed that rats that consumed ramps grown with selenium-containing fertilizer had fewer mammary tumors, suggesting that selenium-enriched ramps may reduce cancer in humans. A 2004 update by the same author reported that out of eight trials on human subjects, seven showed a positive benefit of selenium on cancer reduction or biomarkers for cancer. Consumption of plants from the Allium family, which includes ramps, onions, leeks, and garlic, has been shown to retard the growth of some cancers.

Thanks to a wide range of products made by a few enterprising individuals, ramps can now be consumed year round. I have purchased jelly, mustard, salt, pancake mix, biscuit mix, soup mix, and dip mix, all made with ramps. Pickled ramps have been a popular treat in some areas for decades. More recently, dried ramps have become available. They are convenient to have on hand for cooking. Ramp leaves also freeze well.

15

Ramps Growing Instructions: Methods, Care, Protection, Harvesting, and Marketing

In 1998, when I first started growing ramps, there were only a few individuals in North America producing ramps for commercial purposes. I relied heavily on their experiences to initiate my own successful ramp cultivation efforts and studies. Since that time we have learned a great deal about them, and people are successfully growing ramps all over North America.

Site Selection, Shade, and Moisture

Successful ramp production is possible throughout most of North America except in the Deep South and in hot, dry desert areas where the chilling requirements of the bulbs will not be met or the soil is too dry. They are hardy in zones 3 to 9. When choosing a site for planting your ramps, you should look for a spot where you might find ramps growing naturally. Ramps most often grow under a forest canopy of beech, birch, sugar maple, and/or poplar trees. Ramps will also grow under buckeye, linden (basswood), hickory, and oak. A forested area that includes any of these trees should be right for planting a ramp crop. As with ginseng and goldenseal, you should look for a site with other plants that like similar conditions to the plant you want to grow. Areas that host trillium, toothwort, nettle, black cohosh, ginseng, bloodroot, trout lily, bellwort, or mayapple should be suitable for growing ramps.

Ramps tolerate a wide range of shade levels, except the deepest shade and, in most cases, full sun. When they first emerge in the spring, they should have plenty of sun. Most mixed species forests provide this because the deciduous trees don't have leaves when the ramps are up and growing. After the trees leaf out, they get a dappled light. Dense pine forests are not suitable for ramps because they are too dark. Occasionally, ramps can be found growing in full sun, and a study by Vasseur and Gagnon in Canada using transplants indicated that ramps will grow in full sun if the soil is kept moist enough. We have not had much

Wild ramps growing with foam flowers. Photo by Jackie Greenfield.

Cultivated ramps growing in the woods on the research station. Photo by Jeanine Davis.

success with full sun production in our research studies in western North Carolina, but we were not able to keep the soil consistently moist, because we were in the midst of a severe multi-year drought at the time.

We planted ramp seeds under shadecloth structures providing 30 percent, 47 percent, 63 percent, and 80 percent shade, under a wood lath structure (about 63 percent shade), in full sun, and in the woods. Under the artificial-shade structures, seedling emergence was highest under 30 percent shade (52 percent germination) and decreased as the shade increased. There was only 10 percent germination in the full sun, but 57 percent germination in the woods. As the years went on, the plants grew best under 47 percent to 80 percent shade.

Shade structure suitable for growing ramps. Photo by Jeanine Davis.

Ramps we planted in forest sites, however, grew significantly better than in any of the shade-cloth sites or in the full sun site. The ramps in the full sun site never did size up and eventually all died out. The plants under the shade-cloth structures grew best when the soil was kept moist with overhead irrigation from spring until we had to shut off the irrigation in the fall. Then we put a very heavy layer of mulch on the beds to hold in the moisture. We raked back some of that mulch around the first of March so the plants could come through. The ramps grew fine, but they didn't spread much and rarely flowered and set seed. A few years ago we dug them all up and moved them to the woods. Since then they have grown bigger, some have flowered, and it looks like they are spreading. So, based on these experiences, my suggestion is to plant them in the woods if you can. If you do not have a wooded area available for ramps, you can use an artificially shaded site, but make sure the soil is high in organic matter and that it will stay moist all year round. You can refer to the ginseng and goldenseal chapters to learn how to provide artificial shade for your plants.

Whether in the forest or the field, choose a well-drained site with rich, moist soil high in organic matter. Research by Liette Vasseur of Saint Mary's University in Halifax, Nova Scotia, and Daniel Gagnon of the University of Quebec in Montreal showed that adequate soil moisture is the most important environmental factor influencing ramp seed germination, seedling emergence rates, survival, and growth rate of transplants. Good soil moisture must be maintained throughout all the seasons, not just during the spring growing season when the ramp leaves are visible.

Ramps flourish in forests with rich, moist soils where organic matter from leaves, twigs, and decaying trees builds up year after year. These conditions are usually found in hardwood forests or mixed hardwood and pine forests. If the forest

Susan Bryson raking in limestone for a research study on ramps. Photo by Jeanine Davis.

floor has nothing growing on it, it is not the place to try to grow ramps. Ramps will grow near streams, but do not plant them in swampy areas. If you live in the South and your area is hilly, the direction of the slope to the sun, known as the slope aspect, is very important. North- and east-facing slopes are generally cooler, moister, and shadier than south- or west-facing slopes. The south-facing slopes are usually drier and have poorer soils than the north-facing slopes. Think about how hot and uncomfortable a room with a west-facing window is in the summer.

Site Preparation and Fertilization

Site preparation for ramps is the same as for ginseng or goldenseal, so refer to those sections for detailed information. To plant under a forested canopy, rake back the leaves on the forest floor, removing any unwanted weeds, tree sprouts, or roots. To plant in a pasture or field, turn under the existing vegetation and do what is necessary to control weeds, especially grasses. This might involve growing a cover crop, such as soybeans or clover for a season. Till, or break up the soil by hand, and incorporate organic matter such as composted leaves. You can build raised beds, as described in the goldenseal section, or use the trench system that Scott describes for ginseng. Raised beds are not usually necessary for ramps unless you are planting where there is a chance of soggy conditions. The raised beds will improve drainage.

In the natural environment, studies have shown that ramps grow best in soil with high nutrient levels. A study published in 2013 from Quebec showed that ramps produced bigger bulbs when fertilized with an organic fertilizer. My experience has also been that ramps like a nutrient-rich soil, so I usually run a soil test and ask for recommendations for "native ornamentals." If adjustments are recommended, I use organic fertilizers and mix in some compost just to increase the organic matter.

Jackie Greenfield, a former research associate in my program, collected soil samples from dozens of healthy native and naturalized ramp sites in far western North Carolina to see if there was an ideal soil nutrient condition for ramps in our area. Those soil tests suggested that calcium might be an important nutrient for ramps because calcium levels were high in most of the samples. Earlier studies on ginseng by my predecessor Dr. Tom Konsler, and by me on goldenseal, showed those plants grew faster and larger when the soil pH was increased, so we wondered if ramps would respond similarly to a change in pH.

We designed a set of studies in the woods at our research station looking at the effects of calcium, increased by adding gypsum, and soil pH, increased by adding the mineral olivine, on survival and growth of ramps. After three years, we found that raising the soil calcium had increased the survivability and bulb size of the ramps, but raising the soil pH over 4.9 reduced survivability and bulb size. The Quebec study mentioned above and a 2005 study from West Virginia also got better ramp growth by adding gypsum to the soil. The latter study also showed an improvement from the raising soil pH, but they started with an initial soil pH of 3.83 to 4.28 and the lime only raised it to about 4.5. So, based on these findings, and until we get more information, I suggest that growers add some gypsum when they plant, say about 100 pounds to 1,000 square feet, and don't adjust the soil pH unless it is below 4.5. After explaining all that, I must also say that many growers don't add anything to the soil when they plant ramps and the ramps seem to do just fine.

George Cox planting ramps in a research study on soil pH and calcium. Photo by Jeanine Davis.

Propagation

Direct Seeding

Unlike ginseng and goldenseal seed which must always be kept moist, ramp seed is stored dried and can be held that way for at least a few years. Fall is the best time for seeding ramps, although seeds can be planted almost any time of the year. Moist soil and protection from wildlife must be provided for the one to two years it takes ramp seeds to germinate. Maintaining moist soil through the stratification process can be assured by choosing the proper site,

Immature and mature ramp seeds in seed capsules. Photo by Jeanine Davis.

applying a good layer of mulch, and providing irrigation when necessary. Sowing your seeds in a conveniently located small nursery bed is a good approach. First, find a perfect location as described earlier—one that has dappled sunlight and rich, moist soil. Rake back all the leaves and sticks, loosen the soil on top of the bed, and rake it to prepare a fine seedbed. Next, sow the ramp seeds thinly on top of the ground by broadcasting the seeds or sowing them in rows about 3" apart, both ways. Press the seed gently into the soil with your hand or the flat side of a hoe. Then cover the bed with several inches of dry leaves from the surrounding forest, composted leaves, or high-quality compost. If you have trouble with rodents, turkeys, or other animals disturbing your seed, you might have to build some kind of protection around your bed, such as covering it with a screen held down with big metal pins until the seeds start to germinate.

Seeds sown in the fall can be expected to germinate the second spring after sowing; about 18 months later. Every now and then, a few seeds will germinate the first spring after sowing, and just as often some wait until the third spring after sowing to germinate. In general, expect to take good care of what looks like an empty bed for a year and a half. Keep it moist, mulched, weeded, and protected from wildlife.

Eventually your seedlings will emerge. They won't look like much; you might even need to get down on your hands and knees to see them. They will look like very thin blades of grass. Continue to keep them weeded, mulched, moist, and protected from wildlife. When the bulbs are two to three years old, you can move them in the spring to their final growing area. Production, from sowing the seed to bulb harvest, takes five to seven years.

There are some growers who firmly believe that ramp seeds should be sown where they will stay until harvest. They say they get bigger harvests that way. I prefer to use nursery beds so protection can be provided that would be impractical if the seeds were spread out over a production area. I think more plants survive with my method, and I have moved young plants many times without any problems, but it is a two-step process. Try it both ways and figure out what works best for you.

Transplanting

Since ramp seeds take so long to germinate, many growers choose to start their first plantings with dormant bulbs or young plants. If you plant large bulbs, you can expect to make your first harvest after three years of growth. Dormant bulbs can usually be purchased in February for spring planting. You can dig and transplant bulbs from nursery beds or wild populations (to which you have legal access) anytime between September and March, but February through early March is still the best time in most places. My own studies have demonstrated this to be the case, and Bernatchez and colleagues just confirmed it in a paper published in 2013 from Quebec. You can also transplant young, growing plants. This is most successful when the leaves are just emerging from the bulbs and not fully expanded. Very early spring, middle to late March in the southern Appalachians, is usually the optimum time for transplanting young plants.

Young ramp plant with bulb. Photo by Jackie Greenfield.

Weather conditions greatly determine when you should dig bulbs or young plants.

Ramp seed heads on left and ramp bulbs on right. Photo by Jeanine Davis.

It is not unusual for a large snowfall in late February or early March to delay digging. If you want to dig your own bulbs or plants, wait until the ground has thawed and gently dig into the ramp patch with a fork or spade, taking great care not to damage the bulbs. If you plan to move growing plants, as opposed to dormant bulbs, it is best to leave some of the soil on the roots during the transplanting process.

Choose an appropriate planting site, as described earlier. Rake off the leaf litter, till the soil, add any needed amendments, and work them in. Make raised beds if you plan to do so. To plant your bulbs or transplants, dig little trenches 4–5" deep across your beds or planting area. Spacing these trenches 4–6" apart gives the plants some room to multiply. I make trenches by simply dragging a hand trowel across the bed. Set dormant bulbs approximately 3" apart and 3" deep, with the growing point facing up. Cover with soil so just the very tip of the bulb shows above the soil surface. Transplant leafed-out plants at the same depth they had been growing. Finally, cover both bulbs and transplants with several inches of mulch. Transplants will usually reach harvestable size in four to six years.

Mulching

I have not conducted research on mulches yet, but based on my experience and that of growers I have interviewed, hardwood leaves appear to be the best mulch for ramps. Growers report that hardwood bark and most commercially prepared mulches, such as bark mixes and coconut shells, should be avoided. I have used hardwood bark under shade structures and it worked just fine, but in the woods, leaves are the most practical. I do not know anyone who has tried straw. Cover the planted area with several inches of mulch and check the site throughout the year to make sure the mulch layer remains adequate to hold in moisture and prevent most weed growth. During the prolonged drought we experienced here a few years back, the leaves were so dry they kept blowing away, and it was difficult to keep a good layer on the beds. Ramps do not like dry soil, whether they are actively growing or not, so the leaf mulch had to be replaced several times during the summer. The easiest way I found to do this was to blow them back onto the beds using a gas-powered leaf blower.

Care and Protection

There is little information available on diseases or insects of ramps. Weeds, however, should definitely be controlled, especially until the plants are well established. How much time you will need to spend on weed control is very site specific. I have had some sites where there were almost no weeds; on other sites, we wage a constant battle with poison ivy, grasses, and tree seedlings. A good layer of mulch will always help, so be sure to maintain that all year round.

In 2002, we discovered a disease on ramps in western North Carolina that was identified as Septoria leaf spot. It causes an unsightly spotting on the leaves, but does not seem to have much effect on bulb growth or flowering. However, although

Ramp leaves with Septoria leaf spot. Photo by Jackie Greenfield.

the leaves are not usually the portion of the ramp that is consumed, the foliage is almost always sold with the bulb, so the spotting could reduce the marketability of the ramps.

There is a new threat on the horizon that has many ramp growers and wild-harvesters concerned. It is a small, speckled brown, white, and black moth called a leek moth (*Acrolepiopsis assectella* Zeller). It is native to Europe and feeds exclusively on plants in the Allium family, i.e., onions, leeks, and garlics. It was discovered in Ontario in 1993, and as of 2013 had spread to Quebec and New York. It prefers to eat cultivated leeks and onions, but if those aren't available it will eat other Allium plants. On onions and leeks, the first generation caterpillar feeds on leaves in May through June. The second generation feeds on bulbs in July and August. It overwinters as a pupa in leaf litter, dead plant material, and in the soil. In Europe, a little wasp has been released that parasitizes the caterpillar. Apparently, that has suppressed the populations of the little pest. In North America, there are conventional and organic insecticides that will control it, so the real concern is for wild ramp populations. A study published in 2007 from Cornell indicated that the leek moth likes ramps just as much as it likes cultivated leeks. The field data suggest that native ramps are at risk from attack from this new insect. So monitor your plants for this insect and the damage that it causes. A quick Internet search for "leek moth" will bring up pictures of what to look for.

Seed Collection and Stratification

Ramp seeds have what is called morpho-physiological dormancy. They have to be exposed to warm-moist conditions for a period of time to break dormancy on the seedling root followed by cold-moist conditions for a period of time to break dormancy on the seedling shoot. The whole process is called stratification and takes about 18 months. Many people, including myself, have tried to speed up the process. I just read a 2013 paper about a study by J. DeValue in which ramp seeds, which have a hard seed coat, were scarified (nicked or scratched) and exposed to different temperature regimes. Scarification didn't do much of anything, and the final germination was about 40 percent when the seeds were held at 68°F for ten weeks before sowing. I usually get germination rates around

Jackie collecting ramp seeds and bulbs. Photo by Jeanine Davis.

60 percent by sowing directly outdoors without using any seed treatments.

In 2003, Jackie Greenfield and another graduate student studied the germination requirements of ramp seeds. Their research findings should help us develop a reliable, predictable method of producing seedlings the spring after seed harvest. In the meantime, most growers are using the method I described earlier: sowing dry seed in the fall and waiting one and a half years. There are a few methods that have been described to me that are supposed to speed the germination process, but they require holding the seeds in moist environments and fluctuating the temperatures for three or four months. The objective is to mimic what would happen in nature over several seasons, and it might work, but most of the methods described to me are not very practical and would be difficult to duplicate without incubators. One method that has proven successful for some growers that you might want to try is to harvest fresh seeds and store them in moist sand at room temperature for two months. Then move them to the refrigerator at about 40°F for the rest of the winter and sow them outdoors in the early spring. Growers have reported results ranging from 0 to 90 percent germination with this method. If you do not get good germination the first spring, there will probably be considerably more the second spring, and in rare cases the seed will wait until the third spring to germinate, especially if the soil has been on the dry side.

The following is the easiest procedure that has worked well for me. Collect the mature, black seed from the plants in late August before the seeds fall to the ground; then, either immediately plant the seeds in

Bunches of ramps being sold at the Western North Carolina Farmers' Market in Asheville. Photo by Jackie Greenfield.

a nursery bed or store the seeds in paper envelopes in a cool, dry place until you are ready to plant them in a moist site. Little seedlings should emerge after exposure to one warm season and one cold season in the soil. This usually means after 18 months. If you planted the fresh seed early enough in the season for the seeds to get a long period of warm temperatures before the soil cooled down for the winter, you should obtain good emergence the first spring after sowing—in about seven months.

Harvest and Post-harvest Handling

It is best to wait until a bed of ramps is mature before you begin harvesting. I consider a bed of ramps to be mature when a majority of the plants have flowered several times and there is evidence that the bulbs have divided a few times, i.e., there are many more bulbs there than you planted. Measure off a small area where you know how many plants were originally set and see what kind of increase has occurred in numbers of bulbs. I do not have enough information on ramps to tell you the economic threshold, in other words, what kind of increase you need in order to make a profit, but my guess is you want to have at least three good-sized bulbs for every one that you planted.

Traditional ramp digging tool. Photo by Agatha Kaplan.

There are four basic approaches to harvesting ramps: digging all the ramps from the whole bed or a section of the bed; digging just the largest plants from the bed (known as thinning); digging clumps of ramps from several small areas scattered around the bed; or cutting the leaves off the largest plants in the bed. If either the whole bed or sections of beds will be dug, the recommendation is to have enough bed space to allow for rotations of between five and seven years and to replant areas equivalent in size to what you harvested. In other words, to have continuous harvest year after year, harvest only one-fifth or one-seventh of your planting area each year. If you want to have a bed of ramps in perpetual production, you could harvest clumps of ramps or thin the beds each year.

Research on wild populations indicates that in order for the population to recover you should harvest no more than 10 percent of the ramps at a time. If this selective digging is your harvest method of choice, carefully dig the largest plants in the bed and any that are growing in large clumps and starting to crowd each other. These may be susceptible to "clump death," which I assume is caused by disease and stress when the plants become too dense. When you dig, take care not to damage bulbs of any of the plants that you want to leave in the bed to continue growing.

You can dig as many plants as you want, but if you limit yourself to no more than 10 percent of your plants, you should be able to harvest from that bed every few years. Janet Rock's work in the Smoky Mountain National Park indicated that a 10 percent harvest once every ten years would be sustainable for most ramp populations in the southern Appalachians. It has been my experience, however, that managed beds recover faster. Selective harvesting does not require replanting each year; however, because you have to harvest so carefully, it is more time consuming than digging all the plants in a bed. Harvesting just the leaves of the plant allows the bulb to remain and grow again the next year. Keep in mind that the leaves must be allowed to exist long enough to produce food for the bulb to grow and survive the next year. So do not harvest leaves as soon as they emerge, but they should be harvested before they get large and tough or start to turn yellow.

There are several different kinds of tools used to harvest ramps. The most commonly used ramp digger is a hand tool the size of a hammer, with a head similar to a maddox. Some of the ramps suppliers listed at the end of this book sell these "official" ramp-digging tools. Other tools suitable for digging ramps include a garden hoe, pick, and ginseng hoe. For commercial operations, having a tool that can be used comfortably all day is essential. I do not know of anyone using any kind of mechanized digger at this time. That will certainly change if large-scale operations come into production.

While digging, keep the harvested ramps cool and moist. When harvesting is complete, wash the ramps thoroughly and trim off the roots. Let the ramps air dry un-

Ramps for sale at the ramp festival in Richwood, West Virginia. Photo by Jackie Greenfield.

til the surface moisture is gone, and then pack them in waxed cardboard produce boxes and store in a cool place, preferably a walk-in cooler. Do not store the ramps in airtight containers, as they will break down very quickly. Ramps are usually packed in bundles secured with a rubber band. At this time, there is no standard for how large that bundle should be. I have observed bundles holding from 5 to 15 ramps and selling for between $2.50 and $7.00.

Marketing

If you live in an area where ramps are a part of the culture, you will not have any trouble selling them. In these areas, folks grow up eating ramps and look forward to having some every spring. Many of these people do not have the time, inclination, or ability to gather ramps anymore, so they will gladly pay someone to do it. Where I live in the mountains of western North Carolina, ramps are sold at roadside stands, tailgate markets, farmers' markets, and even some supermarkets. Ramps are highly perishable though, and they should be kept in a cooler or refrigerated unit if they will not be sold within an hour or two.

In areas where tourists come for ramp festivals or to enjoy old time mountain culture, consider producing food products made of ramps. There is no end to the products you can create from the bulbs and tops, such as condiments, baking mixes, dip and

Table 17. Farm Enterprise Budget for One-tenth Acre of Ramps Grown in the Forest

Income	Quantity	Price per Unit	3-Year Totals
Wholesale sales of bulbs	80 lbs	$10.50/lb	$840
Restaurant sales of bulbs	190 lbs	$16.00/lb	$3,040
Gross Income			$3,880
Expenses			
Labor: Site preparation and planting	50 hrs	7.25/hr	$363
Labor: Weeding and mulching	80 hrs	7.25/hr	$580
Labor: Harvest, washing, and packing	60 hrs	7.25/hr	$435
Bulbs for planting	9,600	194/1,000	$1,862
Packaging			$110
Total Expenses			$3,350
Net Income			$530

soup mixes, and seasonings. Put them in attractive packages with cute descriptive names and promote your mountain culture. A good example is the ramp cornmeal mixes sold by the Smoky Mountain Native Plant Association in western North Carolina. They sell the mix in an attractive paper bag with recipes and a short story about their ramp business on the back.

You might also want to take advantage of the growing popularity of ramps among city dwellers. To them, ramps are novelty items, and they are willing to pay premium prices to sample them. I do not envision ramps ever being as popular as broccoli or carrots, but they might rival kumquats someday! To make it in the competitive specialty produce department, it takes creative marketing. One of the easiest ways to get started is to give a free case of ramps to an adventuresome chef at an upscale restaurant.

Budget

This is a completely hypothetical budget. No growers have provided me with reliable figures, and I was not able to continue the research to generate enough data to create budgets, so this is my best guess. Yield estimates are based on the numbers of plants planted per unit area (minus some for expected die-out) and the weight of the average bulb in a purchased bunch. I chose two marketing methods—wholesale and restaurant sales by the pound. Production costs are based on what we have seen with other forest crops. The net return is not very exciting, but a grower could greatly increase the profit margin by using other marketing strategies, making value-added products, and by keeping all the labor within the family. As in many other budgets in this book, the wage rate is the current US federal minimum wage.

16

Ramps Growers' Stories

The Smoky Mountain Native Plants Association

When I wrote the first edition of this book, I briefly mentioned the Smoky Mountain Native Plants Association (SMNPA) and showed a picture of their ramp cornmeal mix. All these years later, that little organization is still alive and well, so I thought it would be fun to tell you a little bit more about them. I've known the president of the association, Beverly Whitehead, for a very long time, so I gave her a call one August morning in 2013 and asked her to tell me their story.

The SMNPA has 60 members, a mixed group of farmers, growers, and wild-harvesters dedicated to earning income through sustainably harvesting, propagating, using, and preserving Appalachian native plants for future generations. Their office is located in the beautiful Stecoah Valley in far western North Carolina. They operate out of a restored historic school building which now houses the Stecoah Valley Cultural Arts Center. The non-profit SMNPA was formed to conduct research and offer educational programming to help families earn year-round income in a rather isolated community with high unemployment and limited job opportunities. Contributing to the lack of work is the fact that 68 percent of the land in the county is federal forest land with an additional 10 percent owned by the Tennessee Valley Authority, Alcoa, Inc., and a land trust. This is great for hunting, fishing, and tourism, but results in mostly seasonal jobs for the residents. Many adults are forced to move elsewhere during the winter to find jobs to support their families, a practice referred to as "working off." The association members initiated several income generating projects, including a seasonal county farmers' market. Their most successful and visible project is the Ramps Project.

The SMNPA is committed to preserving ramps while at the same time making an income from them. They have spent years figuring out the best way to propagate, grow, and harvest ramps. They teach the harvesters sustainable harvesting techniques and Good Agricultural Practices (GAP), which includes special harvesting practices. Members have agreed not to sell fresh ramps until the ramps can be sustainably farm raised. They have developed a line of value-added products made from sustainably harvested ramps. Currently, 50 percent of the ramps used in their products are farm raised, with that percentage increasing every year.

Harvesting and Propagating Ramps

The members practice a ramps-harvesting technique that they learned from Cherokee tribe members. They do not harvest the whole bulb. Instead, they use a knife to cut the bulb on an angle below the soil surface, harvesting the young leaves and the upper portion of the bulb, but leaving the roots in the ground to regrow. Beverly claims the plants grow back stronger and more lush as a result.

Since the goal of the association is to eventually raise all their own ramps, they have conducted studies on the best approach to propagation. They plant seeds, and have had about a 70 percent success rate with them. Ramps are long-lived perennials that love rough terrain and cold weather. From sowing of seed to bloom takes seven years; to first commercial harvest is a total of ten years. Beverly cautions new growers: if you plant ramps seeds, when you check your planting area that first spring, don't pull up anything that looks like grass because it could be a first-year ramp seedling. It is quicker and easier to propagate ramps by planting the roots. This is a good use of the roots that are cut away from ramp bulbs at ramp festivals and would otherwise just be thrown away. To propagate by roots, they suggest harvesting a whole plant and cutting the bulb at an angle so there is at least ⅛" to ¼" of bulb still attached to the root plate (the solid section where the roots are attached) with all the roots intact. The leaves and bulb are kept for eating or processing, and the piece of the bulb with the root plate and roots are used for propagating. Although they have not done a scientific study on it, they have found that if they cut the root plate in half, or just nick it with a knife before planting, it seems to increase the number of root pieces that develop into new plants. They store the root pieces in the refrigerator before planting them in the fall.

Even though the SMNPA is not selling fresh ramps yet, they do keep up with the market. Beverly reported that fresh ramps brought good prices during the 2013 season. A gallon bag of fresh, clean ramps (about 1.5 pounds) sold for $25 to $60 depending on the location and event. They sold for $25 per fresh pound in New York. The association is content to let others sell fresh ramps because they know that adding value to their ramps should increase their profits. As a result, SMNPA has put their efforts into making products from ramps. Every spring, members harvest young ramps. Their method of leaving the roots in the ground is not only good for the plants,

Stone-ground ramp cornmeal sold by the Smoky Mountain Native Plant Association. Photo by Karen Hardy.

since they will regrow, it also reduces production time in the kitchen because they are not bringing dirt inside! They developed a protocol to process and store the ramps so they can maintain a three-year supply as insurance against a bad year for ramp harvesting.

Value-added Products

The first value-added product that the association created and sold was a traditional "Cornmeal with Ramps" with recipes on the back of the package for "Bear Hunter's Cornpone from 1835" and "Grandma Amo's 1894 Ramp Cornbread." It was featured at the Smithsonian Institute's 2003 FolkLife Festival on Southern Appalachia. Since then, 47 local restaurants and tourist shops have carried the ramp cornmeal and other SMNPA products. Initially, association members were making money from the project. It provided sufficient supplemental income for member families to stay home year round instead of "working off." But the recession was hard on the tourist industry in their remote community. Many of the family-owned shops and restaurants went out of business. Only four of those businesses survived the recession and continue to sell the ramp products. This is causing a reevaluation of the project, including the packaging, target markets, and expansion into more urban settings, such as Asheville, which is a two-hour drive away on winding, mountainous roads.

In recent years, the association has expanded the product line to include ramp salt, ramp flakes, and a ramp seasoning. Their latest product is a ramp dip mix that the consumer adds to sour cream or cream cheese. Inside the ramp dip mix package are four additional recipes for using it, including ramp popcorn and ramp butter.

The association has been awarded many grants to help them develop their business model, conduct research, and do educational programs. For example, in 2008 they received a grant to maximize their production system. They learned that they could harvest and clean 50 pounds of ramps in a day. The grants are particularly important right now as the association makes changes in marketing and distribution following the recession. For example, through some market research regarding packaging, they learned that consumers prefer a brown paper bag for cornmeal over the shiny white paper bag they had been using. So they are currently redesigning their packages and labels.

The association has a website, but since they did not get reliable Internet service in their area until 2013, they are just now developing a webstore. Most of their sales take place at festivals and shows where they set up displays that also serve to educate people about ramps and SMNPA's mission. Beverly was excited to report that their products will also be sold in a natural food store opening soon in Asheville that will feature local food products. They are adding UPC codes to their packages to facilitate sales in larger commercial stores such as this one.

Keeping current with state and federal regulations is challenging for a small organization. The association helps members understand and comply with federal and state laws and regulations. SMNPA participated in the North Carolina review and response to the USDA Food and Drug Administration's proposed Food Safety Modernization Act rules to explain why small producers need to be exempt if they are to have a chance to stay economically viable.

The SMNPA members also grow a number of other woodland medicinal plants and have hosted on-farm tests for my program several times. One of these tests included ginseng, which they continued to plant and grow after our trial period was completed. I was saddened to learn that at least one grower had all their ginseng stolen, even the baby plants. Fortunately, their ramps have remained untouched.

Glen Facemire: Commercial Ramp Grower

Glen Facemire is a bona fide commercial ramp grower with, what he claims is, "the only ramp farm in the country." When I became seriously involved in ramp research in 1998, he was one of the first people I contacted. In April of 1999, my associate Jackie Greenfield and I made a special trip to Richwood, West Virginia to visit with Glen and his wife, Norene. We timed our visit to coincide with the annual Feast of the Ramson—the Richwood Ramp Festival. We had a wonderful time touring the G & N Ramp Farm and visiting their little shop in town. I learned more about ramps in those few days than I did during all my reading on the subject during the previous three months! We designed our initial field research studies based mostly on information obtained during that visit. In April of

Glen Facemire, noted West Virginia ramp grower, in the woods. Photo by Jeanine Davis.

2003, I spoke with Glen again specifically to write this story about his ramp business. In September 2013 we talked again. I have included both interviews here.

Interview with Glen in April 2003

Preserving Ramps and the Culture Associated with Them

Glen is a retired United States Postal Service employee who has always been interested in ramps and has a deep appreciation for the tasty little plant. Prior to starting the ramp business, he sowed ramp seeds all over his property for many years. He even sowed ramps along his rural mail route. This turned out to be quite educational, for it allowed him to closely observe ramps growing under all different conditions year after year. Then, when he retired, he decided to start a business that would allow him to help conserve ramps, and at the same time, make a little money. He says that his main interest is still the preservation of ramps, and he feels that now is the time to encourage other people to sow ramps everywhere—before digging of the wild plants is outlawed because their populations are decreasing.

Glen not only wants to preserve the plants in their native habitat but also to preserve the culture of harvesting ramps and holding community festivals for them. He helped influence the United States Forest Service to start rebuilding ramp populations in the national forests instead of banning spring ramp harvests. He thinks people can be taught to replace what they take from the forest by seeding areas where they have been harvested. He did note that this will require sowing a very large number of seeds, because ramps have a high mortality rate in a natural setting, often 95 percent or more. At least part of this is due to the fact that birds and mice seem to be particularly fond of ramp seeds and seedlings.

Glen explained that people in his region are talking about the possibility of seeding ramps in areas where the timber has been clear-cut and replanted in trees. The ramps would grow, mature, and be harvested long before the trees are ready to be cut. This is a form of agroforestry that he thinks could benefit both the timber companies, who own the land, and the local residents, who would plant and harvest the ramps. He envisions people paying the timber companies for the rights to go into these newly reforested areas and grow their ramps. This would provide the timber companies with a small amount of income from the land before the timber is mature—something they do not have now. He thinks that if it works, this idea could be used in many other parts of the country, and for other herbs, too.

In my conversations with Glen, I found that there are big differences in the pressures on wild ramp populations between where he lives in West Virginia and where I live in western North Carolina. Western North Carolina is developing rapidly. It is a very desirable place to live and vacation, which means there are increasing numbers of people. As I mentioned earlier, our spring ramp festivals are quite popular, and the demand for ramps at roadside stands and specialty markets is growing. Many local people have discovered that harvesting and selling ramps for these markets is a good way to make a little extra income in the spring. All of this has put tremendous pressure on wild ramp populations, and United States Forest Service botanists report serious declines in the number of ramps in the national forests. In contrast, Glen lives in a much less populated area, and the tourism industry is smaller. He thinks that is one of the main reasons there is no scarcity of ramps in West Virginia; there just are not as many people collecting ramps in West Virginia as there are in western North Carolina. He also believes that the climate and soils in his area are more conducive to rapid growth and reproduction of ramps than they are in my area. That might well be true, because I am located on the southernmost end of the natural ramp range.

This could be important information for people wanting to grow ramps in other parts of the country. Glen is of the opinion that the very distinct seasons in West Virginia are good for ramps. What he means is: when it is summer in West Virginia, it is hot; and, when it is winter, it is cold. In contrast, in western North Carolina, the winters often consist of alternating periods of cold and warm temperatures. It is not unusual to have shirt-sleeve weather in February followed by a major snowstorm in March. Glen thinks these milder, blended seasons in the southern Appalachians are not optimal for ramps.

Glen noted that there are also differences in how ramps are harvested in West Virginia compared to western North Carolina. In his area, most people who harvest ramps are commercial diggers. They dig ramps to make money, and time is money, so they are very selective about where they dig. They only harvest from large, healthy, mature patches of ramps, otherwise it is just not worth their time to stop and dig. They like to dig where they see from 10 to 15 ramps growing in a "pod." In contrast, in my area there are many casual, recreational diggers. Families go out every year to "gather a mess of ramps," and hikers and fishermen dig ramps to cook with the meals they prepare over a campfire. These people tend not to be as selective as the commercial diggers and will often harvest the closest, easiest plants to reach, no matter how few are in the patch.

Closeup of white ramp flower. Photo by Jackie Greenfield.

I noted at the beginning of the discussion on ramps that there are two kinds of ramps, the common *Allium tricoccum* and the rare, often controversial, *A. tricoccum* var. *burdickii*. Glen is fairly certain that all the ramps growing naturally in his region are the common type. He has some ramps from Michigan, however, that are very small, white-stemmed, with a short flowerstalk. He thinks these might be the *burdickii* type.

Glen's Observations about How Ramps Grow and Reproduce

Glen noticed that, at least in his area, ramp patches do not expand much by seed, and he rarely sees baby ramp plants within a mature population. He assumes that most of the seeds are either carried off by birds and small mammals or fall on top of the leaf mulch and dry up and die because they never make contact with moist soil. The ramp patches grow mostly by division of the bulbs. Therefore, in his estimation, most of the seeds produced by wild ramps are wasted, and people should collect them and sow them properly under the leaf mulch, so they have a chance to grow. Some ramp seeds do germinate and grow in the wild. Sometimes they just happen to fall into a moist, protected spot, but Glen has a theory of how others survive. He has observed that in the fall, wild turkeys scratch around ramp patches, digging up and eating young bulbs. During their scratching, they sometimes bend over and then cover up a ramp seed head with leaves, giving those seeds a chance to grow. He has found clusters of little baby ramps growing in ramp patches with turkey damage. Unfortunately, these new little plants are often food for a flock of turkeys that come through during the next year or two, because turkeys are particularly fond of baby ramps, leaves, and bulbs.

A ramp plant must be five to six years old to produce seed. Glen reports that these mature ramps will usually flower every year, but seed production is very unpredictable. It often occurs in alternate years, or sometimes the plants go three years without producing seed. Ramps bloom in late June to early July. Glen's theory is that if late July is very dry, seeds do not develop. He is not sure what pollinates ramp flowers or if they are self-fertile, but he has observed bumblebees and ants working the flowers. He also sees many little spiders on the seed heads when he collects seeds, though he does not know if they play a role in pollination or seed distribution. Ramp seeds mature within the seed head in late summer to early fall. If they are not carried off by animals, they gradually fall to the ground. Without human intervention, those that make contact with the soil, get covered with leaves, and stay moist will usually take 18 months to germinate.

Turkeys are not the only animals that enjoy the flavor of ramps. Glen has observed deer nibbling on the tops of tender young ramp plants in the early spring. They do not usually pull the plant out of the soil or kill it, so the plant will often regrow, but the nibbling certainly sets the plant back and makes it unmarketable that year.

A Few Specifics about How Glen Grows Ramps

Glen grows his ramps at elevations of 2,400 to 2,900 feet in mixed hardwood forests, and he thinks the elevation may be important. He has found that beech trees are the best to plant ramps under and near, but ramps also grow well under poplar, maple, hickory, and oak. He prefers to plant ramps where there are some of the following growing naturally: nettles, trillium, lady's slippers, wild sarsaparilla, goldenseal, or ginseng. He has observed that ramps do not flourish in a very acidic soil. The best site has moist soil (even when the ramps are dormant), a neutral soil pH, and does not get too much sunlight. Some of this is contradictory to what my research has shown so far and what we find in natural ramp populations in western North Carolina. This is a topic that obviously needs more study.

Glen prefers to grow his plants from seed that he has collected. The seeds ripen at the end of August into September and can often still be found on the seed stalk in November. He likes to plant the seed when it is fresh, but says it can be sown anytime. When he is ready to plant, he just pulls back the leaf mulch, scatters the seed over the soil surface, trying to leave several inches between seeds, and then covers the seeds back up with the mulch.

Over the years, Glen has experimented with different aspects of growing ramps. Trying to get ramps to grow from seed has been a major area of study for him. He observed that some of the seed that drop naturally from the plants germinate very quickly. He has even found baby ramps growing in November, but usually, ramp seeds do not germinate for at least 18 months. He has tried all kinds of treatments to get ramp seeds to germinate earlier. He even tried a clover inoculum that is used on roadside fescue plantings and said that it did speed germination somewhat.

Ramp seeds are not as sensitive to their storage conditions as ginseng and goldenseal. Glen has dried ramp seeds, held them in a jar at room temperature for four years, sowed them, and they grew. Glen does not like to plant ramp bulbs in the fall because he thinks it disturbs the soil too much, resulting in the bulb freezing or being pushed out of the soil. Again, this is contradictory to what my research has shown.

Other than deer and turkey, Glen has experienced few pest problems on his cultivated ramps. He did report that some years a worm eats on some of the big bulbs in July, actually cutting rings into the outer surface of the bulb. When this happens, the bulb turns brown and often dies. He has not had any problems with disease.

The Ramp Business

Glen and Norene own G & N Ramp Farm and run a business called Ramp Farm Specialties. Their farm is located just outside the quaint little town of Richwood at an elevation of 2,400 feet. Over the years, they have carried a wide range of products, operated a cute little shop in town, and developed a successful mail-order business.

Much of Glen's business now consists of selling ramp bulbs and seeds for planting. Until quite recently, most of his customers were ordinary people who wanted to put a little patch of ramps in their backyard, herb garden, or wooded area for their own use. He particularly enjoys working with these hobby growers and finds it gratifying to assist them, but the sudden rise in popularity of ramps has resulted in an increased number of customers wanting to grow ramps commercially.

Glen provides ramps for the Richwood Ramp Festival, which is held every April in Richwood, West Virginia. The festival serves about one ton of ramps each year, and in 2003, Glen provided 26 bushels of ramps for the local fire station to prepare. The economy is depressed in West Virginia, and the money ramps bring to the area is important.

Although Ramp Farm Specialties is a successful small business, Glen says it has been more like an overgrown hobby than a big business venture. One of the unique aspects of Ramp Farm Specialties is the amount of effort Glen and Norene have put into making value-added products. They have sold ramp jewelry, framed ramp photos, ramp postcards, and ramp digging tools. They offer three little booklets that they have written and illustrated themselves: a ramp cookbook, a story of ramps, and a grower's guide. The food products they offer are quite special and very tasty. Over the years, they have offered ramp salt, ramp jelly, ramp mustard, ramp potato biscuit mix, ramp gravy mix, dried ramps, and pickled ramps. They develop all the recipes

Glen and Norene staff their Ramp Farm Specialties booth at the Richwood Ramp Festival. Photo by Jeanine Davis.

in their own kitchen and test them on family and friends. They have a flair for creating products that catch people's attention. The ramp mustard, pickles, and salt have been the most popular products, but Glen thinks their ramp gravy mix is their best product. It is excellent with pork chops and rice or served on biscuits with bacon and eggs. Unfortunately, most of their customers do not want to cook and mostly buy products they can eat right out of the jar. They don't know what they are missing! You can see their present line of products on their website (see Comprehensive Resources Directory).

The Marketing and Economics of Ramps
Glen believes this is a good time to enter the ramp growing business because of the rising interest in ramps. Consumers are learning to appreciate ramps as a delicacy and are willing to pay high prices for them as a seasonal specialty. He predicts that prices are going to increase dramatically over the next few years as demand rises faster than the supply stream.

In 2004, a little bundle of ten ramps for eating was selling at roadside stands in his area for $2 to $5. He was selling eating ramps direct to the customer for $0.15 to $0.20 each or a pound of ramps (50 to 60 plants) for $5, which is very inexpensive. He knows of people selling ramps for $15 per pound. At those prices, even a tiny ramp patch can be profitable!

I asked Glen to share what he could about the economics of growing ramps. As I have found with many of these small, specialty businesses, he does not have detailed budget sheets for his operation. Glen does not plant ramps in distinct, carefully measured beds, and he does not harvest from measured-off areas either. Therefore, it is not possible for him to provide the kind of detailed production information that Scott has for ginseng or that John Kershaw has for goldenseal. When I asked if we could create an estimated budget for an acre of ramps, Glen laughed and said, "Now that would be a lot of ramps!" He obviously was not accustomed to thinking about ramp production on that scale, but I know that he has ramps planted over 35 acres!

Glen did share that they have never lost any money in the ramp business. He has also not had to send any bill collectors out to get payments. In his experience, the kind of people who buy ramps are good, honest, down-to-earth people.

Downsizing

Glen and Norene are downsizing their business because it is going almost too well for them. The value-added part of the business has reached the stage where they either have to expand production, which means upgrading to larger facilities and commercial-scale equipment, or they have cut back. They have chosen the latter because they want to spend more time with their grandchildren and have more fun. After all, they are supposed to be retired.

They will maintain the business, but will concentrate mostly on selling ramp seeds and bulbs. Glen really enjoys that part of the business. He likes teaching new people how to grow his beloved little plant. He also finds that people who purchase plants and seeds really appreciate what he is doing, and he gets a lot of gratification from helping them start their own little ramp patches. This is more important to him than the monetary gain. He also will not let this part of the business get any bigger than he can handle on his own, because he just cannot find the labor to dig and pack the quality product he insists on. He said they will probably exhibit at more festivals than they are doing now. He also wants to write a ramp book, which would document everything he knows about ramps and how to grow them.

Glen does want to encourage other people to make value-added products from ramps. He thinks someone could take that part of the business and make it very profitable. Buyers interested in purchasing large volumes of their products, some for sale overseas have approached them, but Glen and Norene just are not up for that at this stage in their lives. He just wants to sell seed and encourage others to grow ramps.

Advice to New Growers

Glen has some very carefully thought-out advice for new ramp growers. Most important is the geographical location of your future ramp patch. Growers in the north can probably grow ramps just about any place where there is a little shade and good soil moisture. He recommends that you look for an area with some hardwood trees to provide shade, but where the plants will get the morning sun. The further south you are, the more careful you are going to have to be about where you locate the patch. He suggests that you choose a site where the ramps will not get any direct late afternoon sun. One way to identify such a site is to look for the place where the snow stays on the ground the longest, such as on the north side of a bank or behind your house.

Glen recommends that you develop a five- to seven-year production program. Sow seeds and plant some bulbs to get started. Then let them mature so they can produce their own seeds, which you will use to expand your plantings even more. Keep planting at least one new patch every year, and within a few years, you will have patches at all different stages. Eventually, you will have at least one mature patch to harvest every year. Seeds are definitely the most economical way to propagate ramps, but most growers are impatient for their first harvest, so they plant some bulbs, too. Count on three to four years to reach harvestable size if you plant bulbs. Plants started from bulbs will produce their own seed the first or second year of growth. Glen suggests that your first patch of ramps should be no bigger than 50' × 50'. This is a large enough area for you to get some marketable production, but small enough for a novice to manage alone.

Glen says that if he were just getting started at growing ramps, he would not spend the time, energy, or money to get soil tests done or to add fertilizers. He suggests that you pick about a dozen places in the woods that look good and plant 10 to 12 bulbs in each of those spots. Then just watch and see if the plants flourish there with just what nature provided them, or if they appear stressed. Glen did say that stressed ramp plants can actually revert back to looking just like a seedling. That is, if you plant a three- to four-year-old bulb and it is stressed by lack of nutrients, a soil that is too tight, or poor moisture, it will look younger and smaller the next year. Each year it will get smaller until it looks like a seedling, and in three to four years, it will just disappear. It uses up all the reserves it has in the bulb and dies. Scott has told me that ginseng will do the same thing! Glen suggests that you build your production areas around the patches that do well without any special effort. Why make this any harder than it needs to be? Give the initial test plots a year or two before you make a decision on their suitability for ramps, and then start your real production areas from seeds.

Again and again, Glen stressed the importance of testing out areas before doing any major planting. He tells his customers this, and most of them follow his advice and are pleased with the results. He also wants to remind readers to plant in areas where your test plants flourish, not just survive. If you follow that advice, you will be successful with ramp growing.

Glen thinks that ramps are just coming into their own, and that many opportunities exist for people to make money from ramps. He encourages people to be creative and have fun with it, as he and Norene have done.

2013 Update from Glen Facemire

Glen published a book, *Having Your Ramps and Eating Them Too*, in 2008. I was honored to review it for him and highly recommend it. You can order it from his website or from online book sellers such as Amazon

.com. It is a wonderful little book in which Glen, in his own unique style, shares how to grow and use ramps.

I spoke with Glen in September 2013. He and his wife, Norene, are still in the ramps business, and as he described it, "still living out in the woods, sheltered from a lot of the nonsense" going on in the world. They appreciate the little extra income the ramps business brings in to supplement their retirement income. He said the book had done very well and had to be reprinted just a year after it was released. They don't really do anything to promote it except post it on their website, but ramps aficionados don't seem to have any trouble finding it.

Changes in Their Ramp Business

Over the past two to three years, Glen and Norene have stopped selling fresh ramps for eating. Their business is now restricted to selling ramp seeds and bulbs for propagating. That is what they had always wanted their business to be, so they are pleased, and they have noticed their profits have gone up a little each year since they made that change. They are making more money while working less. The majority of their business takes place in February and March. They are thrilled with this because, after all, they are supposed to be retired!

Their Customers

Glen would have been a great county extension agent. It is clear that he really likes helping people, teaching them how to grow ramps, and hearing about their experiences. People get excited about their ramp patches and call or write to tell him how their ramps are doing, often sending pictures. Glen told me that interest in growing ramps is highest in the northern states, from Maine over to Michigan. He has to hold bulbs in the refrigerator for his Wisconsin customers because they still have snow on the ground when Glen is digging bulbs. People with nurseries put in orders for 8,000 to 10,000 bulbs at a time to satisfy their customer demands.

People from around the world, but especially Korea and Japan, are also very interested in ramps and ramp products. Ramps, or plants very similar to them, are native in those countries. Glen had a little entourage from Japan visit him; his state agriculture department sent them. They wanted to buy "gobs and gobs of pickled ramps" to take home with them. They told him, through a translator, that ramps, or a plant just like it, were almost "harvested out" in Japan. They wanted to import all kinds of ramp products.

A woman in California was excited to find his website. She hadn't had any ramps since she was in the mountains of Iraq and wanted to buy some to grow for herself. Glen didn't realize there were ramps in Iraq, but when he looked into it, he discovered that there are ramps all over the world, or something very similar to them.

He has customers from all walks of life. For example, he once supplied ramps for a Cherokee gathering in New Mexico, and a deacon from a church in Tennessee ordered ramps because members in his congregation wanted to start their own ramp patches.

Their Value-added Products

Their pickled ramps continue to be a popular product. There are some fancy pickled ramp products on the market, but Glen does not like to add anything that might take away from the flavor of the ramps. When Glen and Norene pickle their ramps, they don't put in any seasoning; just ramps and vinegar!

A Looming Threat?

Glen did have one concern that he wanted to bring to people's attention. There is a new insect in eastern North America that could pose a threat to ramps, especially wild ramps. It is the Allium moth; also called the leek moth. It is native to Europe where it is a destructive pest that feeds exclusively on Allium species, including onions, leeks, and garlic. It was first found in North America in Ontario, Canada, in 1993. Then it was discovered in New York. Glen said it was tracked from New York down to Pennsylvania in 2013. The Department of Agriculture has not found it in West Virginia yet, but they asked Glen to put some traps in his ramp patches to monitor for them. Fortunately, none were found in his traps. Glen thinks this insect could be devastating to ramps if it gets into our forests. He thinks he has seen evidence of it in the woods, but not on his property. It doesn't kill the ramp, but it makes the plant unmarketable.

Final Thoughts

I enjoyed my conversation with Glen and was pleased to know that he was still the Johnny Appleseed of ramps, spreading seeds and knowledge of how to grow them around the world. As I hung up the phone and realized that call was the last of my update interviews for this book, I reflected on how sharing all these people are. Even though several of them have physical issues to deal with, a few are in their golden years, and the recession set some of them back financially, all are optimistic about their plant-based businesses, opportunities for others, and the future of the industry.

PART FOUR

There Are Many Other Woodland Medicinals You Can Grow

Interest in medicinal herbs and natural forest products continues to grow, and it has extended far beyond ginseng, goldenseal, and ramps. As the popularity of natural products increases, the list of plants that people want to use and grow gets longer. Besides having medicinal value, many of the plants described here are very attractive and grow well in wooded areas. If you are a commercial producer, or are considering becoming one, this presents new opportunities. For example, you can grow planting stock of hard-to-obtain plants for other growers; you can produce nursery plants for the herb gardening enthusiasts to put in their gardens and landscapes; or you can expand the number of different kinds of medicinal herbs you grow for the botanicals industry.

Please keep in mind that there is limited information available on how to cultivate most of the plants described here. These are all shade-loving, herbaceous perennials. The approach I take when starting with a new native forest herb at home or in my research program is to use cultural practices successful for ginseng and goldenseal and then modify them as experience is gained with the new plant. Some of the plants described in this section are rare or threatened. Be sensitive to the conservation of these plants in their native habitats and find out if there are rules or regulations concerning the propagation or sale of any of these plants in your state, province, or country. Please obtain all of your planting stock, roots, and seeds legally!

17

Bethroot

Plant Description and Range

Bethroot (*Trillium erectum*) is a trillium that grows in shady, moist forests in eastern North America, from Canada south to the Carolinas and Georgia, and west to Michigan. It is known by many other names including red trillium, birthroot, purple trillium, squaw root, wake-robin, and ill-scented trillium. This herbaceous perennial grows to about two feet in height, with three broad, almost heart-shaped leaves on a stout stem with a single three-petaled flower that is usually brown, maroon, or purple. It is a slow-growing plant and must be four to seven years old to flower. The flower, although quite attractive, has a disagreeable odor that attracts the flies that pollinate it. Many of the flowers will produce a reddish berry that hangs underneath the foliage. The berry will eventually drop its seeds on the ground. The seeds have fleshy appendages called elaiosomes which are attractive to ants who carry the seeds back to their nests to eat the elaiosomes. Small and large mammals also help disperse the seeds. The underground rhizome is wrinkled, stubby, reddish-brown on the outside, ivory colored on the inside, and covered with fibrous roots.

Uses

The roots and rhizomes of this plant are usually harvested in late summer and early fall and used to treat a wide range of disorders. (In discussing bethroot, "root" will refer to both rhizomes and roots.) Historically, the Abnaki Indians ground bethroot roots and used the powder for many childhood illnesses. The Cherokee used bethroot to treat cancer, coughs, asthma, skin conditions, and gastrointestinal problems. An infusion was used to treat problems with menstruation and menopause, and a poultice was often applied for ulcers, tumors, and inflammation. The Iroquois Indians used an infusion of roots and flowers for pimples and sunburn. Some Southeastern Indians were said to eat pieces of the root

Bethroot in bloom. Photo by Karen Hardy.

in food as an aphrodisiac. A recent reference states that the roots have astringent properties and are useful in treating excessive discharges of the bowels and womb, as well as internal bleeding of all kinds. Some midwives still have bethroot in their apothecaries.

How to Grow

Bethroot can be cultivated in many parts of North America. It is hardy to at least zone 3 and must have enough exposure to cold to satisfy the dormancy requirement of the buds on the rhizome each winter. Thus, I do not suggest trying to grow bethroot in the Deep South. This plant is a little harder to cultivate than goldenseal and ramps. I have had the best success growing bethroot in an area that already has other trilliums present, and it has grown much better for me under a natural forest canopy than under a shade-cloth structure. This could be because many trilliums require special fungal relationships, known as mycorrhizae, which are present in abundance in most forest soils. Mycorrhizal associations are mutually beneficial for the fungus and the plant. For the plant, the benefit is usually improved nutrient uptake from the soil. My shade structures are located in a former grassy field which probably does not have the right mycorrhizae present for trilliums.

Bethroot prefers a cool, moist, shady site, preferably under a canopy of mixed hardwood trees. Bethroot likes a slightly acidic soil, around pH 5.5, and benefits from a light application of fertilizer to the soil each spring in the form of fish emulsion, seaweed, or a balanced synthetic fertilizer. I find that it likes a little more soil

Bethroot foliage damaged by *Colletotrichum* sp. Photo by Karen Hardy.

moisture than ginseng or goldenseal, so I do not recommend trying to grow the three together for commercial production purposes. Bethroot will grow in a site where ginseng and goldenseal thrive, but it might not flourish there. I suggest planting bethroot with other moisture-loving plants such as bloodroot, jack in the pulpit, ramps, or spikenard. Choose a site where the plants will get sunlight during early spring before the trees leaf out. A study by Routhier and Lapointe in Canada showed that almost all of the food reserves the plant needs for the next year are accumulated in the rhizome during those first few weeks of spring when the plants are exposed to full sun.

Bethroot is susceptible to a large number of foliar diseases including blights, spots, and rusts. Most of these do not affect the root or the survivability of the plant, but they do affect the appearance of the foliage and probably reduce root growth that year.

Seed Handling

Bethroot seeds have special requirements that must be met in order to germinate. To propagate bethroot from seeds, harvest the ripe fruits before they drop their seeds, usually in late July. Clean the seed of all fleshy tissue and store them in moist sand, buried in the soil, or kept in the refrigerator, as described for goldenseal. Do not let the seeds dry out. Sow the seed in a well-prepared seedbed in late fall or early spring and mulch heavily with leaves or hardwood bark. In *Growing At-risk Medicinal Herbs* (2002), Richo Cech gives a detailed description of how bethroot seeds must undergo a two-phase germination process. In short, only root growth takes place during

Bethroot root. Photo by Jeanine Davis.

the first year, so you will not see anything happening above ground. Just be patient, keep the site mulched and moist, and do not disturb the soil. Shoot growth should begin the second spring after sowing. The seedling will look like a blade of grass.

Vegetative Propagation

I usually propagate bethroot by planting rhizome cuttings in late fall or early spring. Break or cut a mature rhizome into pieces at least ¾" long, making sure each section contains one or more buds and some roots. Plant the rhizome pieces 12" to 14" apart in a well-prepared bed, as described for goldenseal, and mulch heavily with leaves or hardwood bark.

Harvest and Post-harvest Handling

In my experience, bethroot roots grow slowly and will take five years or more to reach what would be considered an economically feasible size to harvest. A saleable nursery plant can be produced in two to three years. It can take seven years or more to grow a harvestable root from seed. There is not a consensus among buyers as to when is the best time to harvest the roots. Some prefer the roots to be dug when the plants are in flower, whereas others want the roots dug in the fall. Check with your buyer before digging. Roots should be washed, dried, and packaged as recommended for ginseng or goldenseal or as specified by your buyer. Roots to be held as planting stock can be stored in a cooler or root cellar in moist sphagnum moss or sawdust.

Marketing and Economics

The market for bethroot is small but steady. In 2007, local dealers paid wild-harvesters in the southern Appalachians $1 to $3 per pound for dried bethroot, and they were purchasing 10 to 500 pounds at a time. Prices in September of 2013 were exactly the same, although individual sales were small. Growers and companies selling wholesale weren't moving much bethroot in 2013; prices in that category ranged from $5 to $27 per dried pound. Bethroot plants were selling in retail nurseries for $8 apiece, and small volumes of planting stock sold for about $180 for 100 small roots.

I do not know anyone who is growing and selling large quantities of cultivated bethroot, but many are selling it on a small-scale. Bethroot, sold as Red Trillium or Wake-robin, is a very popular plant for the shade garden, and you will often see it for sale by specialty nurseries at plant festivals and through online catalogs. There are also a few growers who manage small, semi-wild patches of bethroot in their woods, harvest it when requested, and sell it as a dried and cut root to individual herbalists. Bethroot is just one of many plants in their product line, and they make some money on it. With demand being as low as it is right now, I would not recommend investing in a large planting, and I do not recommend growing it intensively under artificial shade.

Bethroot takes a very long time to grow, so if the demand for it increases, the only way to take advantage of it is to have the plant growing on your property already. This is true for many of the plants in this book, and the approach some growers take is to grow a little of everything and a lot of just a few.

Do I think you could make any money with bethroot? Well, let's do a little "paper farming" here and see what we might get. Years ago, I sampled some bethroot that had been growing under a polypropylene shade structure on the research station in Mills River, North Carolina. The plants had grown for five years from rhizome cuttings, but four of those were drought years. The harvested roots averaged four ounces fresh. I replanted those, but estimate that they would probably have dried down to about 30 percent of the fresh weight. That would be 2.8 ounces each. In Oregon, Richo Cech found that eight bethroot plants yielded one pound of dried root, or about two ounces each. Let's split the difference and say each plant will yield 2.4 dried ounces of root.

Now, let's say I have an acre of woods and half the area can be put in beds; I would have about 20,909 square feet of bed space. If I give each plant 2¼ square feet of bed space, that would be a total of 9,293 plants. If half of them lived to harvest age, that would leave me with 4,646 roots. At 2.4 dried ounces each, I would end up with 697 dried pounds of bethroot. If I sold half of that to a dealer for $3 per pound and half to herbalists at $27 per pound, I would gross $10,455 per acre.

The Future

Bethroot is considered a plant to be watched, because some herbalists believe it could suddenly be in much greater demand. This is of concern to conservationists since bethroot is already listed as threatened, protected, or endangered in several states and has been considered for inclusion on the CITES listing. United Plant Savers also has it on their "at risk" list. I suggest that some growers start producing this plant for nursery sales. At the same time, they should maintain a small area of plants in the forest that could be used for seed production and then turned into a source of rhizome cuttings if large-scale production is suddenly desired. If the demand for bethroot increases to the point that commercial cultivation is required, the lack of available planting stock is going to be a problem. Tissue culture propagated material is another possible source for large volumes of planting stock, and the protocol has been developed and published, but I don't know how well those plants would grow in a natural setting.

18

Black Cohosh

Uses

Black cohosh (*Actaea racemosa*, formerly, and still often called, *Cimicifuga racemosa*) is a popular women's herb, traditionally used in the treatment of menopause and childbirth. Thus, in some areas it is still known as squaw root or papoose root. In the United States, hormone replacement therapy (HRT) has been the treatment of choice for menopause for many years, but in 2001, a government-sponsored study indicated that HRT was so dangerous that the study was terminated early. Claims that HRT reduced the risk of heart disease and decreased or eliminated bone loss due to osteoporosis were not substantiated. Then, in 2003, another study linked use of HRT with heightened risk of developing dementia. As a result, many women decided that the risks associated with using HRT were just too great and turned to natural remedies for relief from unpleasant menopausal symptoms.

Women around the world have used black cohosh for centuries to treat menopause, and federally-funded clinical trials have been conducted in the United States in an attempt to determine just how effective it is, its mode of action, and possible side effects. To date, those trials, including 16 on over 2,000 women, have shown no effect of black cohosh on undesirable menopausal symptoms, and evidence on safety is inconclusive. Yet, the author of the report on those trials said there is adequate justification for conducting further studies. In fact, many women do report relief with black cohosh, and, in ever-increasing numbers, they are purchasing popular black cohosh products. One of those, Remifemin, is manufactured by Schaper and Brümmer in Germany and is distributed in the United States by the pharmaceutical company GlaxoSmithKline.

Demand

Most of the black cohosh used in commercial products is derived from wild

Freshly dug, wild-simulated black cohosh root. Photo by Jeanine Davis.

populations, but most large, easily harvested populations have already been exhausted, so supply is becoming unstable. CITES has considered listing black cohosh as an Appendix II plant, which would put restrictions on international commerce and probably greatly reduce the volume of wild-harvested material that could be sold. United Plant Savers has already listed black cohosh as an "at risk" plant. In the future, cultivation of black cohosh will be necessary to meet the demand. Fortunately, black cohosh is easy to cultivate.

Plant Description and Range

Black cohosh is an attractive herbaceous perennial that grows in hardwood forests throughout most of eastern North America, except for the Deep South. It usually grows in rich, open, moist forest coves, but it can also be found growing on the edges of woodlots and, sometimes, in full sunlight. The plant can become quite large, ranging from three to eight feet tall, with a canopy about three feet across. The underground rhizome is dark brown to black, and knobby, with large, pinkish buds on the upper surface. Thick, black roots cover the bottom and sides of the thick, horizontal rhizome. The rhizome sends up one or more stems topped by a compound leaf usually composed of three distinct branches of many pointy, serrated leaflets. The terminal leaflet on each of these three branches has three lobes. In late spring to early summer, each plant sends up one or more flower stalks with long spikes of tiny, fuzzy looking, white flowers, which open successively from the base upwards. I think the flowers smell awful, although some people consider them fragrant. They do attract a number of small flying and crawling

Black cohosh flower stalks in bud stage. Photo by Jeanine Davis.

Young black cohosh plants. Photo by Jeanine Davis.

insects that help with pollination. However, some people say black cohosh repels insects and refer to it as bugbane. In late summer, the flowers develop into small oval fruit full of brown seeds. The pods eventually dry, and the seeds rattle around inside, which is the origin of another of its common names: rattlesnake root. The plant parts of economic interest are the rhizomes and roots, which are usually harvested in early fall. (In discussing black cohosh, "root" will refer to both rhizomes and roots.) They contain a wide variety of chemical compounds including triterpene glycosides, salicylates, sterols, and alkaloids. The triterpene glycosides, actein and 27-deoxyactein, and the isoflavone, formonentin, are thought to be the main active constituents.

Guidelines for Growing

I have had excellent success growing black cohosh following the same guidelines provided for goldenseal. They grow under the same conditions, although black cohosh will take more sun than goldenseal. Black cohosh can serve as a shading border around the periphery of an artificial-shade structure or near the forest edge. According to Joe-Ann McCoy, director of the Germplasm Repository at the North Carolina Arboretum in Asheville, NC, the more light the plants receive, the more flowers and seeds they produce. Like all the plants discussed so far, black cohosh has a winter dormancy requirement and needs a certain amount of chilling each winter for the buds to break or the seed to germinate. It will grow in hardiness zones 3 to 8, although it grows best in zones 5 to 7.

Soil

Black cohosh needs a well-drained soil and benefits from being planted in raised beds. If good drainage is not provided, the plants may succumb to a root rot fungus during a wet season. Black cohosh thrives in rich soils with a pH of 5 to 6. If the soil is too dry, the plants will not flourish. Most forest soils will have adequate nutrition for black cohosh, but if the soil is not naturally rich in organic matter, or you are planting in a field site under artificial shade, consider incorporating compost or a green manure crop prior to planting. That, combined with a good layer of organic mulch applied after planting, should supply the plant's nutritional needs. If your soil is poor (not a recommended site for black cohosh), or you have experienced a very wet, cool spring, which may inhibit the release of nutrients from the soil, a small amount of a balanced organic fertilizer, providing 10 to 25 pounds of nitrogen to the acre, can be applied to the soil in the spring.

Black cohosh in flower, growing under a shade house. Photo by Jeanine Davis.

Propagation

Black cohosh is most easily propagated by dividing the rhizomes in spring or fall. Joe-Ann McCoy studied seed and rhizome propagation of black cohosh for her Ph.D. All of those field studies were conducted on the research station where I am located or on private land in far western North Carolina. Her research showed that mature black cohosh rhizomes should be cut into pieces 1½" to 3" long with at least one bud on each segment. A mature rhizome often contains an obvious "old" end that will be black and not contain any buds. Do not try to propagate from that part of the rhizome. Plant the rhizome pieces about two inches deep. Space the pieces 18" to 24" apart, center to center, and cover with a good layer of mulch.

Black cohosh can also be propagated by seed, but as with so many of the woodland botanicals, the embryo inside the seed is dormant and certain temperature conditions must be met for the seed to break that dormancy. Sometimes, especially with older seed, this can take more than 24 months. The seed matures in the fall and must be exposed to a warm period followed by a cold period before it will germinate. Sometimes an additional warm period is needed.

The best germination usually occurs if fresh seeds are sown immediately. The seeds dry naturally within the pod and should be collected just as the pod starts to split open. Sow fresh seed about ½" deep and space them about 1½" apart in a well-prepared seedbed. Cover lightly with soil and with a one-inch layer of mulch, such as hardwood bark or leaves, and keep moist. Most of the seed should germinate the following spring after having been exposed to a few weeks of warm temperatures in the fall and months of cold temperatures in the winter. If the seeds do not emerge, keep the seedbed moist and covered with mulch until the next spring, when most of them should come up. Like ginseng seeds, black cohosh seeds can also be stored in a buried stratification box for a year before sowing, but I have had better success sowing fresh seed. The older the seed is before it is exposed to moisture to start the germination process, the longer it takes to germinate.

A study published in 2013 by Kaur and colleagues describes a method to speed up germination for black cohosh. It might not be practical for most growers, but if you are going to produce a large number of transplants in a nursery, it might be worth the effort. The study also clearly illustrates the temperature needs of black cohosh seeds. The researchers started out with seeds that they carefully cleaned with slightly soapy water, a mild bleach solution, and lots of fresh water. If they did not clean the surface of the seeds first, they lost many of them to rot. But they point out that alcohol should not be used because it apparently killed the embryos. The clean seed were incubated in petri dishes on moist filter paper at 77°F for two weeks (warm period). Then they were held at 39°F for three months (cold period). Seeds that exhibited root and shoot emergence were then held at 77°F until they were big enough to pot up. This would not be hard to test out at home because 77°F is slightly warmer than room temperature. Petri dishes set along the warm front edge of your refrigerator might stay at that temperature. The 39°F is standard refrigerator temperature. Remember to keep the seeds moist during the whole process because this can shorten the germination process from the standard 8 to 24 months to only 3½ months.

Pests and Diseases

A number of different diseases can infect black cohosh, including leaf spots and root rots. Please consult Joe-Ann McCoy's Disease List in the back of this book for all the diseases that have been documented on black cohosh so far. Several insects, including cutworms and a leaf-eating beetle, can also infest black cohosh. Weeds are not usually a problem if a forest site is properly prepared and a good layer of mulch is applied. Black cohosh, planted at the recommended spacing, will quickly provide a dense canopy that smothers out other plants. Weeds may be more of a problem when black cohosh is planted in full sun or on the forest edge, but hand-weeding will probably only be required the first year.

Harvest and Post-harvest Handling

Black cohosh roots should be ready to harvest after three to five years of growth from rhizome cuttings. They are usually harvested in the fall, though a few buyers prefer roots dug in early spring. Dig, wash, and dry the roots as described for ginseng or goldenseal. Black cohosh roots tend to be a little more difficult to clean than most other roots, especially if the soil they grew in had lots of clay. The roots must often be broken apart to clean them adequately. Black cohosh roots may take up to a week to dry in a forced-air dryer at 80° to 95°F. Some growers start at the lower temperature for a day or two and then turn the temperature up until the roots are dry enough to snap when broken. If you live in an area with high humidity, you might need to start at 95° to 100°F to get the roots dry. Just remember to always have lots of moving air.

Dried black cohosh roots. Photo by Jeanine Davis.

Package the roots as your buyer specifies, often in large polypropylene sacks (called poly sacks) or cardboard barrels or boxes. Use of burlap sacks is no longer recommended. Store in a cool, dry place, protected from rodents and insects. Reported yields of dried root range anywhere from 750 to 2,500 pounds per acre.

If you are harvesting black cohosh roots for use as planting stock, they must be treated with some care or they will quickly mold and rot. This is especially true if you wash them prior to storage. Either keep the soil the plants were growing in on the roots during storage, or wash them and store them in moist sphagnum moss. The sphagnum moss has natural antifungal properties that inhibit growth of fungi, mold, mildew, and algae. Store the roots in large plastic bags that are loosely gathered at thc top (not closed tight with a twist-tie or knot) set in cardboard boxes and put into a cooler or refrigerator at about 40°F. Check frequently for mold and rot, stirring up the roots with your hands in the process. Some growers have also reported success with wiping off the excess soil and storing the roots in moist sawdust in a root cellar.

Beautiful bed of woods-grown black cohosh. Photo by Jeanine Davis.

Black cohosh in bloom. Photo by Jeanine Davis.

Supply and Demand

Black cohosh demand is only going to increase. There have been several articles stating that even short-term use of HRT is dangerous. Whether this turns out to be accurate or not, many women want to use more natural remedies, leaving millions of women worldwide looking for alternative methods of alleviating sometimes painful and disturbing menopausal symptoms. Sales of black cohosh products increased 14 percent from 2009 to 2010. It was the 8th-top-selling herbal supplement in the United States in 2010 with sales of $9,308,047. Buyers indicate that supplies of wild-harvested material are becoming scarce, and that growers should definitely start cultivating this material.

Apparently, there are only a few people growing black cohosh on a large scale right now. There is some commercial cultivation in Germany, and a few large artificial-shade plantings exist in Canada, but other than that, I think most plantings are less than one-half acre in size. Growers are being cautious about entering the black cohosh market because prices paid for the root are still quite low. Growers who have entered into production are counting on prices increasing as the supplies of black cohosh growing in the wild continue to decrease.

Economics of Production

Prices

Even though demand for black cohosh is strong and steady and has been for years, prices paid to growers and wild-harvesters are not. In 2001, dealers paid wild-harvesters $1.50 to $2.00 per dried pound of root. No one could afford to grow black cohosh at that price. By 2003, the price had risen to $3 to $5 per pound, and I was optimistic that prices would soon reach a level that would support cultivation. Unfortunately, prices paid to the grower or wild-harvester dropped back down to $2.50 to $4.00 a pound and stayed in that range for many years. In 2011, buyers were again paying $3 to $5 per pound, but in 2013, they were back down to $3 to $4 per pound. The last figures I saw indicated that only about five percent of the black cohosh traded globally is cultivated. That will not change if prices stay so low. In May 2013 an article was published by the American Botanical Council that explained that some manufacturers were selling black cohosh products that were actually made with closely related species that were imported and sold as inexpensive black cohosh. The US Good Manufacturing Practices for Dietary Food Supplements, which required positive identification of all ingredients, along with the exposure of the problem, should help eliminate that raw material product stream. That, combined with the rising demand for a more consistent product, may drive the market towards more cultivated black cohosh.

In 2013, growers selling wholesale direct to consumers received an average of $19.50 per pound for organic dried root and $15.35 for conventionally grown root. Retail sales were unusual in 2013; conventional black cohosh root brought an average price of $31.50 per dried pound whereas organic root only averaged $28.00 per dried pound. In the nursery category, the average wholesale price was $125 for 100 plants, and a single plant sold retail for an average of $4.25.

Artificial-shade Production

To the best of my knowledge, there are only a few growers in North America producing black cohosh on a large scale under artificial shade. They can produce 2,000 to 3,000 dried pounds per acre in three years. This is about four times the yield predicted for that production system in 2003. Still, at $3.50 per pound, even though that translates into a gross return of $10,500 per acre, the costs for planting stock, production, harvest, and drying consume most of that. As John Kershaw, the Canadian grower whose story is included in the next chapter explained, if they can't make at least $10 per pound, it doesn't pay to grow the crop.

Forest versus Artificial Shade

Joe-Ann McCoy conducted her Ph.D. research on black cohosh. She was a Clemson University student, but I served on her committee, and her field research was done in North Carolina. Joe-Ann grew black cohosh for three years in two forest sites and in one artificial-shade site (78 percent shade) in western North Carolina under extreme drought conditions. The forest sites were not irrigated, but the shade structure site was. The plants grown under the shade structure exhibited a nine- to 12-fold increase in root weight. In contrast, roots from the forest sites increased from only one- to three-fold. Roots from all three sites dried down to 37 percent of their fresh weight. Using her best figures from the artificial-shade site, if she had planted 500 pounds of rhizome cuttings and harvested the roots three years later, she would have 6,000 pounds of fresh roots, which would dry down to 2,220 pounds. This would translate into a gross return of $6,660 or $8,880 depending on whether she received $3 or $4 per pound, respectively.

This sounded really good until I plugged these figures into the budget John Kershaw provided for goldenseal production under artificial shade in chapter 13. Assuming production costs are the same and figuring $3,500 for planting stock ($7 per fresh pound), the total costs of production, not including the shade structure, were $22,190. This translates into a net loss of between $15,530 and $13,310 for $3 and $4 per dried pound root, respectively. However, if she grew the black cohosh organically, and got $19.50 per pound for the dried root by selling wholesale direct to consumers, her gross returns would have been $43,290. Even if her production costs increased 20 percent for being organic, her net profit would still have been $16,662 for that acre of black cohosh.

So what if you grew black cohosh in the woods in a wild-simulated system? Again doing some "paper farming" and using the research figures, 250 pounds of planting stock at $7 per pound, grown in the woods for three years, should yield 750 pounds fresh (three-fold increase), or 277.5 pounds dry (37 percent of the fresh weight). Since this is wild-simulated, there won't be much in the way of material or equipment costs, but let's figure $715 in labor costs over the three years. Total production costs would be $2,465 per acre. If the crop is sold to a dealer for the 2013 price of $4 per dried pound, there would be a $1,355 loss. If a crop is grown organically, expect to spend about $100 more per year in certification costs (assuming total certification costs are spread out across other crops on the farm). At the 2013 price of $19.50 per pound for wholesale organic black cohosh, there could be a net return of $2,646. If the grower had an online marketplace and could sell the 277.5 pounds organically to retail customers, assuming that added another 15 percent in production costs, there is still a potential return of $7,226.

None of these scenarios include taxes, packaging, fuel, and dozens of other costs. Those need to be figured in. And if you sell retail, you will have to advertise, store the product, and maintain an online store. But regardless, this clearly illustrates that growing and selling black cohosh into the raw botanicals market at current prices is not a money-making deal. Prices will have to exceed $10 per dried pound before any

Table 18. Farm Enterprise Budget for One Acre of Black Cohosh Grown Under Artificial Shade

Expenses		3-Year Totals
Labor (site prep, planting, mulching, weeding, harvesting) 780 hrs at $7.25/hr		$5,655
Planting stock (rhizomes) 500 lbs at $7/lb		$3,500
Fertilizer and lime		$204
Plant protection		$272
Fuel, oil, lubrication		$136
Mulch		$2,480
Utilities		$273
Packaging		$310
Total Expenses		$12,830
Income Scenarios based on different prices received per pound	**Gross Income**	**Net Income**
Dried root sales: 3,000 lbs at $4/lb	$12,000	−$830
Dried root sales: 3,000 lbs at $5/lb	$15,000	$2,170
Dried root sales: 3,000 lbs at $6/lb	$18,000	$5,170
Dried root sales: 3,000 lbs at $15/lb	$45,000	$32,170

profit can be gained. But, there are markets willing to pay higher prices for small volumes of high quality black cohosh. If you have multiple woodland herb crops and are good at marketing, or have the right connections with some small manufacturers or herbalists, a profit can be made. Even then, production costs must be kept low. Until you know if your roots are going to increase three-fold or 12-fold in weight, I would not suggest investing in an artificial-shade structure. If, however, you already have a structure, or are going to grow it in the forest, it is probably a reasonable risk. My recommendation for anyone wanting to attempt black cohosh for the first time is to grow certified organic black cohosh, keep production costs as low as possible, and try to develop direct marketing channels to get the highest price possible.

Selling Black Cohosh with High Triterpene Glycoside Levels

Many of the black cohosh products on the market are standardized and have a guaranteed level of triterpene glycosides. Although I have not yet heard of anyone being paid a premium for black cohosh with high triterpene glycoside levels, it can be assumed that this will be desirable in the future. There is probably public and private research being done on this right now, but at the time of this publication, I could not find any results. Growers who are interested in trying to increase the value of their black cohosh in this manner, should keep close records on how changes in fertilizer, soil pH, light, and time of harvest influence the triterpene glycoside levels in their roots. There are commercial laboratories that will analyze your roots for you. Be sure to contact them early to find out how they want samples prepared and how large a sample they need.

Nursery Plants

Because black cohosh is such a pretty and popular plant, many people want to have some growing in the woods behind their homes. This can be a lucrative market for some growers. In 2007, single black cohosh plants sold in retail nurseries for $6 apiece. Wholesale nursery stock brought about $1.25 per plant. Currently, most growers who want to buy enough roots to plant an acre or more must purchase them from buyers who source the material from wild-harvesters. Being a source of high-quality, cultivated planting stock will be a good niche for a few growers.

Production Budget

As I explained in the goldenseal chapter, my research associate Jackie Greenfield and I were involved in a special medicinal herb project for two years, which included development of a number of production budgets for medicinal herbs. Because I consider these the best budgets currently available, I am including a modified version of the one on black cohosh here. At current dealer prices, even if a grower already owns a shade structure, the profit potential for black cohosh is questionable. This is especially true if the grower has to hire labor. However, for every dollar that the price per pound rises, profit increases by roughly $3,000 per acre, so unit price is critical.

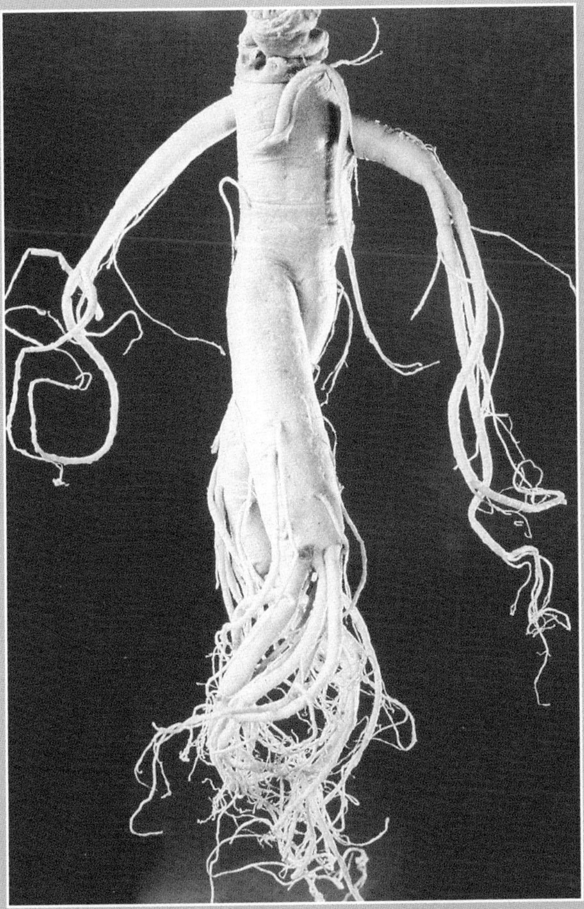

Top: The name ginseng means "man root" or "man essence" in Chinese, and sometimes the root does take the shape of a man. Photo courtesy of General Nutrition Corporation.

Bottom left: Berries ripening on a woodland planting of American ginseng. Photo by Crede Calhoun, courtesy of Larry Harding.

Below: A head of scarlet ginseng berries. Photo, Jeanine Davis.

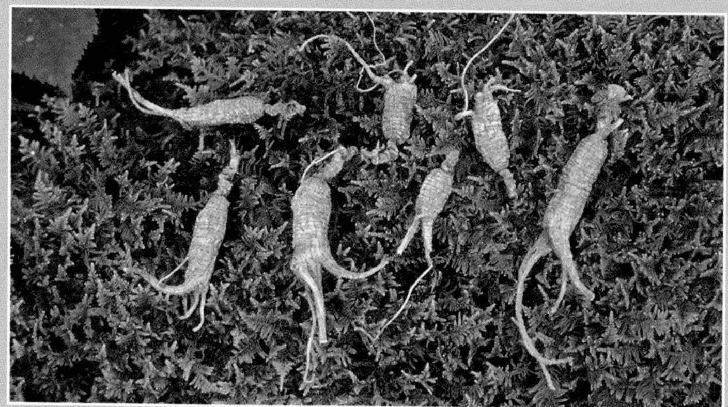

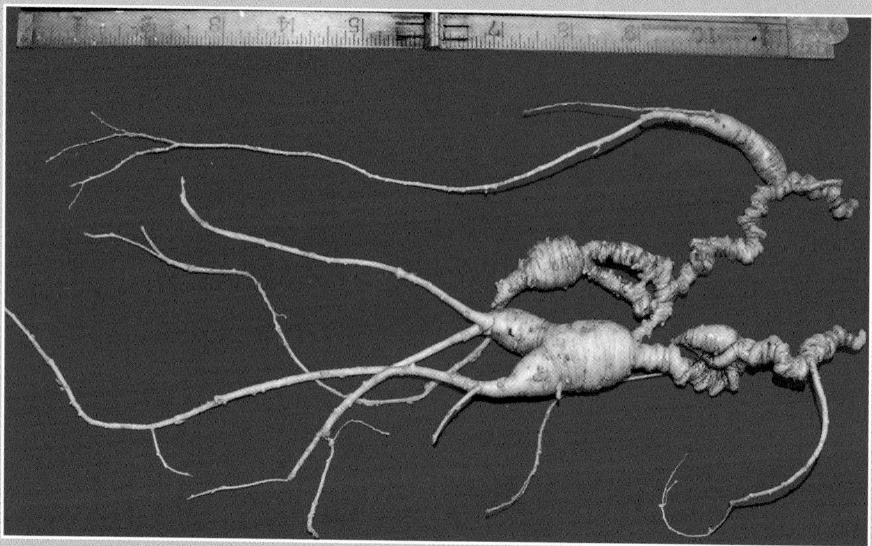

Top: Highest grade eight-year-old woods-cultivated American ginseng roots that might be mixed with wild. Photo, Mark Haskett.

Middle: Polypropylene shade structure over mature ginseng beds in British Columbia. Photo, Al Oliver.

Bottom: Wild 132-year-old (at least) American ginseng root found in the Catskill Mountains of New York. This root sold in 2003 for $2,800.00. Photo, Scott Harris of Sylvan Botanicals.

Top: Ginseng and goldenseal can be interplanted together in a wild-simulated planting. Photo, Jeanine Davis.

Above: Some growers cut the foliage off their ginseng plants prior to the opening of ginseng harvest season to prevent theft. Photo, Jeanine Davis.

Left: American ginseng production under tree ferns in Tasmania, Australia. Photo courtesy of Graeme and Wendy Elphinstone of Tasmanian Organic Ginseng.

Above: Full-color ginseng tattoo on Cornell Cooperative Extension Agent Bob Beyfuss' arm.
Photo courtesy of Bob Beyfuss.

Top: Wild 'sang thriving next to its common companion, maidenhair fern.
Photo, Robert Beyfuss.

Above: Future Tuckasegee 'sang farmer inspects a sample of newly washed roots.
Photo, Mark Haskett.

Top: Jeanine grew woodland botanicals under this same polypropylene shade structure for over 20 years.
Photo, Jeanine Davis.

Left: Slug damage to this ripe goldenseal fruit reveals the small black seeds inside.
Photo, Jeanine Davis.

Right: The typical bullseye pattern of Alternaria on a goldenseal leaf.
Photo, Jeanine Davis.

Top: Wild-simulated goldenseal in Kentucky. Photo, Jeanine Davis.

Middle: Woods-grown goldenseal root. Photo, Jeanine Davis.

Bottom: Young goldenseal plants being sold as planting stock by Kitchen Sink Farm at a conference in North Carolina. Photo, Jeanine Davis.

Top: Ramp seed heads holding green capsules and mature black seeds (left), with fresh bulbs (right).
Photo Jackie Greenfield.

Middle: A nice harvest of freshly dug ramps.
Photo Jackie Greenfield.

Bottom: Occasionally the flower bud of the ramp, known as a scape, emerges in the late spring, before the leaves have died back.
Photo, Jackie Greenfield.

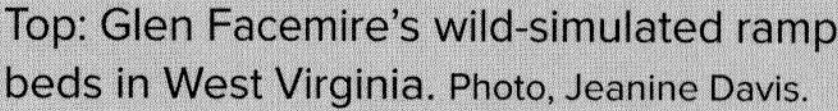

Top: Glen Facemire's wild-simulated ramp beds in West Virginia. Photo, Jeanine Davis.

Above: Bethroot has many names: wake-robin, purple trillium and stinking trillium, among others. Photo, Jeanine Davis.

Right: Black cohosh flowers with pollinators.
Photo, Jeanine Davis.

Above: Bloodroot seedlings are single-lobed round leaves.
Photo, Jeanine Davis.

Top: A good stand of black cohosh growing in the woodland gardens at the research station.
Photo, Jeanine Davis.

Above: Freshly dug black cohosh roots.
Photo, Jeanine Davis.

Top left: Bloodroot seeds with fleshy appendages called elaiosomes.
Photo, Karen Hardy

Top right: Slicing a root reveals the internal color that gives bloodroot its name.
Photo, Jeanine Davis.

Bottom: Emerging bloodroot leaves still wound around the unexpanded flower bud.
Photo, Agatha Kaplan.

Top left: Blue cohosh flower buds and flower in early spring.
Photo, Karen Hardy.

Top right: Male false unicorn plant in bloom.
Photo, Agatha Kaplan.

Bottom left: False unicorn with long seedstalk and seed pods in the early fall.
Photo, Jeanine Davis.

Top: Freshly dug false unicorn root.
Photo, Jeanine Davis.

Middle: The shiny green leaves and delicate white flower spikes of mature galax.
Photo, Virginia Lindsey.

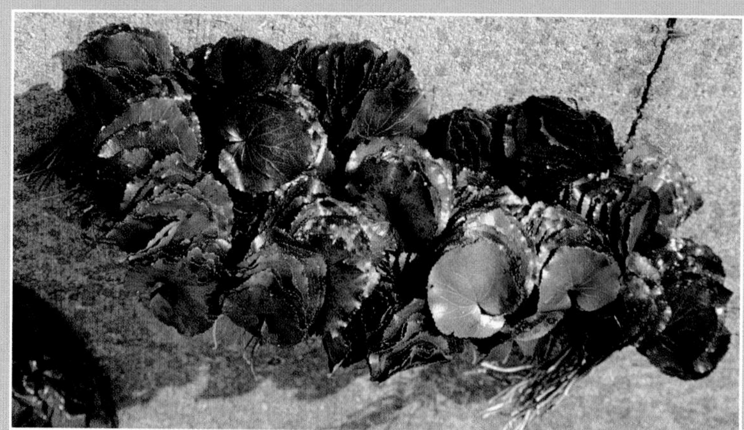

Bottom: Galax leaves are more valuable when they turn red in the fall.
Photo, Jackie Greenfield.

Top: The blossom of the mayapple.
Photo, Agatha Kaplan.

Below left: An example of rust on the mayapple leaf.
Photo, Jeanine Davis.

Below right: Pinkroot flowers.
Photo, Karen Hardy.

Top left: Pinkroot growing under artificial shade.
Photo, Tyler Nance.

Top right: The beautiful magenta berries of the spikenard plant.
Photo, Jeanine Davis.

Bottom: Spikenard's foliage and flowers make it an attractive landscape plant. Photo, Tyler Nance.

Top: Spikenard roots can grow quite large..
Photo, Jeanine Davis.

Left: Freshly harvested wild ginger roots.
Photo, Jeanine Davis.

Right: The reddish-brown blossom of the wild ginger nestles close to the ground under its leaves.
Photo, Agatha Kaplan.

Top left: Wild ginger makes an attractive groundcover. Photo, Jeanine Davis.

Top right: Delicate yellow flowers on wild indigo plants. Photo, Jeanine Davis.

Bottom: Visitors touring the woodland gardens at the research station. Photo, Jeanine Davis.

19

Bloodroot

Plant Range and Uses

Bloodroot (*Sanguinaria canadensis*) is an attractive, low-growing plant with large waxy leaves and pretty, white flowers that is among the first blooms on the forest floor each spring. Bloodroot has a wide range and grows in forests throughout eastern North America and as far west as the Rocky Mountains. It grows within the forest and along the forest edges. Depending on the location, bloodroot is known by many other names, including red Indian paint, red root, and puccoon. The underground rhizome of bloodroot, which "bleeds" a red sap when cut, has a long history of use by Native Americans as a clothing and basket dye, skin paint, and medicine. It has proven antimicrobial activities and has long been used to treat skin lesions.

The primary active component in bloodroot is the alkaloid sanguinarine. It is red-orange and water soluble. Sanguinarine inhibits the bacteria that cause tooth decay and has been used to treat gingivitis. For a long time, it was the active ingredient in Viadent toothpaste. A 2003 study indicated that bloodroot root extract inhibited growth of the *Helicobacter pylori*, the bacteria responsible for stomach ulcers. Bloodroot, however, may be toxic when taken internally and should not be used casually. Caution should also be exercised when handling bloodroot. I do not react to it, but many people experience rashes or burns from the sap in the rhizomes, roots, and stems. Some people in my program just have to wash their hands after handling bloodroot, but others must wear gloves to avoid an unpleasant reaction. My former graduate student, Maria Onofrietti, who worked on bloodroot nonstop for several years, advises that you should never touch your face or rub your eyes when working around bloodroot, because extreme irritation can occur.

Mature bloodroot leaf. Photo by Jeanine Davis.

Bloodroot root. Photo by Jeanine Davis.

Plant Description

Bloodroot has a small, reddish-brown rhizome covered with a number of black fibrous roots. Very early in the spring, a tightly curled leaf emerges from the soil. A white, showy flower pushes up out of that little tube and blooms before the leaf completely expands. These are always among the earliest of flowers in the forest, although peak flowering varies depending on temperatures. The rhizome slowly grows and branches and, in time, may produce over ten leaves and flowers. After the flower blooms, the leaves continue to grow and expand for several weeks and produce a dense canopy about eight inches tall. The leaves are large, flat, waxy, and lobed. There may be variation in the number of lobes on the leaves. The flowers are cross-pollinated by bees but can also self-pollinate. Once the flowers are spent, a long green pod quickly develops, filled with many shiny, brown-black, round seeds. Each seed has a little white fleshy attachment, known as an elaiosome, which looks like a tiny worm. In early June in western North Carolina, the seedpods split open, and the seeds fall to the ground. If left there, ants will usually carry the seeds into their nests and eat the elaiosomes. This is nature's way of dispersing the seeds. Fire ants also collect and disperse the seeds, but move more of the seeds to the forest edge than native ants do. Rodents can damage and destroy seeds, both in the pods and after they have been carried off by ants. Many of the intact seeds that find their way into the soil will germinate the next year, each producing a single little, round, and waxy leaf. We find many seedlings growing among the mature plants in our production beds.

Shade

Bloodroot grows under all levels of shade in the forest, although the biggest, densest patches are often in the higher light areas.

Bloodroot flower. Photo by Karen Hardy.

Robin Suggs, of Moon Branch Botanicals in far western North Carolina, put in a bloodroot planting study that stretched from the forest edge to deep within the forest. When he harvested the roots (in discussing bloodroot, "root" will refer to both rhizomes and roots) after one season of growth, he noted that the rhizomes from the forest edge had many more buds than did the rhizomes from deeper in the forest. In a study by Marino and colleagues looking specifically at light and fertilization effects on bloodroot growing in the forest, bloodroot produced more leaves in response to higher levels of light and nutrients. Those researchers assumed that the plants with the most leaves also had the largest rhizomes. I have consistently grown excellent bloodroot under 78 percent shade cloth and in mixed hardwood forests where some sunlight gets through the canopy. Plants growing under artificial shade often die by middle to late July, whereas in the forest, the plants retain their foliage until fall. I do not know the reason for this, although I have observed the same phenomenon on goldenseal and mayapple. It is not a concern, though, because root growth under artificial shade is greater than in the forest. At this time, I recommend planting bloodroot under 70 to 80 percent shade provided by either a natural forest canopy or shade cloth, but you may also want to experiment with growing it under higher light conditions.

How to Grow

Bloodroot is hardy to zone 3 and will grow throughout much of North America. It likes a moist soil, so have irrigation available if the soil tends to get dry late in the summer season. This adaptable herb will grow in dry sites, but will often go down early in the season, which probably reduces root growth. Bloodroot is an easy herb to grow. Prepare raised beds and adjust the soil pH to 5.5 to 6.5. We do not fertilize our bloodroot, relying on the natural fertility of our rich mountain soils to meet the plants' nutritional needs. Several studies have shown, however, that fertilization can increase leaf and root growth, but the sanguinarine concentration, which can affect price, will probably be reduced. If your site is not naturally fertile, consider adding a small amount of a balanced fertilizer.

Vegetative Propagation

Bloodroot is easily propagated by rhizome cuttings. This is usually done in the fall, but I have also done it successfully in the spring

Bloodroot being planted for research at the Mountain Research Station in Waynesville, North Carolina. Photo by David Danehower.

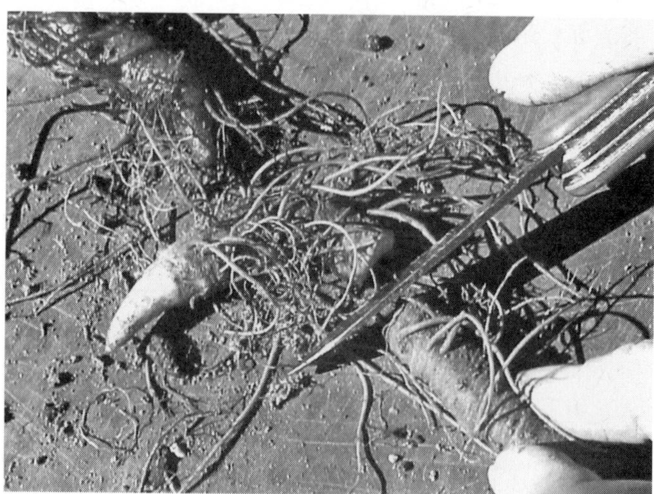

Bloodroot roots being cut for planting. Photo by Jeanine Davis.

from rhizomes stored in the cooler over the winter. Cut the rhizomes into 2" pieces, including at least one bud on each piece. Plant the rhizome pieces in well-prepared raised beds in the same manner as goldenseal, but only under about one inch of soil. Use a 6" × 6" spacing, and cover with a good layer of mulch, such as hardwood bark or the natural forest leaf mulch.

Cold Requirement

Although bloodroot will grow throughout most of the South, the rhizomes do have a chilling requirement that must be met for the buds to break in the spring. As a result, it is not recommended for commercial production any further south than zone 8. As part of her master's degree research, Maria Onofrietti determined how much chilling the rhizomes need to break dormancy. In her first study, she found that about half of the rhizomes broke dormancy with as little as four weeks storage at 40°F, but 12 weeks were required for full emergence. She did not look at a range of temperatures, so we do not know just what the critical low temperature is. Until that is figured out, you should probably not try to produce bloodroot commercially in an area where you get less than a couple of months of cold temperatures.

Seed Handling

Bloodroot can be propagated by seeds, but they can be a little tricky to handle. First, do not let the seed dry out. As with many of the plants discussed in this book, the best germination results from seeds that are sown as soon after maturing on the plant as possible. Sow in a well-prepared seedbed and cover lightly with about an inch of soil and about three inches of mulch. We have also had success with putting the freshly harvested seed into a screen-mesh bag, burying it under a couple of inches of soil and mulch in the shade, and sowing the seeds in the late fall or very early spring. In our experience, most of the seeds handled either way will usually germinate the first spring after harvest, but many growers report that the majority of their seeds do not germinate until the second spring. My former technician, Karen Hardy, had that experience. She put fresh bloodroot seeds in mesh bags and buried them under our 78 percent shade cloth structure, about five inches deep in the soil, with three inches of mulch on top. Late in the spring, she sowed

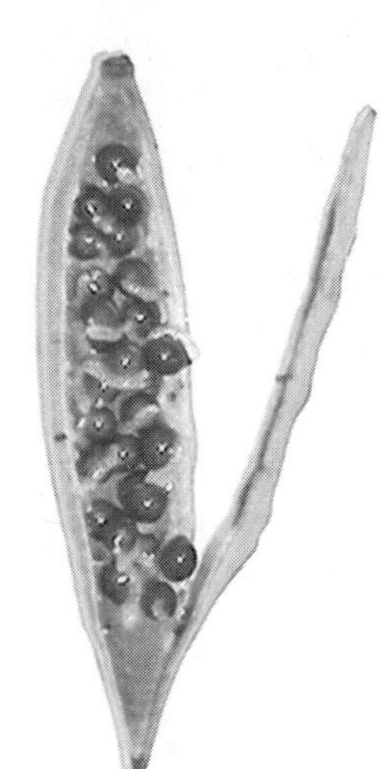

Bloodroot seed pod with immature seeds inside (*left*), with mature seeds (*middle*), and with elaiosomes clearly visible (*right*). Photos by Karen Hardy.

Bloodroot seed pods in mesh collection bags. Photo by Karen Hardy.

the seeds as described above, but got no emergence. We suspect the seeds did not germinate because she buried them too deeply, and as a result, they were not exposed to an adequate warm treatment that first summer. The research done by Dr. Jim Affolter and his students at the University of Georgia in Athens, who have been studying bloodroot seed propagation for many years, support this theory. Their studies show that the seedling root (hypocotyl) and seedling shoot (epicotyl) require different temperatures to break dormancy. The hypocotyl dormancy can easily be broken with a warm treatment, which must be done before the cold treatment is imposed. This occurs naturally outside if the seeds fall to the forest floor and make soil contact, because the seeds mature early enough in the season to get lots of warm temperature exposure before winter sets in. We now leave our seeds in the seedbed, buried under no more than an inch of soil and three inches of mulch, and get good germination the next spring.

The trickiest part of handling bloodroot seeds is collecting them. The seedpods pop open suddenly, scattering the little brown-black seeds all over the mulch where they are almost impossible to find. Some people monitor their plants daily and harvest the pods when they are swollen and just starting to turn yellow. We cannot watch the plants that closely during the busy spring season, so we use little mesh bags to capture the seeds. We cut out 6" circles of wedding veil fabric, run a thread around the perimeter, and pull the thread ends to gather up the edges and make a little bag. The bag is fitted over the immature pod and tied loosely around the flower-stalk. We then wait until all the pods have ruptured and pinch off the flower-stalk near the soil line and gather all the little bags in a bucket. We

Leaf lesions at the end of the growing season. Photo by Karen Hardy.

sort the seeds out over a tray where there is no risk of losing any. The seeds are sown into a seedbed immediately and the little bags are saved to be used again the next year. This is actually not as slow a practice as it might sound, and you do not have to bag all the flowers. Just bag as many as you think you will need to gather seed. The rest will self-sow throughout your beds.

Pests and Diseases

In the wild, we rarely observe any disease or insect problems on bloodroot. Now that we are growing bloodroot in concentrated plantings, however, some pest problems have materialized. In 2003, we had our first case of Alternaria leaf blight. This may be an isolated incident, since spring 2003 was the wettest on record in our area (up till that time), but now we know the plant is susceptible. In late June and early July, plants growing under the artificial-shade structure will sometimes develop big black lesions on the leaves. The pathology clinic has never been able to definitively diagnose the disorder. Since it does not appear to hinder root growth, and the plant comes up healthy the following spring, we have not tried to prevent or control it. At the University of Georgia, they routinely sprayed fungicides on the bloodroot in their studies to prevent Alternaria and Botrytis. We have not observed either problem in our forest plantings. Slugs will occasionally nibble on the leaves, but they do not usually cause serious damage. In 2003, we also had our first experiences with "critter" damage in our forest plantings. Turkey, groundhogs, and apparently deer, caused significant damage in some of our beds. A study by Rockwood and Lobstein showed that groundhog and deer grazing can reduce or eliminate seed production, especially in the year after the damage has occurred.

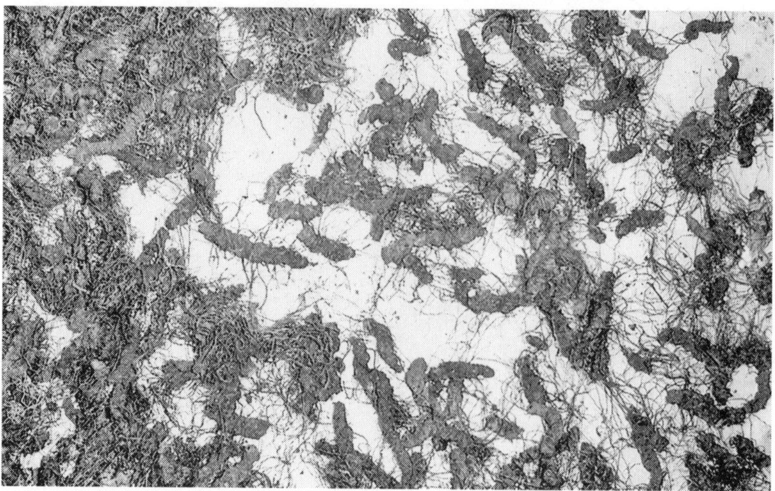

Dried bloodroot that will be used for making dye. Photo by Michael Bruckner and courtesy of Qualia Arts & Crafts Mutual.

The most disturbing problem we have had with bloodroot is the mysterious disappearance of the root. This has happened to us twice now, both at the research station and in on-farm trials. It has also happened in grower fields in the United States and Canada. In some cases, the bloodroot grows just fine for several years, and then starts slowly disappearing, a little more each year. Only seedlings remain. In spring of 2013, we gave bloodroot planting kits to a group of people and planted some ourselves at three locations. None of it came up. When we investigated, we couldn't find even a scrap of root to send to the pathologists. At the research station, we planted that new bloodroot near where other bloodroot had been growing for years and those older plants are fine. In western North Carolina, 2013 was a season for the record books; it was the coolest and wettest in 14 years. The disease Pythium is a suspect, but we don't have any evidence and only suggest it because of the unusually wet soil conditions. I don't know if the loss was due to excessive wetness or not, but why did the other bloodroot do fine? A quick search of the Internet revealed that this has happened to others, including home gardeners. My suggestion is to watch your bloodroot closely, and if you notice any disappearing, get a soil and root sample from that patch to a plant disease clinic as soon as possible. We need to identify what this mystery phenomenon is.

Harvest and Post-harvest Handling

Bloodroot roots are usually harvested during the growing season because the plant tops often die back by late summer, making it difficult to find the roots in the fall. Some buyers do want fall-dug roots, in which case the areas to be dug should be flagged while the foliage is still alive. We dig bloodroot with forks, but large-scale growers producing under shade cloth will want to use ginseng diggers or something similar. Wash and dry the roots as recommended for ginseng or goldenseal. It takes bloodroot three to seven days to dry at about 95°F with high airflow. It can be dried at lower temperatures, but be sure to maintain high air movement with a fan and watch the roots closely because bloodroot molds quickly. Package as the buyer specifies, usually in polypropylene sacks or cardboard barrels. Store in a cool, dry place, protected from rodents and insects. If harvesting for use as planting stock, leave some soil on the roots or pack them in moist sphagnum moss. Store in cardboard boxes or in boxes lined with unsealed plastic bags in a cooler at about 40°F until planting in the spring or fall. Check the roots every few weeks to make sure they are moist and not getting moldy.

Yield and Gross Return Estimates

How long bloodroot needs to grow before producing economically feasible root yields is still not known. The data gathered from our research plots, demonstration gardens, on-farm tests, and information given by wild-harvesters suggest that plants need to grow between four and five years from rhizome cuttings and at least six years from seed. I also do not have reliable yield estimates. Growers in Ontario expect to harvest 1,000 to 1,500 pounds of dried root per acre after three years of growth under artificial shade, but I do not know anyone who has cultivated enough of it yet to really know. Richo Cech reported in *Growing At-risk Medicinal Herbs*, that the average fresh weight of 300 three-year-old forest-grown bloodroot plants was 20 grams (0.7 ounces). Multiplying that figure by 139,200 plants per acre (6" spacing, 4'-wide beds on 5' centers) and accounting for a water content of 76 percent, the yield per acre would be 1,462 pounds of dried root. That is right in line with what the Ontario growers estimate. At $8 to $16 per dried pound, that would translate into a gross return of $11,696 to $23,392 per acre after three years. Keep in mind, that at a six-inch spacing under shade cloth, you will need to plant over 1,000 pounds of rhizomes pieces per acre. How much the planting stock costs will have a big effect on how much your net returns per acre will be. In contrast, in the woods, you would probably plant and harvest only about half as much because of the area taken up by trees, but production costs would be much less.

Markets and Prices

In 2003, the market for bloodroot was increasing rapidly. The demand was fueled by Phytobiotics, a German company that was using bloodroot in an animal feed product. The sanguinarine in bloodroot is an appetite stimulant that helps the animals gain weight faster. For that company, it was important for the bloodroot to have high sanguinarine content. Research at the University of Georgia showed that alkaloid concentrations in bloodroot increase with decreasing light intensity and fertilizer levels. In related studies, they also found that alkaloid content in bloodroot declines with elevation, fluctuates seasonally as the plant grows, increases with rhizome water content, and varies from site to site. Although these studies were not designed to develop commercial production guidelines, they do show that the chemical composition of bloodroot can be manipulated by changing the growing conditions. In studies I was involved in, we found that alkaloid levels varied seasonally in wild bloodroot and were highest in the spring. Bloodroot roots grown under artificial shade exhibited much less variability in alkaloid level and were much larger than wild-harvested roots.

Unfortunately, growing bloodroot for Phytobiotics did not turn out the way we anticipated. By 2005 it was obvious to the company that bloodroot grew very slowly—too slowly for their purposes. The price they were willing to pay, $9 per dried pound, was too low for growers producing bloodroot in the woods to make any money. Then we learned that the company could extract the sanguinarine they needed from plume poppy (*Macleaya cordata* and *M. microcarpa*), both of which grow much faster and easier than bloodroot. So that burst the bloodroot bubble.

For many years, the market for bloodroot was low, as were the prices, but that has changed once again. In 2013 the demand for bloodroot far exceeded supply, and raw botanical dealers were paying up to $16 per dried pound. I have not been able to find out what is driving the demand, but the rumor is that there are new skin cancer products being developed. There is also a small demand for the root as a source of dye for crafts. I am collaborating on a bloodroot-growing project with Cherokee basket makers in western North Carolina.

If demand for bloodroot continues to grow, wild populations will not support the quantities required, and United Plant Savers has already listed bloodroot on its "at risk" list. The current prices being paid by dealers are $8 to $16 per dried pound. I don't think you can make money at $8 a pound, but you can at $16.

Nursery and Planting Stock Production

There are nurseries selling small volumes of cultivated planting stock, but I am not aware of any nurseries selling large volumes. Growers wanting to plant large areas in bloodroot must purchase wild-harvested roots. This is not the best situation because you never know where those plants were dug, and there can be a great deal of variability among the plants. An order of several hundred pounds of bloodroot planting stock may contain roots gathered from several states. A few years back, growers in Canada purchased roots from forests in Tennessee, Kentucky, and North Carolina. They found the "Southern roots" did not always adapt well to their northern climate, and high mortality rates were common. If the demand for cultivated bloodroot continues to rise, sources of high quality, cultivated planting stock will be in greater demand. If a grower could produce 4,500 pounds of fresh root per acre and sell it for planting stock at $5 per pound, this would translate into a gross return of $22,500 per acre. Bloodroot is also an attractive forest wildflower that blooms very early. Currently, wholesale nurseries are selling bloodroot in volume for $40 for 100 roots. Individual plants are selling retail for $2.50 to $10.00 apiece. This could offer a new grower a very good starting point.

Production Budget

Jackie Greenfield and I were involved in a special medicinal herb project that resulted in the development of a number of production budgets for medicinal herbs. I am including a modified budget for bloodroot here. This is completely hypothetical. I used the estimated yields from those who are growing it and production costs based on other similar crops.

Table 19. Farm Enterprise Budget for One Acre of Bloodroot Grown under Artificial Shade*

Expenses	3-Year Total
Labor (planting, mulching, weeding, harvesting): 700 hrs at $7.25/hr	$5,075
Planting stock (rhizomes): 1,000 lbs at $5/lb	$5,000
Fertilizer and lime	$204
Plant protection	$136
Fuel, oil, lubrication	$136
Mulch	$2,232
Utilities	$273
Packaging	$372
Total Expenses	$13,428
Gross Income (dried root sales: 1,500 lbs at $16/lb)	$24,000
Net Profit (after three years)	$10,572

*It is assumed that this grower used an existing polypropylene shade structure and already owned the irrigation and drying facilities. Wage rate is the U.S. federal minimum wage (2013).

20

Blue Cohosh

Range and Types

Blue cohosh (*Caulophyllum thalictroides*) is another beautiful, useful medicinal plant that is commonly found in hardwood forests throughout eastern North America, particularly in New England. A larger version of blue cohosh, known as giant blue cohosh, is a separate species (*Caulophyllum giganteum*). Giant blue cohosh is a more northern species than the common blue cohosh and does not grow in abundance anywhere. The two can be difficult to distinguish, and giant blue cohosh is sometimes mistakenly harvested as common blue cohosh. Indeed, the two have similar medicinal uses. There is very little information available on giant blue cohosh, and I have no experience growing it. This chapter addresses growing and marketing common blue cohosh, but I understand that medicinally, the two can be used interchangeably and both have similar propagation and cultivation requirements. The giant blue cohosh might be particularly well suited as a landscape plant.

Plant Description

Blue cohosh is a deciduous perennial. The common blue cohosh grows to about three feet in height; the giant blue cohosh is larger. The leaves of both are compound and composed of three to five leaflets with two to three lobes. They are waxy on the surface and smooth on the edges, not serrated like black cohosh. The plants begin growing in early spring, usually by April in the southern Appalachians. When the leaves first emerge, the foliage is blue to almost purple, with a bluish, waxy film on the surface. As the leaves expand, clusters of little flowers open. *C. thalictroides* has yellowish flowers, and *C. giganteum* has purple flowers. As the season progresses, the blue color of the foliage fades, and the leaves become greener. The flowers turn into clusters of blue-purple berries that are

Blue cohosh. Photo by Agatha Kaplan.

held above the foliage. The rhizome, which is the plant part of economic interest, is thick, brown, knotty, and covered with many fibrous roots. Blue cohosh is a beautiful herb that has a market as a landscape plant for use in wooded areas.

Uses

Blue cohosh is known by many names, including papoose root, squaw root, and blue ginseng. Like black cohosh, it is considered an important woman's herb and has been shown to have estrogen-like activity. An important difference between the two plants, according to the literature, is that blue cohosh raises blood pressure, while black cohosh tends to lower blood pressure. Traditionally, blue cohosh was valued as a childbirth aid. This has been attributed to the presence of the glycoside caulosaponin, which is believed to stimulate uterine contractions. A study published in 2012 from the National Center for Natural Products Research in Mississippi raised serious concerns about the safety of blue cohosh, particularly for pregnant women, because of potential adverse effects on mitochondrial function. Many Native American tribes, such as the Meskwaki, Menominee, Ojibwa, Potawatomi, and the Omaha, took blue cohosh for several gynecological problems, including treatment of menstrual problems and for difficult labor. The Cherokee valued blue cohosh as an anticonvulsant, a sedative, and an aid for dermatological and gastrointestinal conditions. The Chip-

Blue cohosh cultivated in the woods. Photo by Tyler Nance.

pewa used it as an analgesic. Blue cohosh extract has exhibited antimicrobial and anti-inflammatory properties, which may explain its reported effectiveness as a pain reliever and as a treatment for poison ivy rash.

How to Grow

Blue cohosh is hardy to zone 3 and can easily be grown as far south as Georgia. It prefers a slightly moister soil than many of the plants discussed in this book, and in its natural habitat, it often grows in cool, moist rich coves along stream banks. It has been my experience that the roots will rot in soggy conditions. (In discussing blue cohosh, "root" refers to both rhizomes and roots.) I suggest that you select a site to

Blue cohosh fruit. Photo by Tyler Nance.

Blue cohosh flowers and flower buds. Photo by Agatha Kaplan.

cultivate blue cohosh where the soil stays moist but not waterlogged. Other than that, growing blue cohosh requires most of the same conditions as goldenseal.

Choose a site for blue cohosh with at least 75 percent shade. It can probably take more or less shade, but that research has not yet been done. Blue cohosh grows naturally in soils with a pH ranging from 4.5 to 7.0. I have obtained the best growth in my location with a soil pH of 5.5 to 6.0. Blue cohosh likes a rich soil, high in organic matter. If your soil does not match this description, incorporate compost or leaves into the soil if it is practical to do so. Build raised beds, and after planting, cover with organic mulch. Young blue cohosh plants grow slowly and do not compete well with fast-growing weeds, so keep the beds well weeded for at the least the first growing season.

So far, I have not had any pest or disease problems on blue cohosh. We have occasionally lost plants over the winter, but the number has been very small. Because of its need for moisture, you should definitely have the ability to irrigate if you intend to plant a large amount of blue cohosh. You might never need it, but you should be able to get water to blue cohosh if a time of extended drought occurs.

Vegetative Propagation

Blue cohosh is easily propagated by rhizome pieces. Cut the rhizomes into 3" to 4" pieces in the early spring or late fall and plant between three and four inches deep in the soil. If possible, cut the pieces so there is a big bud on each one, but do not worry if there is no large bud. Like goldenseal, blue cohosh rhizomes are covered with tiny buds that will sprout even if a large bud is

Blue cohosh roots and tops. Photo by Agatha Kaplan.

damaged or missing. Space the rhizome pieces 18–24" apart and keep the soil moist. Most of these pieces will sprout the first growing season, but be patient, because if a large bud is not present, they might not emerge until the second season.

Seed Handling

Blue cohosh can also be propagated by seed, but the seeds are slow to germinate, and germination rates are often very low. Blue cohosh seeds undergo a two-part germination process, and it may take up to three years before a shoot emerges from the soil. Richo Cech describes this in some detail in his book *Growing At-risk Medicinal Herbs.* As with so many of these woodland plants, it is best to collect the seeds right after they mature in middle to late summer and sow them as soon as possible. Remove the flesh of the fruit from around the seed, using the method described for ginseng or goldenseal. Keep the seeds moist until planting by holding them in moist sand or covered with damp paper towels. Because the seedlings grow very slowly, it is best to start them in a nursery bed and move them to a production area after the second or third year of growth. Sow the seeds several inches apart and about ½ inch deep in a well-prepared seedbed in the shade. Choose a site that will remain moist all year round. Cover with a good, thick layer of mulch, preferably hardwood leaves. The seeds should germinate the first or second spring after sowing.

Harvest and Post-harvest Handling

In the wild, blue cohosh is usually harvested in the fall after the seeds have matured. There is no evidence to suggest that it would be any different under cultivation. Dig the roots carefully, trying to keep the fibrous roots intact. Digging can easily be done with a fork or spade. If you intend to use your harvested roots as planting stock, leave some soil on the roots or pack them in moist sphagnum moss. Store them in cardboard boxes, or in boxes lined with plastic bags, but do not seal the plastic bags. Store the roots in a cooler at about 40°F until planting in the spring or fall.

Blue cohosh roots tend to be twisted, knotty, and matted. This makes them very difficult to clean; you might have to cut some of the roots to remove all the dirt. A root-washer or high-pressure stream of water can be used, but expect to spend more time cleaning these roots than most others. Blue cohosh should be dried in the same manner as ginseng or goldenseal. Use a dryer with low temperatures (80° to 95°F) and high airflow. Package the roots for delivery as specified by the buyer, usually in polypropylene sacks or cardboard cartons (do not use plastic bags), and store them in a cool, dry place, protected from rodents and insects, until delivery.

Economics of Production

There are very few people who grow blue cohosh commercially, so little has been reported about yields, costs of production, or financial returns. Because blue cohosh grows so slowly, it may take six years or more of growth from rhizome cuttings—and even longer from seeds—before the roots are large enough to harvest. Richo Cech reports that the average fresh weight of a mature blue cohosh root is about 1.5 ounces and it is about 70 percent water. Using his figures and my estimate of 4,800 plants per acre in the woods (18" spacing, 4' beds on 8' centers), I estimate a potential yield of 135 dried pounds of blue cohosh roots per acre. Under artificial shade, at 11,600 plants per acre (18" spacing, 4' beds on 5' centers), estimated potential yield is 326 dried pounds per acre. Please keep in mind that these are all paper calculations. No one has shared any commercial yield figures with me yet, and my plots are too small to extrapolate data from reliably.

Markets

There has always been some market for blue cohosh, but the volume sold every year is small and prices are low. Prices paid by raw botanical buyers of roots and herbs from 2002–2013 have been in the range of $0.80 to $2.25 per dried pound. In September 2013, the price paid to wild-harvesters was $1 per dried pound. There is no way that cultivation can be profitable at those prices. Wholesale prices, from the grower to the consumer, ranged from $7.90 to $10.30 per dried pound. With yields of 135 dried pounds per acre in the forest, this would gross $1,067 to $1,391 per acre. That is not much money after six or more years of growth. Herbalists have told me that there will always be some demand for blue cohosh, and United Plant Savers includes it on their "at risk" plant list, which is why I have included it here. Several prominent buyers in the industry told me that the market for blue cohosh is steady, but they only buy about 500 pounds of dried root each year. This is an example of a plant that a property owner could manage wild populations for, harvesting small amounts when necessary.

There is, however, a slightly larger market for blue cohosh landscape plants. Blue cohosh makes a beautiful addition to forest gardens, and many homeowners would purchase blue cohosh if it were made more available to them. Blue cohosh retails in local garden nurseries for $5.00 to $7.95 for a 4" potted plant. One hundred plants can be purchased through several mail order nurseries for $90 to $420. That is the market that it makes sense to concentrate on at the present time. If blue cohosh does suddenly increase in demand, planting stock will also be needed. A nursery growing for the landscape industry could maintain several propagation beds of blue cohosh that could quickly be divided and grown for planting stock if the market changed.

21

False Unicorn

Names and Range

False unicorn, also known as star grubroot, star root, grub root, devil's bit, helonias, starwort, and fairywand (*Chamaelirium luteum*) is an herbaceous perennial, which, if you look closely, can be found growing in forests, bogs, and meadows throughout much of eastern North America from Ontario, south to Florida, and west to Texas. It is a scarce plant that brings a high per-pound price and could quickly be threatened if it were suddenly discovered to have a popular use. United Plant Savers includes it on its "at risk" plant list and strongly encourages its cultivation.

Plant Description

False unicorn is a slow-growing, unobtrusive little plant. Its foliage consists of a low, rather compact rosette of long leaves that range in length from two to six inches. The leaves closest to the soil are broad and large and get smaller and narrower as they advance up the flowering stalk. There are separate male and female plants that each send up tall slender flower stalks with white (male) or greenish (female) flowers. The flowers bloom in spring and are small, delicate, lacy looking, and arranged in a long spike. The female flower spike is upright and three to four feet tall. By late summer, the end of the flower spike is covered with small, light-green, three-celled pods full of many little seeds. The male flower-stalk is only 18 inches tall, less showy than the female stalk, and slightly curved. Several references mention that the plants do not flower every year, but this has not been my experience.

The underground rhizome, which is the plant part of economic interest, is usually between ½ and 2 inches long, hard, light gray to dark brown, rough, and often shaped somewhat like a little curved carrot. It usually has some long, thick roots near the broad stem end and fibrous little roots along the rest of its length. The rhizome has an interesting identifying characteristic.

Male false unicorn flower. Photo by Agatha Kaplan.

When cut crosswise, one can see that the roots enter the rhizome through little holes arranged in a ring just inside the outer edge of the rhizome. C. F. Millspaugh, a well-known 19th-century author on medicinal plants, described the little roots running through these holes as "freely movable, like thread in the eye of a needle."

In discussing false unicorn, "root" will refer to both roots and rhizomes.

Origin of Names and a Similar Plant

This plant has some unusual names. According to Millspaugh, the name "devil's bit" originated in a Native American story in which the devil bites off a portion of the rhizome in order to curtail its medicinal utility. The name unicorn root and grub-root clearly refer to the shape of the rhizome. The old literature mentions that *Chamaelirium luteum* and *Alteris farinosa* were often confused, probably because of similar common names and the fact that both have basal rosettes of leaves and long flower spikes. *Chamaelirium luteum* is commonly called starwort and false unicorn, whereas *Alteris farinosa* is often called stargrass and true unicorn.

Uses

Native Americans traditionally used false unicorn to treat coughs and morning sickness and to prevent miscarriages. The medicinal properties of false unicorn are attributed to steroidal saponins found in the root. In 2011, a paper was published reporting the discovery of two previously unknown saponins in the roots. Early settlers used the roots of false unicorn to treat a variety of disorders, including headaches, depression, colic, and worms. The roots were also used to remedy vitamin C deficiency and to induce the flow of certain body fluids, including urine and saliva. False unicorn is now used primarily as a woman's herb.

How to Grow and Harvest

False unicorn plants may be grown in a shaded landscape, but most of us who have tried to grow it on a large scale have found it challenging. It can be cultivated in zones 5

False unicorn thriving in a woodland garden. Photo by Jeanine Davis.

to 8, but the plant is slow to grow, and mortality rates can be high. In its native habitat, false unicorn prefers a rich soil that is high in organic matter and slightly acidic, in the pH range of 4.5 to 6.0. It grows best in moist but well-drained soil. Although it often grows in open woods, it is not unusual to find it in dense forests. I have grown false unicorn for over 20 years under a polypropylene shade structure alongside all of the other plants discussed in this book. False unicorn has been the least successful of all under the shade cloth. It lives, but it grows very slowly, and the leaves do not get as large as those that I have observed in the wild. I saw beautiful false unicorn growing in forest test plots at the Goldenseal Sanctuary in southern Ohio. The plants were large, with big, dark green leaves. It could be that the soil under my shade structure, which contains a lot of clay, was not the best for false unicorn. I also suspect that we kept the hardwood bark mulch too deep for false unicorn. Plants that I have growing in the forest, where the soil is loamier and the mulch consists of only a couple inches of leaves, look much healthier than those

Young false unicorn plant. Photo by Agatha Kaplan.

Very young false unicorn plant. Photo by Jeanine Davis.

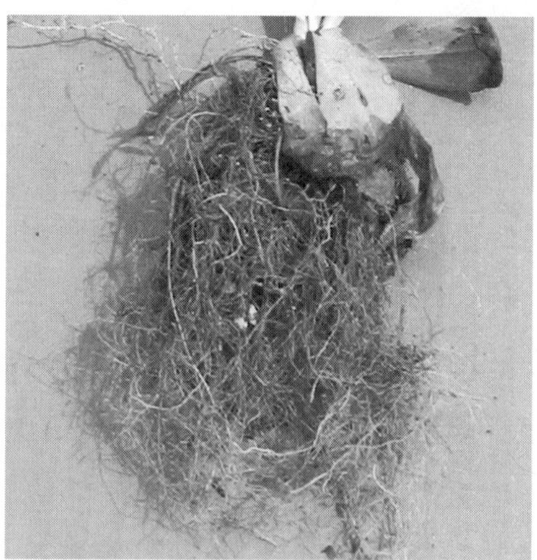

Young false unicorn plant with mass of fibrous roots. Photo by Agatha Kaplan.

A mica washer converted to a root washer. Photo by Jeanine Davis.

that were under the shade structure. The ones in the woods also flower and set seed regularly.

Other than a little study done years ago at what is now the Goldenseal Sanctuary in southern Ohio, I am not aware of any research being done on the cultivation of false unicorn. I have never had the funding to conduct replicated studies on how such factors as soil pH, soil fertility, plant spacing, and mulch influence growth and yield. Until that work is done, I suggest that growers follow the basic recommendations for growing goldenseal in the forest, with the following modifications. Set plants 12" apart, apply a thin layer of mulch for the growing season, keep the beds well weeded, and, in areas where the winters are very cold, add several inches of extra mulch for the winter. It will probably take about six years to grow false unicorn to an economically feasible harvest stage. When they finally reach that size, harvest the roots in the fall using a fork or spade. Wash carefully and dry at 80° to 95°F with lots of air movement. Package the roots as directed by the buyer.

Seed Handling

It has been my experience that false unicorn is one of the easiest of the woodland botanical seeds to germinate. Unlike most of the other seeds discussed in this book, false unicorn seeds are dry, and we have successfully used seeds that were stored in paper envelopes in a seed cabinet for up to two years. An intern in my program compared a large number of germination treatments on false unicorn seeds and found that they all worked almost equally well. We sow the dry seed directly into a shaded nursery bed in the fall or spring, cover with a mere inch of soil and about an inch of mulch, and keep it moist. Usually 80 percent or more of the seeds will germinate the first spring after sowing. We have also started seedlings in flats of a soilless potting mix in the greenhouse during the winter and grown them for one to two years before transplanting into production beds.

Our experience with false unicorn seeds appears to be an anomaly. Most growers and researchers report that false unicorn seeds are difficult to germinate. Richo Cech explains in his book that false unicorn seeds have inhibiting factors that must be overcome before germination occurs. This takes place naturally if seeds are sown outside; but, according to research by Dr. Carol Baskin, the cold treatment must be provided artificially if seeds are sown indoors. If sowing outside in fall or early winter is not possible, Richo suggests storing the seed in a moist medium, such as peat moss or sand, in a refrigerator at 40°F for 8 to 12 weeks before sowing into flats and

False unicorn seed head. Photo by Agatha Kaplan.

setting in a shade house or shade garden. In my program, we have not provided a cold treatment before sowing seed in the greenhouse and have obtained an average germination rate of 80 to 90 percent. Dr. Baskin and Richo Cech have much more experience with these seeds than I do, so I cannot explain the differences in our experiences with false unicorn seed. My recommendation is to try both methods: with and without a cold treatment. It will probably take from six to eight years for false unicorn to reach harvestable size when grown from seeds.

Vegetative Propagation

False unicorn is easily propagated from rhizome cuttings. The rhizomes are covered with little buds that sprout readily. Unfortunately, the rhizome of a mature plant is very small to begin with. To get more than two pieces from one plant, the cuttings must be very small. I suggest that these tiny cuttings be grown in nursery beds or in 4" pots that are kept in a shaded area for several years before planting into production beds. The estimated time from cuttings to root harvest is five to six years.

Marketing

The market for cultivated false unicorn is not large, but marketing studies and reports continue to predict a rising demand for this herb. Currently, most of the market is for dried root, but there is also some demand for false unicorn planting stock for growers trying to meet this predicted demand. The plant can also be sold in flats of 100 plants or in 4" pots for home landscapes.

False unicorn has been one of the most valuable herbs in the raw botanicals market for many years. The volumes sold are not large, but the prices are high. In 2004, raw material buyers (dealers) paid wild-harvesters between $27.50 and $45.00 per dried pound of root. In 2007, prices rose to $30 to $85 per pound. In 2013, the prices were $50 to $65 per dried pound. On the retail market, dried root was selling for about $21 per dried ounce in 2013, and seeds were selling for over $200 per ounce.

False unicorn is definitely a plant that bears watching. There is a lot of talk about it in the industry. If the market for this plant suddenly takes off, those growers who have it in the field will be the ones who will get the premium prices for it because there will be so little available. If you have the space to grow it and are willing to invest in the planting stock, you might want to put a small area in on speculation.

22

Galax

Galax (*Galax urceolata*) is a low-growing plant with shiny, durable, long-lasting leaves that are highly valued by florists. It is not currently cultivated, and as a result, huge volumes of leaves are harvested from wild populations in the forests of southern Appalachia and sent to florists around the world. Galax also has a history of use for medicinal purposes. An infusion of the root was used to treat kidney dysfunction and nervous conditions. It was also used as a remedy for skin problems and wound healing. Usually referred to as galax, it does have a few other common names, including beetleweed, coltsfoot, and wandflower.

Plant Description and Range

Galax is native to the forests of the southern Appalachian mountains. It grows from as far north as New York and Ohio, and south to Georgia and Alabama. It is hardy in zones 5 through 8. Galax is an evergreen herbaceous perennial with leaves that vary from round to heart-shaped and range in diameter from two to five inches. Occasionally, plants can be found with extremely large leaves reaching nearly ten inches in diameter. The leaves have very small, rounded teeth along their perimeters. Most of the year, the leaves are a very glossy dark green, but when exposed to light during late fall through winter, they turn dark red to almost purple. New leaves begin growing in early spring and normally live two years. The plant is usually between 6 and 12 inches tall and spreads, very much like wild ginger. In late spring to early summer, the mature plants put out 10-inch-long spikes of little white flowers.

The Galax Market

The florist industry seems to have an insatiable demand for galax leaves, which are versatile and can be stored for weeks without any deterioration. Galax leaves are used in floral arrangements, in table decorations at weddings, and as plate garnishes in upscale restaurants. The red leaves are even

Population of wild galax. Photo by Jackie Greenfield.

Galax in bloom. Photo by Dick Bir.

more desirable than the green and bring higher prices, although they do not store as long as the green leaves, and when put into a dark cooler, revert to green.

The galax industry is well organized, and product flow from forest to florist is through well-established distribution channels. The only flaw in the system is that all the galax that moves through the industry is wild-harvested. In 2007 it was estimated that over two billion leaves were wild-harvested each year in western North Carolina at a value of over $20 million. In 2013, there was an estimated three billion leaves harvested in the Southern Appalachian mountains, returning $30 to $60 million to the wild-harvesters. As the demand for galax leaves rises, concern over wild populations also increases. In 2001, the United States Forest Service instituted a ban on the harvesting of galax from May 1 through June 15 to help insure the long-term sustainability of native populations. Galax is actively growing during these months, making them more susceptible to damage by foot traffic and harvesting. The new leaves are too small for the market at this stage, so the theory is that if the plants are left alone during this time, they will all be healthier. It is too soon to tell whether or

Galax will spread out over a wide area if left to grow naturally. Photo by Jackie Greenfield.

not this approach will have a positive impact on wild populations.

In 2003, Jackie Greenfield and I published a study on the non-timber forest products industry in western North Carolina that contained an extensive report on the galax industry. Jackie obtained details on the packaging and distribution of galax, which are described here. Twenty-five galax leaves are bundled together with a rubber band, twist tie, or twine. The bundles are then packed into waxed cardboard boxes sized to accommodate 2,500 or 5,000 leaves. Sales are usually done by 2,500 or 5,000 leaf counts. These boxes are stacked on pallets and shipped by the tractor-trailer load to United States markets and ports. Galax is exported to many countries, including Belgium, China, England, France, Germany, Holland, Japan, and the Netherlands.

Galax prices fluctuate with the season, being higher in fall and winter than in spring and summer. Since 2004, the price paid to the harvester has ranged from $20 to over $100 per box of 5,000 leaves. In 2013, the average price paid to the harvester was $100 for a box of 5,000 leaves. Some dealers pay according to a grading system that includes leaf size, color, and quality. Dealers want leaves to be at least 3½ inches in diameter, dark green to dark red, glossy, and blemish-free.

The florist trade often receives boxes of 100 leaves in bundles of 10 or 25. In 2013, the wholesale price from the harvester directly into the florist trade was $14.50 for a box of 100 leaves. The wholesale price to the florists was $92 to $169 for 100 leaves. Retail customers were paying $36 to $42.50 for 25 leaves.

In 2013, there were very few nurseries advertising galax plants for sale. The prices were $72 for a flat of 100 plants and $5.50 to $6.99 for a single plant.

How to Grow

Due to the extreme popularity of galax, it is important to start cultivating this plant. Wild-harvesters who pull up galax, rather than simply cutting the leaves and allowing the plants to survive, have contributed to diminishing wild populations. There is very limited information available on how to grow galax, but there are a few nurseries doing it, and research is being conducted. Some of the following observations and recommendations provided here come from the research conducted by Dick Bir, retired extension specialist in the Department of Horticultural Science at North Carolina State University; Joe Conner, retired technician in that same department; and Claude Deyton, retired extension technician with North Carolina Agricultural and Technical State University, whose office was located in Yancey County, North Carolina.

Galax prefers moist, acidic soils (pH 4.2 to 4.5) that are high in organic matter. If the soil is not naturally high in organic matter, a high-quality compost should be incorporated prior to planting. It is assumed that mycorrhizae (soil fungi that develop symbiotic relationships with plants) are essential for galax to prosper. Galax will grow in full-to-partial shade in hardwood forests or mixed hardwood and pine forests. To the best of my knowledge, successful cultivation in full sun has not been accomplished. If natural shade is not available, it can be grown under 50 to 80 percent shade provided by polypropylene shade cloth. Galax will tolerate dry conditions and a wide range of soil types. The buds on the rhizomes do have a chilling requirement that must be met each winter to resume growth. Tests on seedlings revealed that 30 days at 45°F was adequate to stimulate bud break on 95 percent of the plants.

Propagation and Production

Plants can be propagated from seed by sowing them in a well-prepared nursery bed in autumn, early winter, or very early spring. Seeds can also be started in flats or pots placed in a shaded greenhouse or cold frame. Dick Bir, Joe Conner, and Claude Deyton reported that no scarification or stratification of the seeds was necessary for germination. They obtained good germination within 14 to 21 days by sowing seeds in sphagnum moss in a greenhouse during the winter with temperatures set at 65°F at night and 75°F during the day. Seeds were kept moist with an automatic mist system or by holding the pots in a plastic humidity chamber.

Plants can also be propagated by rhizome division, preferably in early spring. Dick, Joe, and Claude found that bud break was dramatically increased by soaking the rhizome pieces in water at room temperature overnight before planting them. They

This is what a wild-simulated galax production area might look like. Photo by Jackie Greenfield.

concluded that seed propagation into a nursery bed, followed by transplanting into production beds several years later, was the most practical method for growing galax. They do caution that galax is very slow growing, no matter which method is used. It is estimated that it takes at least four years for galax to reach maturity from seed.

When the galax plants are ready to be set into a production bed, they should be spaced 18" to 24" apart in beds and mulched with several inches of leaves, composted sawdust, or hardwood bark. Once established, the plants should be left undisturbed for at least four years. The only pest problems encountered so far are slugs, whiteflies, and a leaf spot.

Potential for Profitability

I do not have estimates for yields or economic returns for cultivated galax. There are growers interested in producing galax to sell as a finished product (leaves) and as planting stock. As the populations of native galax decrease, we can expect to see a rising demand for leaves from cultivated plants. Whether galax can be produced profitably for this market is yet to be seen, but clearly, the first growers to have galax available for the market will have the greatest chances for economic gain.

23

Mayapple

Plant Description and Range

Mayapple (*Podophyllum peltatum*) is an attractive and unusual plant that many gardeners want in their wooded areas. It emerges in early spring and produces a one- to two-foot-tall stalk that forks at the top to bear two large, shiny, umbrella-like leaves. The leaves are 6 to 12 inches in diameter, round in outline, with usually five to seven deep lobes on each leaf. Hidden underneath these leaves, a single white flower appears between April and May. The flower develops into a small "apple-like" fruit, which starts out green and turns yellow as it ripens. Usually, a few plants in each population of mayapple produce only one leaf and do not flower. The underground, horizontal rhizome of the mayapple, which is the part of the plant that is used medicinally, is reddish brown and can grow very long. Rhizomes of wild plants are often two to three feet long, and there have been reports of rhizomes over six feet long. The rhizome is generally smooth and flexible, round, and about the thickness of a pencil. The rhizome is dark brown on the outside and white on the inside. Along the length of the rhizome are thickened areas with stem scars on the top and masses of fibrous roots on the bottom. In the discussion of mayapple, the rhizome and roots will be referred to as the root.

Mayapple is known by various names in different parts of the country. Some of the more common names include mandrake, umbrella plant, Indian apple, and devil's apple. The mayapple fruits are sometimes called raccoon berries. The unripe, green fruit of the mayapple is poisonous, but the ripe yellow fruit is edible and has a pleasant, strawberry-like taste. In some areas, ripe mayapple fruit is made into jam and preserves, but, according to Stephen Foster (who has written widely on medicinals), all other parts of the plant can be fatally toxic.

Mayapple grows in moist hardwood forests, along forest edges, in open areas in the woods, and in moist meadows. Very large

The mayapple fruit hangs below the foliage. It is edible. Photo by Tyler Nance.

patches can be found on steep hillsides in a mountain forest or near streams. Mayapple grows naturally throughout most of eastern North America, west to Minnesota and Texas, and south to Florida. Although this plant appears to be abundant at this time, there is concern about the renewed interest in mayapple; United Plant Savers has listed it on its "plants to watch" list.

Traditional Uses

Native Americans traditionally used mayapple to expel parasitic worms, as a laxative, and to treat skin tumors and warts. Early settlers used it to treat typhoid fever, dysentery, hepatitis, and cholera. Non-medicinal uses for mayapple are as a poison and an insecticide. For example, the Cherokee soaked corn seed in an extract from mayapple roots before planting the seed to keep crows and insects away. Similarly, a decoction of the whole mayapple plant was sprinkled on potato plants to kill potato bugs. Many herbalists do not use mayapple because of its toxic properties. Dried mayapple root powder is a very powerful laxative and should not be taken except under the direction of an herbalist experienced with its use.

New Market Opportunities

Although mayapple is rarely used by herbalists today, it is now of interest to some researchers. According to H. Becker with the USDA, *Podophyllum* species contain a compound called podophyllotoxin, which is used for making cancer-fighting chemicals. Etoposide is a semisynthetic antitumor agent derived from podophyllotoxin and is the active ingredient in a number of chemotherapy drugs, e.g., Toposar, VP-16, and VePeside, used for the treatment of lung, breast, ovarian, brain, stomach, pancreatic, lymphoid, and testicular cancer. C. T. Lee and her colleagues at the Natural Products Laboratory at the University of North Carolina at Chapel Hill have reported that modified derivatives of podophyllotoxin had anti-HIV activity.

Previously the major source for podophyllotoxin was the Asian mayapple (*Podopyllum emodi*). This species has now been over-collected, is considered endangered in India, and export of the plant has been restricted. As a result, attention has turned to the abundant wild populations of the American mayapple (*P. peltatum*). In the past, the American mayapple was considered a less desirable source of podophyllotoxin. Scientists working at the University of Mississippi in Oxford, however, developed a method for extracting podophyllotoxin and found our native North American mayapple can yield more podophyllotoxin than any other source. Dr. Rita Moraes and colleagues at the University of Mississippi discovered that the leaves of mayapple contained higher concentrations of podophyllotoxin than the rhizomes. They concluded that this makes American mayapple an excellent candidate for agricultural production of the antitumor compound.

Growing Mayapple for Leaves

It would be desirable to quickly develop a cultivated source of American mayapple rather than risk destroying the natural populations of yet another native medicinal plant. The late Dr. Kent Cushman, and Dr. Muhammad Maqbool, both formerly at Mississippi State University, created a cultivation system for mayapple, particularly for leaf production. In one study, they looked at the effect of shade on mayapple growth and leaf yield. They grew mayapple under full sun or under little polypropylene-covered tunnels that provided shade from 30 to 80 percent. Results showed that the more shade there was, the larger the leaves grew. Unfortunately, the podophyllotoxin content in leaves decreased with increasing shade. The leaves of the plants in full sun died back much earlier than those growing under shade. They did not look at the effects on root growth. There was also a study examining the influence of different kinds of mulches on mayapple. They found that use of bark mulch resulted in better plant growth than straw mulch.

General Production Guidelines

Mayapple can be easily cultivated in zones 3 to 9. I have been quite successful cultivating mayapple following the recommendations for goldenseal. We produce it in raised beds under about 78 percent shade in moist, fertile, well-drained soil with a soil pH around 5.8, covered with a hardwood bark mulch. Mayapple is easy to propagate by dividing the rhizomes into 2" to 3" pieces, each containing a node, and spacing them 12" × 12" apart. We have not attempted seed propagation, and I do not know anyone who has done that on a commercial scale. Mayapple is one of the earliest plants to die back in the summer, so you should not be alarmed

Freshly dug mayapple roots. Photo by Jeanine Davis.

if the foliage starts to turn yellow and die back in late July or August.

I do not have any data on how long mayapple has to grow before it yields economically feasible roots. Based on our preliminary studies, I'd guess it will take about four years from planting the rhizome pieces until harvest. The roots of mature mayapple plants are harvested in the late summer to early fall. The roots are washed and can usually be dried whole, in the same manner as goldenseal, at 80° to 95°F with lots of airflow. It takes about 3½ pounds of fresh mayapple roots to make one pound of dried root. The only data on leaf yields are too preliminary to estimate what they would be on any kind of commercial scale.

Disease

In 2002, I encountered my first experience with disease on mayapple. It expressed itself as clusters of bright orange spots on the undersides of the leaves and was diagnosed as a rust specific to mayapple (*Puccinia podophylli*). Unlike most other rusts, it does not have another host. It does not appear to cause any real harm to the plants, but it would lower the value of leaves and might eventually reduce root growth. The pathologist recommended that we remove and destroy all affected leaves. I reasoned that the disease organisms were probably also in the mulch. So in the fall, we dug up all the plants in the affected beds and replanted them in a new area. We destroyed the mulch in the diseased beds and allowed those beds to lie fallow for a year before replanting another plant species in them. Rust symptoms did not appear on the transplanted mayapple the next year. But it did reappear the following year. Since then, the rust appears sporadically in the mayapple beds, both in the woods and under artificial shade, infecting 10 to 80 percent of the leaves. But the plants continue to grow, flower, set fruit, and put on good rhizome growth as evidenced by the increasing size of each plot.

Markets

The popularity of mayapple in the marketplace has gone up and down over the years. Prices paid by dealers to harvesters for dried mayapple root were only $0.10 per pound in 2004. The prices increased substantially to $2 to $5 per dried pound by 2013. A recent quote I found on a trapper forum about wild-harvesting mayapple is "You would have to pick dump truck loads of mayapple to make much money."

A harvest of mayapple roots. Photo by Agatha Kaplan.

Some farmers, however, sold small volumes of mayapple root directly to processors and small herbalists, and got upwards of $7.25 per dried pound wholesale in 2007. In 2013, this wholesale price had risen to $8.50 to $10.50 per dried pound. Although the demand is small, dried mayapple root sells in the retail market for up to $16.50 per pound.

While waiting for the large wholesale market for dried mayapple root or leaves to develop, growers can sell mayapple through other channels. Mayapple is an attractive forest plant, and a single plant retails at local nurseries for up to $7.95 per plant. There is room to develop this plant as a landscape plant for forested areas, and with some promotion and consumer education, it could become a popular nursery plant.

My suggestion for growers interested in producing mayapple on a large scale is to start small and learn how to grow the plant efficiently. Estimate what your yields and costs of production will be so you can calculate what price per pound you will need to obtain to make a profit. Then keep a close watch on the research being conducted in Mississippi. In particular, try to stay informed of announcements of any new advancements, patents, or interest by large pharmaceutical companies. That way, perhaps you can be one of the first commercial producers of mayapple for any new products that might develop from the ongoing research.

24

Pinkroot

Plant Description and Range

Pinkroot (*Spigelia marilandica*) is also known as Carolina pinkroot, Carolina pink, Indian pink, American wormroot, and wormgrass. It is a beautiful plant, which usually reaches a height of one to two feet with angular stems arising from a perennial rootstock. The two- to four-inch long leaves are ovate in shape, positioned opposite each other on the stem, and are joined right to the stem without a petiole. The plant has very showy flowers that bloom from early spring through late summer. They are produced on the ends of the stems in a kind of one-sided spike. The flowers are one to two inches long and tube-shaped. They are a striking red on the outside and yellow on the inside. The root is very small, only between one and two inches long and usually less than ¼ inch thick. It is dark brown and rather twisted, with bud scars on the top side and long, light-colored fibrous roots on the underside. In discussing pinkroot, "root" will refer to both roots and rhizomes. Pinkroot is scarce, but can be found growing naturally in woods and thickets throughout the eastern United States, mostly in the southern states as far west as Texas. It is on the United Plant Savers "plants to watch" list and is on the protected plant list in several states, including North Carolina.

Uses

Pinkroot has a long history of use as a medicinal herb among Native Americans. The Cherokees made a decoction of the roots as a treatment for worms. The colonists in the South learned how to use the plant medicinally from the Cherokees and the Osages. The early settlers made a tea from the leaves to aid digestion and from the roots to treat malaria. Its most important use, however, has always been to rid the intestines of worms. The literature indicates that there was a fair amount of controversy over the effectiveness of pinkroot. Some physicians reported that it did not work.

Cultivated pinkroot. Photo by Agatha Kaplan.

Others claimed that only fresh root was effective. There were also reports that due to adulteration and substitution with other species, pinkroot preparations were often ineffective. Whatever the reason, pinkroot use has been minimal since the early 1800s. Presently, pinkroot is getting some renewed attention, and a tea made from the root is sometimes used to aid digestion.

How to Grow

There is little information published on how to grow pinkroot as a forest plant. It is considered hardy in zones 5 to 10, and I have found it easy and satisfying to grow. It is simple to propagate by root divisions and grows well following the same cultivation recommendations provided for goldenseal. I do not know the optimum soil pH or plant spacing as yet, but we have been successful with a soil pH of about 6.0 and a plant spacing of 12" to 18". The roots are harvested in the fall. They are washed, dried, and packaged like ginseng or goldenseal.

There is a little more information on how to grow pinkroot as a nursery plant. Interestingly, it is considered difficult to propagate. I guess that is because in the nursery business stem cuttings are more commonly used than rhizome divisions. A study published in 2012 showed that stem-tip cuttings had an 85 percent success rate if bottom heat was supplied below the flats of cuttings.

Markets and Profit Potential

Currently, there is no information available on root yields or length of time required to grow the plant before it is ready to be harvested. Considering how small the rhizome is, I question whether it would ever be profitable to grow it for dried root sales. Dealers paid between $5 and $7 per dried pound for the root in 2007 and it was still $5 per pound in 2013. Some harvesters are selling it wholesale direct to herbalists for up to $57 per dried pound. It can retail directly to the consumer for $9.95 to $32 per dried ounce. The foliage of pinkroot is quite abundant, and there is an occasional market for the herb. If the market grows, production might be economically feasible. The herb should be harvested when the plant is in full bloom.

Early spring growth on cultivated pinkroot. Photo by Agatha Kaplan.

Pinkroot roots. Photo by Jeanine Davis.

Pinkroot flower. Photo by Agatha Kaplan.

In my opinion, the real market for this medicinal herb is as a nursery plant. In a woodland setting, pinkroot is stunning. Not many nurseries sell it, leaving room on the market for more growers to offer it. I have found potted plants being sold for $16 apiece and flats of 100 plants for $175 to $325. Pinkroot is so easy to divide and it grows so quickly that a grower should be able to produce large numbers of plants with little effort.

25

Spikenard

Plant Description and Range

Spikenard (*Aralia racemosa*) is also known as American spikenard, spignet, spiceberry, Indian-root, fleabane, and old-man's-root. This many-branched perennial shrub grows to between one and six feet tall. It has large, pinnately compound leaves, consisting of oval leaflets with doubly serrated edges. The plants bear numerous elongated flower stalks that can grow up to ten feet tall. Clusters of yellow-green flowers develop into round red-to-purple berries in late July and August. The rhizome is large and thick, with big bud scars and many long thin roots. In discussing spikenard, "root" will refer to both roots and rhizomes. A distinguishing characteristic of the root is a strong spicy odor. Spikenard grows naturally in woods along streams and rivers. It prefers a rich soil but also grows in very rocky soils. It is native to Canada and the United States and can found as far south as Georgia and west to the Rocky Mountains. It is included on the United Plant Savers "plants to watch" list.

Uses

Many Native American tribes have an extensive history of using spikenard. The Algonquins used it to treat diabetes and tuberculosis. The Cherokees used it to treat burns, wounds, coughs, asthma, and many kinds of menstrual problems. They also used it to stimulate labor. The Chippewas applied it to muscle aches and sprains. The Malecites used it for headaches. Many reports exist about its use for the treatment of backaches, colds, and skin diseases.

In western North Carolina, people refer to it as the backache plant. Interestingly, a 2011 study demonstrated the analgesic activity of spikenard. A 2009 study in western North Carolina reported the spikenard root extract has antitumor activity. Spikenard is an ingredient in several commercial homeopathic remedies.

Spikenard foliage. Photo by Agatha Kaplan.

Spikenard has attractive foliage, making it a useful landscape plant. Photo by Jeanine Davis.

How to Grow

Spikenard is easy to cultivate and can be grown in zones 3 to 7. Although I have found no information on producing it commercially, we grow it in our demonstration gardens and research plots the same way that we grow goldenseal and black cohosh. We divide the rhizomes in late fall or very early spring and plant them on an 18" × 18" spacing in raised beds. We adjust the soil pH to between 5.8 and 6.5 and incorporate high-quality compost.

A young spikenard plant sprouting from its large lateral rhizome. Photo by Agatha Kaplan.

Then we mulch it with bark or leaves and let it grow. It is usually a very carefree plant and quickly shades out any weeds. We have not seen any insects on it; however, 2003 and 2004 were extremely wet years in western North Carolina, and some of the spikenard went down with what appeared to be a leaf blight. We cut back the diseased plants and removed the affected plant material from the area. The disease did not return the following year.

Spikenard roots are traditionally harvested in summer and fall and can be used fresh or dried. They are easy to wash and can be dried in a forced-air dryer at 90° to 95°F. The berries, which are used in some traditional preparations, should be collected when ripe.

From the flower spike emerge bright magenta berries in the fall. Photo by Agatha Kaplan.

Markets

The current market demand for dried spikenard root is small but steady. Prices, however, had been volatile. In 2007, dealers paid harvesters $1.50 to $2.50 per dried pound. The price rose to $4 per pound in 2011 but dropped dramatically to $1 per pound by September 2013. Wholesale prices in 2013 were $18.30 per dried pound and spikenard retailed for over $47 per pound. This attractive plant creates a tall, lacy screen with beautiful tall flower spikes. It is not a well-known plant, and only a few nurseries carry it. Individual plants retail for $6 to $10 each, but with some promotion, demand and prices could probably be increased.

26

Wild Ginger

Plant Description and Range

Wild ginger (*Asarum canadense*) is a low-growing, perennial plant that is usually only between six and twelve inches tall. The plants grow in dense groups with an extensive rhizome system. Two leaves arise on long, slender petioles from nodes on the rhizomes. The leaves are dark green, broadly heart-shaped, and usually two to four inches across. Occasionally, plants with very large leaves—up to eight inches across—can be found. The leaves release a wonderful ginger odor when brushed or crushed. Purple-to-brown, bell-shaped flowers develop close to the ground and are usually hidden under the leaves. The flowers look like deep little bowls, one to three inches in diameter, with three long, pointy curled lobes. The self-pollinated flowers eventually develop into leathery six-celled seed capsules. The seeds of wild ginger are naturally dispersed by ants. The rhizome is rather small, about ¼-inch thick, yellowish in color, with roots growing from nodes spaced closely down its length. The dried root has a spicy odor very similar to the ginger root you buy in the grocery store, but it is not related.

Wild ginger is a common plant that grows in the fertile soils of forests and forest roadsides in eastern North America from Canada, south to North Carolina, and west to Kansas. It prefers dark, cool, moist woods with rich soils high in organic matter.

Names and Uses

Wild ginger is known by many different names, depending on the region. They include Indian ginger, Canada snakeroot, Southern snakeroot, black snakeroot, heart snakeroot, snakeroot, coltsfoot, false coltsfoot, catfoot, colic root, and little brown jug.

Native Americans have used wild ginger medicinally for a long time. There are records of the Cherokees employing it to treat menstrual problems, stomach pain, and swollen breasts. The Chippewas

Wild ginger leaves. Photo by Agatha Kaplan.

applied wild ginger roots to bruises and contusions. The Iroquois took it for urinary disorders, and, according to one report, to prevent bad dreams caused by the deceased. If the Meskwakis found a dead animal they wanted to eat, they cooked it with wild ginger root to prevent ptomaine poisoning. Some of the traditional folk applications for wild ginger were for gas, heart palpitations, fevers, and nerves.

Present-day uses are as a peripheral vasodilator and as an indigestion reliever. Wild ginger contains aristolochic acid, which clinical trials have shown promotes healing and prevents infection. The plant has also been used to reduce the toxic effects of commercial antibiotics. However, Richo Cech informed me that some people experience nausea and vomiting after ingesting wild ginger root or root tea. He also said that wild ginger is a renal irritant, potentially toxic, and regulated by the FDA.

How to Grow

There is little information available on how to grow wild ginger, and none that I can find on commercial cultivation. It can be grown in zones 3 to 8. It likes deep shade, at least 80 percent. Plant it in a moist site with acidic soil high in organic matter. It propagates easily by rhizome cuttings. Use pieces between 2" and 3" long, ideally with each containing a node. Space the plants about 6" apart. Given the right conditions, wild ginger will quickly occupy a bed. Wild ginger seeds need a cold stratification period of three weeks or longer. Sow fresh seed in a nursery bed so it can be exposed

Wild ginger flower on the ground at the base of the ginger leaves. Photo by Agatha Kaplan.

to the warm temperatures that season and the cold of winter, and expect to get about 50 percent germination the following spring. I have not harvested enough roots yet to know what the yields might be, but our plants are ready to be divided every three to four years. The roots are harvested in the fall and are used fresh or dried. Some herbalists prefer to use the leaves, claiming they are more potent than the roots. The leaves can be harvested anytime.

Markets

The market for wild ginger root is small and steady, but prices are low compared to most of the plants in this book. Harvesters received about $0.50 per dried pound of root in 2004, and that only increased to $1 per pound by 2013. Wholesale prices in 2013 were $5.75 to $12.25 per dried pound, and retail prices were $17 to $19 per pound.

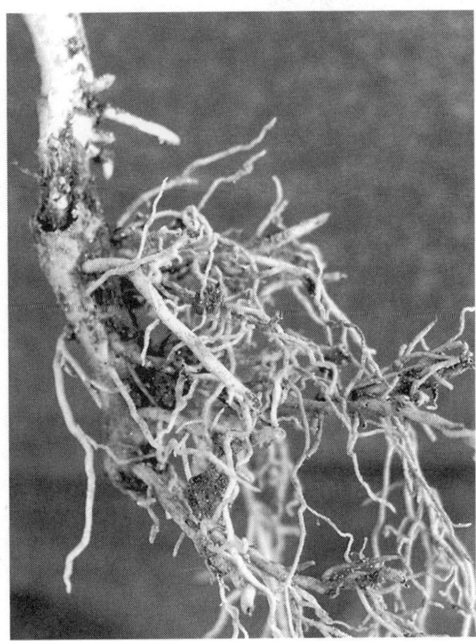

Wild ginger root. Photo by Agatha Kaplan.

I think that this is another plant where the profit potential is in nursery production. Wild ginger plants are popular among home gardeners because they grow under dense shade in moist areas and form a beautiful carpet of green very quickly. The unusual little flowers add interest and value. Individual potted plants retail for $4. Wholesale, bare root plants sell for $36 or more per hundred.

27

Wild Indigo

Plant Description and Range

Wild indigo (*Baptisia tinctoria*) is a bushy little herbaceous perennial that grows to two to four feet in height and about two feet across. The plant has many branching stems that contain small, pea-like leaflets, about ½-inch long, that are grouped in threes on delicate stems. On the ends of most of these stems, small yellow flowers bloom for a long time in the summer. The flowers eventually develop into attractive brown to black pods. The root is large, rough, and irregular, with a thick black bark on the outside and a yellowish interior. It has many long fibrous roots and is bitter to the taste.

Wild indigo can be found growing naturally in the eastern part of North America, from Canada to Florida, and west to the Mississippi River. It is most plentiful on sandy soils, particularly near the coast. It has a few other common names, such as yellow indigo, false indigo, dyer's baptisia, and rattle bush.

Uses

Wild indigo has a long and, literally, colorful history, since it was once used to make a blue dye. According to Millspaugh, "The young shoots of this plant resemble, in form and general appearance, those of asparagus, and are used, especially in New England, in lieu of that herb for a pottage." Harding explains that in Virginia and Maryland, farmers tied bunches of wild indigo to horses' harnesses to keep flies away. However, wild indigo is mostly known as a treatment for fever, sore throats, and typhus; as an antiseptic to treat wounds; and as an immune system enhancer. The Native Americans made good use of this plant. The Cherokees employed it to treat toothaches and as an emetic and purgative. The Micmacs collected the root to treat gonorrhea and kidney trouble. There has been a small surge in popularity for wild indigo for the prevention of colds and the flu. Wild indigo contains certain polysaccharides and

Wild indigo has small, pea-like leaflets. Photo by Jeanine Davis.

The small yellow blossoms of the wild indigo. Photo by Agatha Kaplan.

proteins that some believe are responsible for the immune system's response. Modern herbalists report that the bark of the root has antiseptic properties, which makes it useful as a wash for skin ulcerations or a gargle for mouth sores.

How to Grow

Wild indigo grows in zones 3 to 9 in partial shade to full sun. I have grown it successfully with other woodland botanicals under 80 percent shade, but the excessive shade does slow down its growth. Most impor-

Wild indigo root. Photo by Jeanine Davis.

tantly, it prefers a drier soil than the other plants described in this book. If grown in soil that stays moist for long periods, wild indigo will develop root rot that will quickly kill the plant. A sandy soil is best. If that is not available, do everything you can to improve water drainage. Work organic matter, and perhaps some sand, into the soil; build extra high beds; and choose a site with a slope.

Propagate wild indigo by dividing the root into 2" to 3" pieces in the fall or spring. Plant these root pieces about 18" × 18" apart in raised beds, mulch them heavily with bark or leaves, and keep them weeded, especially the first year. I have not experienced any insect or disease problems on wild indigo so far. The plant can also be propagated by stem cuttings in early spring before the stems get woody.

I have successfully grown some wild indigo from seed by sowing seed into a well-prepared seedbed in the late summer, when the seed pods are dry. Percent germination and survival have been low. My former technician obtained much better results by chilling the seeds at 40°F for three months, scarifying the seeds by scratching them against sandpaper, soaking them in warm water for 24 hours, and sowing them

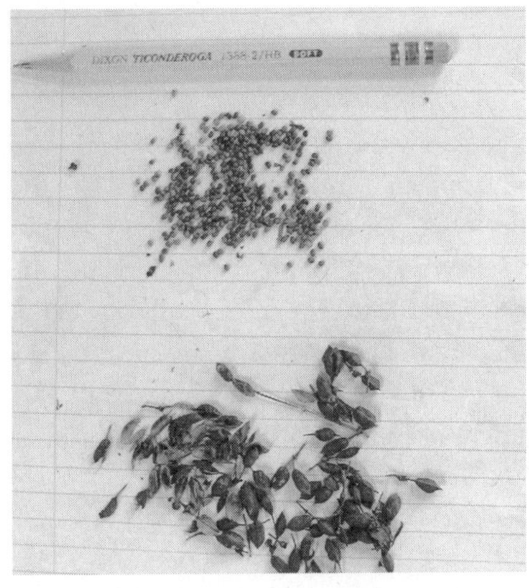

Wild indigo seeds. Photo by Karen Hardy.

Wild indigo transplants grown from seed. Photo by Karen Hardy.

into flats with bottom heat. Scarification or soaking alone did nothing, but the two together significantly improved germination.

Roots are harvested after three to four years of growth. They are usually dug in the fall, although a few buyers request spring roots. The roots can be quite large and might need to be cut to facilitate washing and drying.

Markets

The market for wild indigo is small but steady. Interest in it as an immune system enhancer could mean higher volumes will be required in the future. In the meantime, it can be sold as a very attractive ornamental plant for shaded gardens. Prices paid to harvesters have changed dramatically over the years. Prices paid for dried root were $2 to $4 per pound in 2004, $16 per pound in 2010, and $8 to $12 per pound in 2013. Selling wholesale direct brought $42.75 per dried pound in 2013, and retail prices ranged from $49 to $71.93 per pound. Attractive in the shade garden, a single plant sold for $8 to $9 in 2013.

28

Other Forest Botanicals Growers' Stories

Joe Hollis: Small-scale Specialty Nursery

I first interviewed Joe for this book in 2003. I called him ten years later, in August of 2013, to get an update. The original interview is included here in its entirety followed by that update. I hope you enjoy them both.

My First Interview with Joe Hollis in 2003

Joe and I have known each other for many years. He lives only a few hours from me, and we are involved in many of the same herb, organic, and permaculture activities. Permaculture is a simple lifestyle in which the gardener strives to create a landscape that is self-sustaining, supplying much of the food, fuel, and materials for shelter and clothing that the gardener needs. Joe is truly knowledgeable about native plants and medicinal herbs and is one of the few growers I know who makes all his own herbal preparations. Joe also grows Chinese medicinal herbs and was one of the first to do so in this country. I want to share Joe's story because he grows all the plants I have discussed and because he is someone who really enjoys what he does. His permaculture approach to life means his needs are simple. As a result, he does not have the constant concerns about making money that most of the rest of us have. This is not a story about how to get rich growing medicinal herbs, but I think his approach to life and plants is unique and interesting.

Joe's Place

Joe lives in the tiny community of Celo, not far from Asheville, North Carolina. This community is comprised of people who try to live in harmony with their environment. They tend to be creative, intelligent, rather eccentric people who take an active role in trying to bring about change in the world. Joe is one of these people, and he takes a quiet, gentle approach to life and to educating people about the plants around them.

Joe Hollis at his farm in Celo, North Carolina. Photo by Karen Hardy.

I visited Joe on a bright, fresh, spring morning in 2003. When I asked my employees if anyone wanted to come along for the ride, three of them quickly piled into my pick-up truck with big smiles on their faces. They didn't want to miss the opportunity to see what delightful spring plants were emerging in Joe's famous gardens. Yancey County, where Joe lives, is rural, mountainous and, at this particular time of year, very green. We had a pleasant drive over winding country roads, leaving the noise and activity of Asheville behind us, as we headed deep into the quiet green haven that Joe calls home.

Joe owns about three acres of heavily wooded land on a steep hillside. His property butts right up against the Pisgah National Forest at an elevation of about 3,500 feet. I asked him if the forest rangers

Interesting architecture on Joe's farm. Photo by Karen Hardy.

Joe's apothecary. Photo by Karen Hardy.

minded that his hammock actually hung in the National Forest! We parked the truck on the narrow road below his property and walked the steep trail up past his gardens to his greenhouses and shop. Along the way, we passed a quaint, hand-hewn log cabin, a twisted vine arbor in terraced gardens, a permanent yurt, several glass cold frames, and an attractive wood-frame home. My technicians toured the gardens and greenhouses while Joe and I made ourselves comfortable in his library/shop/apothecary.

The Permaculture Lifestyle

Something you need to keep in mind as you read about Joe and his business is that he is a permaculturist and a homesteader. He lives very simply, and his costs of living are nominal. He supplies his own power and water. As a result, the only monthly bill he receives is for his telephone and Internet service. He raises and processes much of his own food. This kind of lifestyle is certainly not for everyone, but he is content with it, and his home and gardens are comfortable. He has almost everything he needs right there on his three little acres. He does not make much money, but he doesn't require much.

How He Got Started

Joe bought his little piece of heaven about 30 years ago when he returned to the United States after a stint in the Peace Corps. Knowing he wanted to make a living off the land, he tried growing organic fruits and vegetables, but soon determined that was not his passion. Joe is a voracious reader, as evidenced by his extensive library, and he started reading about herbal medicine.

Joe's library. Photo by Karen Hardy.

He soon found himself fascinated by medicinal plants. He learned everything he could about them and how they are used. He visited knowledgeable people and started studying the plants themselves in cultivated gardens and in the wild. Soon he began collecting and growing native medicinal herbs. Then he studied to be an herbalist. He was really enjoying himself and had a good feel for the plants and how to use them. His friends encouraged him to start a commercial medicinal herb farm because there was no one else around doing it. Finally, he decided to give it a try. In time, he became interested in Chinese medicinals and started growing those, too. He was one of the earliest cooperators in Chinese medicinal herb trials coordinated by Jean Giblette of High Falls Gardens in Philmont, New York.

Mountain Gardens

Joe named his new business "Mountain Gardens." It is a diverse business through which he sells plants directly from his farm and at local festivals and herb events, runs a mail-order seed company through a paper catalog and a website, teaches workshops on growing and using herbs, makes and sells tinctures and Chinese prescriptions, and wild-harvests herbs to order. He has a shop and apothecary on-site and welcomes visitors to wander his gardens, greenhouses, and library. He also supplies wildflowers and native medicinal plants to local garden centers. He has a little over one acre of land in production in addition to a variety of greenhouses, overwintering structures, and cold frames.

Mountain Gardens is in an ideal location. It is within an easy drive of Asheville,

North Carolina, where there is a high concentration of people interested in alternative lifestyles, herbal medicines, vegetarian diets, sustainable agriculture, and natural health care. This community is very supportive of his business, and he thinks this is an important point to make. If he were located outside of a similar-sized, conservative, traditional city, he probably would not have as many customers.

How to Make Money from Native Medicinal Plants

Joe is still trying to figure out what is profitable for him to do and what is not. For example, he was asked to wild-harvest five pounds of ramps for $5 per pound. It sounded like a good deal at the time, because Joe knew where he could find a ready supply of ramps. However, it took many hours to harvest, wash, pack, and deliver the ramps, and in the end, he realized it was not worth the amount of time it took. In contrast, shortly thereafter a European company making homeopathic remedies asked Joe to harvest about six and a half pounds of blue cohosh for $38.60 per pound. For about the same amount of time and effort, this was a much more lucrative project.

The Internet has introduced Joe and Mountain Gardens to a large number of people who might never have known about his business otherwise. Native plant chat rooms and online discussion groups have really helped to build his reputation. He is encouraged by the contacts he has made online and redesigned his website to attract more business and allow him to take orders online. His website highlights his most unusual and specialty items, such as wasabi (Japanese horseradish), Dong Quai, false unicorn, Asian ginseng, ramps, and small-flowered willow herb.

Presently, Mountain Gardens is not what Joe would consider profitable by most people's standards. He reinvests almost everything he makes back into the business. He is able to do this because his own living expenses are so low, but he has plans to expand and to make Mountain Gardens profitable. For example, he intends to include more agri-tourism and educational activities. He has found that people like to make the drive through the mountains to his place and spend some time in his quiet gardens surrounded by the forest. Joe is having a professionally designed brochure made and intends to do more advertising. He plans to start concentrating on a few specialty plants and really develop a reputation for being the best source for them. He also wants to do more teaching and increase the Chinese medicine portion of his business.

One example of a specialty plant that Joe is going to develop is what he calls "Southern ginseng" (*Gynostemma pentaphyllum*). It is a rampant, perennial vine in the Cucurbit family that the Japanese use. It is very high in ginsenosides and has properties similar to ginseng. The leaves and stems are dried and made into a tea that tastes like ginseng. Currently, he grows it for a company and thinks there will be an expanding market for it.

Shoppers at one of Joe's greenhouses. Photo by Karen Hardy.

Apprentices' Help

One of Joe's methods for keeping operating expenses low is to accept apprentices who help with all the hard work, such as planting, weeding, and harvesting. He provides room and board in return for giving his apprentices the opportunity to work and live with a very knowledgeable herbalist and plantsman. They help him accomplish things that he could not afford to do otherwise, such as topping all his ginseng to prevent theft. He likes having five people working for him at any one time but is finding it harder and harder to get good workers. He thinks this is because so many farms in the country are now using apprentices. When he first started doing it many years ago, this was a new concept. Now there is competition. He is trying to figure out a way to provide a stipend to make his operation more attractive to potential apprentices.

Advice to New Growers

Joe grows more than a hundred different kinds of plants and has thousands of plants to tend. He loves caring for all of them and considers it great fun. His advice to others, though, is that focusing on a much smaller number of crops and building your business around them is really the way to make money. Learn how to grow quality plants efficiently, and then market them aggressively. The plants you choose to grow should be appropriate for your site and fit well with the native vegetation in your area. Look at the plants growing around you. Which ones are useful? Are they ornamental, medicinal, or a source of nutrition or fiber? Are there plants from similar bioregions in the world that you could grow? For example, there are many plants in East Asia that are adaptable to eastern North America. One is balloon flower (*Platycodon grandiflorum*).

It is a Chinese medicinal herb with a flower that looks like a purple balloon in the bud stage and opens up into a beautiful star. The plant is already used as an ornamental in North America. In China, it is used as a medicinal to warm the body, and in Korea the roots are eaten.

In the end, I do not really expect Joe to take his own good advice, because he is having too much fun growing all the different kinds of plants he does. Since he does not owe anybody anything and lives simply, he will probably just keep doing what he's doing. It is a wonderful lifestyle that he savors.

Follow-up Interview with Joe Hollis in 2013

It was fun talking to Joe again. Although we see each other frequently, we don't often have time to just relax and talk. It was interesting to read over the interview from ten years ago and talk about what has changed in Joe's life and at Mountain Gardens.

Joe has now been on his little piece of paradise for 41 years. His Mountain Gardens business is growing and evolving as any good business should. He still greatly enjoys what he does, although he doesn't do everything the same way he did ten years ago. One big change he has made is that he doesn't travel for speaking engagements as much as he used to. He prefers to have the students come to him so they can learn with the plants all around him. He does still teach one class a semester at the Daoist Traditions School in Asheville, where he also put in a Chinese herb garden, but don't expect to see Joe at all the conferences that he used to attend.

I asked Joe if the phone and Internet were still his only bills. He said that no, he made a mistake when he told me that ten years ago. He actually has two more bills that he has to pay: insurance for his truck and taxes. That's still not much compared to most of the rest of us!

Changes at Mountain Gardens

One of the biggest and most positive changes for Joe with his business was getting rid of the paper catalog and going completely electronic with his mail-order business. He says that has been a really good thing; it is cheaper, less wasteful of natural resources, and he doesn't have to store all those catalogs. He has fully embraced the Internet business model.

Ten years ago, he had a little over an acre in production. That has expanded to one and a half acres now, with most of the expansion going to food production. Wild foods, wild mountain vegetables, and perennial vegetables in general are now bigger parts of his nursery business, gardens, and educational programs. He is excited about this new aspect to his business and will be offering more seeds and plants for them in the coming years.

He had been very excited about Southern ginseng (*Gynostemma pentaphyllum*) when we did the interview, so I asked how that had worked out. He was happy to report that it has done well. He sells lots of it, and it is flourishing—maybe a bit too much, because it is taking over parts of his

Joe's living, working, and growing spaces are intermingled in a comfortable manner. Photo by Karen Hardy.

garden! He knows he could do more with this favorite little plant if he could just find the time.

Chinese herbs continue to grow in popularity, and he is expanding what he grows and provides each year. He will put several dozen new herbs in beds this year. Longevity herbs are his current interest, and he is looking for longevity herbs that are hardy in his environment.

There has been interest in wasabi production in western North Carolina for many years, and at least three other farms have tried to make a business out of it, but Joe is really the only one in the Southeast region doing anything with it. Because it is so popular, it brings in a nice little income for him. He can make $3,000 to $4,000 a year from wasabi without much effort. It is one of his most lucrative crops. But before all my readers run out to start a wasabi business, you might want to try to grow it on your own for a few years. I happen to know why two of those other wasabi businesses did not make it, and it wasn't from lack of trying. Wasabi is a finicky grower; but Joe has figured it out.

Apprentices Are Still Important

We also talked about his apprentices because that was a topic of great concern to him ten years ago. He currently has seven

Joe has a dedicated seed storage cabinet where everything is carefully marked and kept dark and dry. Photo by Karen Hardy.

apprentices working with him, and he now takes a different approach with them. He is trying to set up little businesses for each of them. For example, he has developed a nice little business selling bare root plants. They are easy and inexpensive to produce and ship, and his customers are pleased with them. So he wants to turn that business over to one or two of his apprentices to manage for themselves and make their own money from. Another business idea for an apprentice is to make a longevity tonic with domestically grown, organic herbs. He thinks he could convince more apprentices to stay around for a second year if they had that kind of vested interest. Second-year interns are important to Joe because they know the gardens, the seasonal tasks, and the rhythm of the business. They can help train the new apprentices in the basics, leaving Joe more time to teach the second year apprentices more advanced topics. It also helps build a community when there are overlapping generations of apprentices in the gardens.

Problems, Challenges, and Profitability

We talked about problems and challenges he has encountered over the years. There haven't been many, but I think a lot of that has to do with the fact that Joe is so laid back; he rides the ups and downs of life pretty smoothly. His biggest complaint during our interview was that his solar dryer didn't work too well this year, and that was because it rained almost the entire 2013 growing season. We didn't see the sun for two days in a row until September.

I reminded Joe that he said he was going to try to improve the profitability of his business and asked how that had worked out. He said he has done a little better on that account, but he knows his weak points and has not done much to improve them yet. Number one, he says, is he is not good at promoting the events he conducts. He offers lots of them, but they are not well-attended. His gardens make for such a great classroom, and the people that do come love it. So he needs to improve how he promotes them.

Finally, I asked if he had anything special he wanted to share with my readers, he said "I wish more people would raise wild-simulated ginseng!" But then he added, "Of course, it can really ruin your whole day/week/month when your ginseng gets ripped off. So, think about keeping a big dog, for starters."

John Kershaw: Ontario Farmer Producing Herbs on a Large Scale

In the goldenseal section of this book, there is a story about John Kershaw, a goldenseal grower in Ontario, Canada. When I called him to update his story in September 2013, I learned that he was growing much more than goldenseal. So here I've included what he shared about growing some of his other plants, from both the 2003 and 2013 interviews.

2003 Interview with John Kershaw about Diversifying His Farm with Other Woodland Botanicals

John started in the medicinal herb production business by growing goldenseal, but has now branched out into many different herbs, some of which are quite obscure. He grows these other herbs because the companies he regularly sells to tell him they have a market for them. Since there is so little information available, he often tries these crops without knowing very much about them except for what locals tell him about where they grow in the wild. To use his words, he "goes into it pretty blind." However, he enjoys the challenge and is willing to try to grow just about anything if he knows a potential profitable market exists for it.

In 2002, John started growing false unicorn root, black cohosh, bloodroot, and *Echinacea angustifolia*. He was still evaluating these new plants when I spoke to him and reported it has not all gone smoothly. For example, the first bloodroot planting stock he bought came from wild populations in North Carolina; it rotted in the ground. Since there is still quite a lot of bloodroot growing wild in his area, his most recent planting came from material he collected locally, and it has done much better.

He says there is a strong market for black cohosh, and it grows well for him; roots increase in weight five-fold in three years, so he has expanded those plantings. He was also told there is a growing market for false unicorn root, but he's moving into that crop with caution because of cost and because he is not as confident about its future as a cultivated crop as he is about some other crops. False unicorn root grows slowly, and he will be able to estimate the economic potential of the plant once he harvests his first crop after three years of growth.

John Kershaw grows black cohosh under artificial shade. Photo by John Kershaw.

2013 Interview with John Kershaw about Growing Other Woodland Botanicals

During the ten years since the first interview, John had expanded into other medicinal herbs. He has since quit growing some of those because he found that buyers didn't want to pay a reasonable price for cultivated herbs; they could get wild-harvested material so much cheaper. Considering the risks he had to take to cultivate these herbs, it just wasn't worth it. Over the past ten years, the industry has been very up and down. Demand for medicinal herbs is picking back up, which means he will probably expand some again, but he thinks he will be one of the few in his area to do so. Most of the farmers he knew growing medicinal herbs ten years ago have left the industry and returned to growing tobacco. Many of them now grow larger acres of tobacco and ship their crops directly to processors. They are making good money doing so. Those still growing medicinal herbs produce mostly ginseng because it is reliable and can bring a reasonable return per pound. The growers have improved

Beautiful cultivated black cohosh root. Photo by John Kershaw.

their production systems and can bring in yields of 4,000 dried pounds of ginseng per acre. There are also some farmers growing Echinacea, including John. A new crop in the area is sedum used for living roofs on buildings.

A major challenge for John since 2002 has been the dollar exchange; the United States and Canadian dollar are now almost equal. Back in 2002, there was a 30 percent exchange difference in the two dollars. This has really decreased the profitability of some of the herbal crops. Much of the larger volume herbs are now coming from eastern Europe where labor costs are much lower.

Black Cohosh

At one time, black cohosh was John's biggest medicinal herb crop. He produced an astonishing 21,000 pounds of dried root in one harvest, but he only received $3.50 per pound Canadian for it. It is hard to make money at that price. It is a difficult and time consuming root to wash because it has a very fibrous root system that hangs onto the sandy soil it is grown in. That large crop he harvested had over 2,500 hours of labor in it just to wash the root. This doesn't include the hours for garden teardown or the costs of drying and packaging. Even though black cohosh grows extremely well, with flower spikes reaching up to touch the shade cloth ten feet above them, and it yields an impressive 2,000 to 3,000 dried pounds per acre, if it doesn't bring at least $10 per dried pound, it doesn't pay for John to grow it. I asked John whether he had considered growing it in the open instead of under shade cloth. I thought that might make it cheaper to produce, and some of my researcher friends report that black cohosh will grow just as well in the open sun as under shade. But John reports that they have to grow it under shade in Ontario. If grown in the sun, they have to irrigate it constantly, so there is no savings. The shade structure, which costs $10,000 to $12,000 per acre, can be depreciated over time; irrigation water and management cannot.

He did mention that black cohosh and ginseng grow well together under shade. We have noticed that in our woods plots in North Carolina, too. We see less disease in both species when we grow them together.

Bloodroot

Even though bloodroot demand and prices are up, it is not John's favorite plant to grow or wild-harvest. He collects bloodroot for planting stock from wild populations and finds it almost always growing with poison oak and poison ivy. We have noticed the same thing in western North Carolina. The plants seem to attract each other. This is a real problem for diggers, however, and it can make bloodroot an unpleasant plant for employees to work with.

Years ago, John purchased a large amount of bloodroot rootstock from the United States, but it had all rotted after planting. As I explained in the bloodroot chapter in this book, I experienced similar losses myself, most recently in 2013. John's material had been of good quality when he planted it. He split the roots into pieces for planting and planted it the way he does

Sack full of bloodroot planting stock. Photo by Jeanine Davis.

goldenseal. The plants were completely gone the following spring. John doesn't have an explanation for this phenomenon, but if you want to grow it, John suggests you should always use local bloodroot, and you should plant whole roots instead of splitting them.

John's solution to dealing with poison ivy and dying rhizomes is to propagate bloodroot mostly by seed. In his area, seed is ready to collect just as he is making his first cutting of hay. It is like clockwork. He puts the fresh seed in a jar and keeps it in the refrigerator. He sows the seed in the fall and usually has a good stand the following spring. Unfortunately, sometimes he loses a large percentage of those plants over the next season. Still, it can be a profitable crop. He expects a yield potential of 1,000 pounds of dried root per acre, but at $15 per pound, the return is questionable in cultivation under shade.

False Unicorn

John likes to grow false unicorn. It's a challenging little plant that should be grown commercially before it is endangered in the wild. Dried root prices are worth around US$65 per dried pound. The soil on his farm is sandy and false unicorn thrives in it. Although he hasn't tested it yet, he has read that false unicorn likes the soil slightly more acidic than some of his other herbs. His method of propagation is very different from mine. I do not plant rhizome pieces; I only propagate from seeds. John has had no success with seeds and always uses rhizome cuttings.

He noted that frost can heave false unicorn out of the soil. The smooth hairs on the root cause it to be easily heaved. It needs a summer to set its feet in the ground before it starts to grow. This tells me that it should be planted a little deeper and mulched heavier. But John says if it is mulched too heavily, the plant sits in the straw mulch struggling to see the light of day. It will survive, but it's a struggle, so it really needs to have most of the mulch lifted off in the spring. Other than that, it is a very tough plant for its size and competes with other plants quite well.

We both laughed when he told me that he had talked to Joe Hollis in North Carolina about false unicorn earlier that day. Joe lives about an hour north of me and is another one of the growers in this section of the book. The medicinal herb world is a small one!

PART FIVE

Growing Woodland Medicinals in the Home Garden

An increasing number of the presentations I give on growing woodland botanicals at conferences and festivals are for home gardeners. Hands-on propagation workshops fill up quickly, and we now offer special tours for home gardeners of our woodland medicinal herb gardens at the research station. Usually, someone at these events has a copy of one of the older editions of this book that they brought along for an author signature. Even though the original book was written specifically for the commercial grower, it is being used by gardeners. Clearly there was a niche to be filled. So when it came time to revise the book again, I decided to add a section for home gardeners because there are some big differences in approach and expected outcomes between them and commercial growers. The biggest difference is there is no concern for profitability; gardeners can grow these plants for the sheer joy of it.

When I first moved to western North Carolina to take my position with North Carolina State University, we lived in a little house nestled in the woods on an oversized lot in a small neighborhood. For ten years we tried to grow vegetables in a small clearing near the house without much success, but we also planted native medicinal herbs and ornamentals throughout the woods, and they flourished. Over the years we built walking paths through the woods and created habitat for different kinds of plants. There was an area under the dogwoods near the "forest edge" planted with galax and shortia. Behind it grew a patch of mayapple that quadrupled in size while we lived there. We planted bloodroot, ginseng, goldenseal, wild ginger, blue cohosh, ramps, sassafras (*Sassafras albidum*), and Solomon's seal (*Polygonatum biflorum*). And I will never forget the excitement I felt when we discovered two pink lady's slippers (*Cypripedium acaule*) growing in a corner of the lot far from our paths. We carefully guarded that area and watched for them to emerge each year. Which they did!

My objective with this section is to help others create woodland medicinal herb gardens to enjoy with their families like we did. Woodland gardens fit well into busy lives. When properly established they will take care of themselves for the most part. Weeding is minimal and watering is usually only required in a drought. I have fond memories of walking the garden trails with my children looking for what had emerged or bloomed since we had last been through. My daughter, as young as four years old, always looked forward to finding the first "little brown jug" on the wild ginger plants each spring.

Keeping with the design of the previous two sections of the book, I visited and interviewed six amazing gardeners. What a joy it was to spend a few hours with each of them. And then I was delighted when Scott said he wanted to add a chapter to this section for growing ginseng in the home garden.

29

Making the Perfect Woodland Garden Site

You've decided you want a woodland garden, but you don't know how to start. Don't be intimidated by the prospect. A shade garden can be as simple or complex as you want it to be. You can plan it all out in advance or let it evolve naturally as time and money allow. What I describe in this chapter is the ideal situation, but if that doesn't fit into your life right now, don't worry about it. When my husband and I started our woodland gardens, we had a baby and a toddler at home, we both had demanding jobs with long hours, and money and time were in short supply. In the beginning, we just cut some paths through our woods and started putting in plants as they entered our lives (we were both extension horticulturists at the time, so bringing free plants home was a common occurrence). It wasn't until the kids were in school that we were able to build retaining walls and stairs, bring in big loads of mulch, and invest in a few specimen trees. So what I've described here is the ideal situation, but it is certainly not the only way to do it. Take what you can from this part of the book and design your own little woodland paradise.

Whether you own ten acres or a tenth of an acre, you can have a woodland medicinal herb garden. This can be a quiet, tranquil place that changes with the seasons and is a little different every year. Some woodland gardens have almost a fairy-tale character to them. My woodland garden at the research station provides that ambiance, especially in the early morning or late afternoon. We walk quietly down the path, enjoying the peacefulness, the rich earthy smells, and the special quality of the light filtering through the trees. You can create a place like that right in your own backyard.

Get Inspired

If you don't have a clear picture in your mind of what you want to create, gain inspiration by visiting other gardens, both private and public. Check with a local garden club to see if there is an upcoming tour

A sea of foam flowers in a woodland garden. Photo by Margaret Bloomquist.

Many people visit our woodland medicinal garden at the research station to get ideas for their own gardens. Photo by Jeanine Davis.

Could new or existing structures near your gardens be a problem for you or for the neighbors? Photo by Jeanine Davis.

of members' gardens or if there is a special shade garden in your area to visit. There are also many pictures on the Internet of beautiful gardens that you might want to emulate. Consider what you want in the short term and the long term. Do you want a tiny jewel of a garden that will stay that way forever? Do you want to start out with a little garden that can grow and evolve as time and resources allow? Or do you want to plan and establish a big shade garden right now?

Jot It Down

Now is a good time to start a garden journal, in either electronic or paper format. Use this journal to write down ideas and sketch out plans. If it is a paper journal, cut pictures from magazines and print pictures from the Internet and tape them in the journal. If it is an electronic file, scan in pictures you find in magazines and sketches you've made. Start a Pinterest page. And keep all your ideas; don't throw them out as you move along. You might reject an idea now, but in a few years that idea might be just what you want. The journal is also an excellent place to record sources of plants, seeds, mulch, compost, and accessories. Two years ago, I was at a home show and found the perfect chair for our garden patio, but there wasn't money in the budget at the time. I was forward thinking enough to pick up a business card but not smart enough to scan it into my electronic garden journal. Now we can't find the business card or remember the name of the business.

Take a Serious Look at Your Property

How big is the area you have to work with? Is your plan for that area realistic? How will it affect the use and appearance of the rest of your property? Are there restrictions

This is a wooded area with light shade. Photo by Jeanine Davis.

on what you can do on your property because of neighborhood covenants, or city or county regulations? Could your plans possibly have a negative impact on any of your neighbors? Do you know where your septic field, sewer pipe, and other utilities are? If the area where you want to locate your garden is on a hill, what direction does it face? South-facing slopes will be hotter and drier than north-facing slopes. How steep is the slope?

Shade

Do you have a naturally shaded area to plant your garden in? The ideal site has a hardwood canopy with trees such as poplar, sugar maple, white ash, oak, beech, maple, and birch, and an understory of trees and shrubs such as dogwoods, elderberry, and witch hazel. Is it all deep shade? Are there open areas? Are there different degrees of shade, e.g., dense, light, or partial shade? Partial shade means the area has direct sunlight for three to six hours each day. Pay attention to when the area gets that sun. Three hours of direct late afternoon sun can be hot and damaging to shade plants. Light shade or dappled light is filtered light that works its way through the canopy of deciduous trees. There are patches of direct sunlight that get through at different times of the day, but they are small. This is a preferred state for many of the plants covered in this book. Areas of dense or full shade get less than three hours of direct sunlight each day, with filtered light the rest of the day. Dense shade is not totally dark. Few plants grow on the forest floor under dense

We marked this area of bloodroot with flags so we wouldn't accidently run a path through it later in the season. Photo by Kelly Gaskill.

shade. If you have full shade, you will probably have to open up the tree canopy a bit through some judicious pruning or tree removal. If your canopy is very open and there is too much light, now is the time to think about planting some trees that will add more shade in the coming years.

If you don't have any shade right now, you can create some that is temporary until the trees you plant fill in, or it can be the permanent shade. Construct an arbor or pergola and let vines cover it to simulate the natural opening and closing of a tree canopy that occurs when leaves come out in the spring and fall off in autumn. Use lattice fencing to protect an area from late afternoon sun or put up a shade sail or canopy made of woven polypropylene shade cloth. If you can't afford to build shade right now, consider planting on the north or east side of your house.

Water Access

Can you irrigate the part of your property where you want your shade garden to be? This will affect what you plant if you have dry areas. It will also determine whether you can add a water feature or not. If you really want water and don't have it, now is the time to run the water lines.

Clear Out the Area and Take Inventory

Your first task is to remove what you know needs to go, such as weeds, dead trees and shrubs, and any junk that might have accumulated in that part of the yard. Be cautious. If this is an overgrown area, it could be home to animals and insects that don't like to be disturbed. Wear thick leather gloves, poke around with a tool to give animals a fair warning, and don't stick your hands into places where you don't have a clear view.

Once you can see what you have, take an inventory. Get a good plant identification book and key out what is growing there. If you don't know what something is, take a picture and look it up on the Internet. There are many online forums that will help you identify plants. You don't want to tear something out just because you don't know what it is. It might turn out to be something very special. Keep in mind that some plants that would be considered weeds in a vegetable garden might look beautiful in your shade garden.

Don't forget the trees. Some trees are not desirable because they are trashy, spread too easily, have poor form, or you just don't like them. Identify invasive plants such as privet, multiflora rose, bittersweet, Japanese honeysuckle, garlic mustard, and whatever else is invasive in your area.

When you are finished identifying everything, sketch out where the plants and trees you want to keep are located. Mark them physically with flags and flagging tape. Then get rid of the undesirables. This can be done mechanically with a chainsaw, weedeater, machete, hand pruners, hoe, or shovel, or you can do it chemically through the careful use of herbicides. I only resort to herbicides when encountered with a big area of poison ivy. If you do find poison ivy that is more than a single plant or two, make note of where it is and be careful when planting and working the area. Even after you get rid of the above-ground portion of the plant, the roots can cover a large area, and they also contain the oil that causes the rash.

Get your soil tested for pH and nutrients.
Photo by Alison Dressler.

Evaluate the Soil

Not all woods have the deep, fertile, humus-filled soils characteristic of a rich, temperate hardwood forest cove. Matter of fact, some wooded areas have very dry, poor soil. Determine what you are dealing with by first digging a hole. Is the soil hard, dry, and compacted or is it dark, moist, crumbly, and earthy smelling? There could be a variety of soil types and conditions in your garden area. Sketch those areas out in your journal. You don't want to plant a moisture-loving plant in an area that is always dry by early August. You can improve the soil in small areas by adding compost and chopped leaves, but if it is really dry because it is a south-facing slope, you are also going to need to build or grow a screen to shield the area from late afternoon sun and heat. Your other choice is to work with what you have and plant that area accordingly.

Get the soil tested so you know what the soil pH and nutrient level is (more on that in chapter 31, "How to Grow a Garden in the Woods."). Where I live in western North Carolina, forest soils are very acidic. This is fine for many plants but others will grow better if the soil pH is increased through the addition of lime. You can apply lime and fertilizer just where needed

Check Conditions at the Base of Trees

Often times, the soil right around the base of a tree is sparse and full of tree roots. It is tempting to want to just pile up soil all around the trunk of the tree to create a planting area. This is not recommended because tree roots need to breathe. Putting even a few inches of soil over the root area of some tree species can smother the roots. In these situations, it is probably best for the health of the tree to put down a layer of mulch around the base of the tree and add an art piece or an interesting container with plants growing in it. Some people recommend chopping out some of the large roots to make room to plant. I do not recommend doing this unless you have a tree expert who understands the root structure of that kind of tree and can tell you if removal of that root might be damaging to the tree or not. Also, if the tree roots are shallow, which is often true for smaller trees, the tree will compete with your plants for water and nutrients. In this case, keep the new plants outside the drip line of that tree.

Make an inviting path that winds through your woods garden. Photo by Jeanine Davis.

Installing Your Garden

Put In a Path

Now we get to one of my favorite parts of the process: deciding where to put the paths. Design your paths so they will wind through the garden; this slows people down and encourages them to look around. Include a hidden dead end or two where a bench, art piece, or special plant will reside. You can cut the first paths in with a weedeater and a machete. Once you have your paths the way you want them, start defining the edges and getting rid of the plant material in the paths. In my case, this happened naturally as we moved up and down the paths, but sometimes there was

Wild ginger growing along the garden path. Photo by Jeanine Davis.

some stubborn plant material that had to be cut out. If there are weeds in the path, you can smother them with newspaper or cardboard and cover with mulch. Trim branches that are hanging into the path so you can walk comfortably underneath them and don't risk getting a branch in the eye or caught on clothes.

Decide what you want your paths constructed of. Trails are usually mulched with bark or wood chips, but there might be areas where brick, gravel, or flat stones would be more appropriate and add interest. If there is a wet area, you might do a little boardwalk.

Framing the paths adds a nice finishing touch. You don't have to do this, but I am always impressed with the difference it makes in the overall feel of the garden when it is done. This doesn't have to cost anything. We use branches and small logs in our gardens at the research station. A local mushroom grower inoculated some of the logs lining the paths in his garden with shiitake mushrooms.

Erosion Control

I can't overemphasize how important it is to plan for excessive water events. Look at the lay of your land and imagine what would

Heavy rains washed out a part of our gardens. Afterward, we installed branches and logs which helped slow the flow of water. Photo by Margaret Bloomquist.

happen if six inches of rain fell in one day or there was a sudden heavy downpour during a summer thunderstorm. Is there an area up hill and outside of your garden that could send sheets or rivers of water rushing through a section of your garden? I have experienced this twice and it was devastating to my gardens and my spirit. The first time was totally unexpected and we really couldn't have prepared for it. We went on vacation for a week and during that time the new owners of the lot above us (we lived on a steep hill) had all the trees removed to prepare for construction. A thunderstorm came through and washed out a large part of our gardens and the gravel driveway, filled in all our newly laid drain tile, and brought mud into the house. The second time was the summer of 2013 when we had many weeks of heavy rain, sometimes two to four inches at a time. Houses had recently been constructed just uphill of our gardens at the research station, but there were lawns and buffer zones of shrubs that we thought were adequate. Unfortunately they were not enough to handle that much rain at one time and a large section of our garden washed away. So I strongly suggest that you plan for the worst. Put in berms and direct water away from your gardens. If you don't understand how to control water on the landscape, seek advice from someone who does.

Provide a relaxing place to sit and enjoy your gardens. This is one of the benches in Joe Hollis' garden. Photo by Karen Hardy.

Provide a Relaxing Sitting Area

If you would like to provide a place or two to sit, relax, and enjoy your garden, be on the lookout for an interesting bench, Adirondack chair, hammock, glider, or swing. It can be placed in one of those little dead ends you built into your pathway or in a location with a view of a special place in the garden. Antique stores, craft fairs, garden festivals, flea markets, and yard sales are fun places to look for just the right piece for your garden.

Add a Water Feature

Maybe you are fortunate enough to have a stream, pond, or spring on your property. If you do, be sure to incorporate it into your design. If nature didn't provide

Joe Hollis built these delightful, simple arbors for his gardens. Photo by Karen Hardy.

this for you, don't worry. It isn't necessary, or in some cases even practical, to have a water feature. But imagine enjoying a glass of wine while sitting on a bench next to a small pond, waterfall, or fountain in a little alcove in your shade garden. Sounds lovely, doesn't it? If you can get water out to the place where you want the water feature to be, consider adding one. Kits complete with liners, plumbing, and pumps are available at garden nurseries and home improvement stores. In the course of a weekend you can put in a little pond with a pump to keep water moving (you don't want to create a mosquito nursery). Lined with big, mossy rocks and stocked with some fish, it would be a wonderful addition to your garden. It will encourage more wildlife to visit your garden, too.

Add an Art Piece

While you are shopping for your garden seating, be on the lookout for some interesting art to add to your garden. This could be a gazing globe, statuary, metal sculpture, wind chimes, mobile, or anything decorative or whimsical to add interest and make the garden uniquely yours.

Think Layers

As you plan your garden and choose plant material, think in layers. Plant tall trees, small understory trees, shrubs, and herbaceous plants to provide interesting changes in height throughout the garden. This can be done in the traditional manner, with low plants near the path and increasingly taller plants as you move away from the path. I prefer a more mixed design with groups of

Add some whimsy with a small statue or two as Mossin' Annie did in her garden.
Photo by Jeanine Davis.

Notice the layers of tall trees, medium-sized shrubs, and herbaceous perennials on the forest floor in the Woodland Gardens at the Bullington Gardens. Photo by Jeanine Davis.

Professional-style labels can be used to identify your plants, as Bonnie Arbuckle does in her gardens. Photo by Jeanine Davis.

plants of different heights scattered about in a pleasing design. Also think texture, foliage color, and the timing of when plants emerge and bloom. This is where studying other gardens and pictures of gardens can be extremely helpful in achieving the garden you are dreaming of.

Plant Labels

Whether to include plant labels or not is a personal decision. Some gardeners just have to have them; others think they make the private garden look like a public botanical garden. It's your choice, but if you don't add labels, it is wise to have a good map of your plants. (Put the map in a plastic sleeve because your gardening friends are going to want to carry it with them when they walk through your garden.) If you use labels, consider using the small, discreet, permanent ones that botanical gardens use. These can be ordered online.

30

Choosing the Plants to Grow in Your Garden

As you make your shade garden plans, consider what kind of plants you want in it. Do you want to grow only native plants? Do you want mostly native plants with some exotics thrown in for color and interest? Do you want to plant only medicinal plants? What about designing a garden of medicinal and edible plants? Are you trying to create a particular garden that will need some specific trees or plants, such as Spanish moss hanging from tree limbs for a Southern garden ambiance?

Don't take this all too seriously. Have fun. Collect plants from friends, festivals, and herbalists. If a plant is not a medicinal herb but you really like it, so what? Put it in your garden. And don't be afraid to make mistakes. You can always move plants.

Medicinal Herbs

All the plants in this book are candidates for your garden. Be sure to check out the extensive list in Appendix 1, Forest Botanicals Bought and Sold in the United States and Canada. There is a medicinal herb for every soil and light condition in your garden. Many of these plants are also beautiful ornamentals. Plant bloodroot for showy early spring blooms and interesting foliage through mid-summer or later, if there is enough shade. The native bloodroot has simple white flowers, and double-flowered varieties are available in the nursery trade. "Multiplex," for example, is a bloodroot variety with big, snowball-like flowers. Black cohosh is a favorite plant of mine because of its feathery foliage, tall spikes of white flowers, and the way it quickly fills in an area. It is also more tolerant of sun than many of the other herbs. Galax, with its glossy round leaves that turn red as the weather cools, makes a beautiful ground cover. When planted in little sunny spots in early fall, queen of the meadow (*Eupatorium purpureum*), also known as Joe-Pye weed, will tower above all the other plants with its tuffs of smoky pink flowers. And one of my all-time favorites is the native

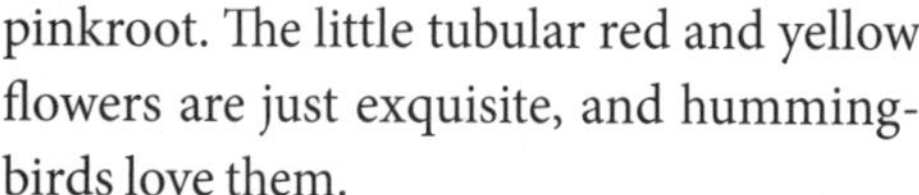

The first bloodroot blooms are always a welcome sight after a long winter. Photo by Kelly Gaskill.

Partridge berry is a delicate little vine with bright red berries. Photo by Jeanine Davis.

pinkroot. The little tubular red and yellow flowers are just exquisite, and hummingbirds love them.

What medicinal plants do you like to use? Be sure to include those in your garden. Goldenseal is one of my "go-to" herbs for boosting the immune system, treating mouth sores, and for use as a topical antiseptic. It is one of the first plants I put in a shade garden. Add new herbs as you learn about them. A plant that I want to add to my gardens is wahoo (*Euonymus atropurpureus*), also known as burning bush. The bark of this deciduous native shrub has medicinal value, but I want to grow it for its fall color and bright red berries.

Caution When Using Ornamental Varieties of Medicinal Plants

When you look at ornamental varieties of your favorite medicinal herbs, pay close attention to the genus and species. I have found many plants being sold as a particular medicinal herb that upon examination were not even the same species. This is not a reason to eliminate a plant from your garden, but label it as an ornamental, particularly if you or anyone else plans to make medicine from plants in your garden.

Vines

Go vertical in your garden by adding some perennial vines that will climb up trees or arbors. Passionflower (*Passiflora incarnata*) grows at the edge of the woods and easily climbs over fences and latticework. The flowers are amazing purple and white confections that turn into little egg-shaped fruits that make the best custard pies I've ever had. Wild yam (*Dioscorea villosa*) is another climbing, medicinal vine; it has beautiful heart-shaped leaves. If you are looking for a trailing vine, consider partridge berry (*Mitchella repens*), with its little green and gray striped leaves and tiny red berries.

The partridge berry winds through the leaves, eventually forming a lacey carpet. Photo by Jeanine Davis.

Shrubs and Small Trees

An aesthetically pleasing shade garden is composed of layers. The middle layer, between the tall trees and low-growing herbaceous plants, can be partially filled with shrubs and small trees. There are a surprising number of medicinal shrubs for you to choose from, including balmony (*Chelone glabra*), bayberry (*Myrica cerifera*), cramp bark (*Viburnum opulus*), New Jersey tea (*Ceanothus americanus*), Oregon grape (*Berberis* [*Mahonia*] species), sumac (*Rhus glabra*), wahoo, and wild hydrangea (*Hydrangea arborescens*). Small medicinal trees include black haw (*Viburnum prunifolium*), fringe tree (*Chionanthus virginicus*), mountain maple (*Acer spicatum*), prickly ash (*Zanthoxylum americanum* and *Z. clava-herculis*), and witch hazel (*Hamamelis virginiana*). Plant a few that are not commonly found in gardens. New Jersey tea, for example, is a wonderful tree that is rarely planted; it's about three feet tall, with grayish foliage and spring clusters of white flowers that butterflies are attracted to.

Witch hazel is a pretty, small tree that is also medicinal. Photo by Jeanine Davis.

Tall trees provide the perfect canopy for the woodland garden. Photo by Jeanine Davis.

Trees

Even if you already have tall trees creating shade for your garden, you might want—or need—to add some more in strategic places. Consider filling empty niches with medicinal trees such as balm of Gilead (*Populus balsamifera*), butternut (*Juglans cinerea*), sassafras (*Sassafras albidum*), slippery elm (*Ulmus rubra*), or white willow (*Salix alba*). Because these are large trees, study their growth habits and location requirements so you can place them appropriately.

Ferns

Ferns are a nice addition to any woodland garden, and most thrive in moist areas with partial shade. There are also dryland ferns that grow in full sun. Maidenhair ferns (*Adiantum pedatum*) are medicinal herbs that add a "fairyland" effect to a garden; they come in ornamental varieties that are dwarf, cold-hardy, and drought tolerant. Consider adding other ferns to fill in areas that might be too moist for most other plants.

Ferns help add that fairyland touch to the woodland garden. Photo by Jeanine Davis.

Use mosses to add softness to your garden as Mossin' Annie did here. Photo by Jeanine Davis.

Berries

Including berries in your woodland herb garden can provide good tasting medicine and food for your family, as well as food for birds and other wildlife. Blackberries (*Rubus villosus*) and elderberries (*Sambucus nigra*) are the two most popular medicinal berries. Making a big batch of elderberry syrup in late summer is a tradition in many of my friends' households.

Mosses

Cultivated mosses are becoming more popular and more available, and they open up a whole new world of possibilities to gardeners. Mosses add a softness to the garden that contributes to the "fairyland" appearance. Spreading across rocks and logs or dripping from trees, they come in a wide range of colors and textures. For our purposes here, there are basically two kinds of moss; tree moss and ground moss. Some mosses are medicinal, including sphagnum moss (*Sphagnum* spp.), Spanish moss (*Tillandsia usneoides*), and club moss (*Lycopodium* spp). Cultivated mosses can now be purchased at specialty nurseries and online. They often come with instructions on how to get them established and keep them growing. Annie Martin runs a small moss nursery and landscaping business in Pisgah Forest, North Carolina, just a few miles from where I live. She is writing a book on moss gardening which should be available in 2015.

Mushrooms

Mushrooms, medicinal and edible, are a fun and easy addition to your shade garden. Shiitake mushrooms (*Lentinula edodes*) are

Add mushrooms to your garden for a gourmet treat. Photo by Jeanine Davis.

the easiest ones to incorporate. Inoculating a stack of oak logs with easily obtainable shiitake spawn is a fun late winter activity for the family. If this is more than you want to tackle, you can pick up a few pre-inoculated logs at many gardening, organic, and homestead-oriented festivals. Shiitake mushrooms are medicinal and tasty. Other native medicinal mushrooms that you can cultivate or encourage to grow in your shade garden are reishi (*Ganoderma* spp.), lion's mane (*Hericium erinaceus*), maitake (hen of the woods) (*Grifola frondosa*), and turkey tail (*Trametes versicolor*). Other edible mushrooms that can be cultivated include oyster (*Pleurotus* spp.), chicken of the woods (*Laetiporus* spp.), and if you are up for a challenge, the morel (*Morchella esculenta*).

Non-medicinal Plants to Consider

If you decide to include non-native and non-medicinal plants in your garden, the possibilities are endless. If you looking to add more early spring color, consider planting spring-flowering bulbs such as trout lilies (*Erythronium* spp.) or Spanish bluebells (*Hyacinthoides hispanica*). You might want to intersperse those among hostas (*Hosta* spp.) that will come up and fill in the area as the spring flowers die down. Hostas are available in many handsome variations, and the graceful spikes of purple or white flowers in mid-summer are a nice contrast against all the green that time of year.

Deer-resistant Plants

If deer are a problem in your area, consider adding some plants that are less attractive

Trout lily (*Trillium sessile*) is a beautiful non-medicinal native plant that looks good in the woodland garden. Photo by Alison Dressler.

to deer than others. Keeping in mind that deer will eat almost anything if they are hungry enough, there are some plants that they often leave alone. In areas with high deer pressure I've observed that ramps, mayapple, black cohosh, spikenard, Indian turnip (*Arisaema triphyllum*), and wild ginger are usually not on the menu.

Plant Conservation Issues

A few of the plants suggested in this book are protected plant species in certain areas. Some of them may require a permit to be cultivated. Others might be restricted from cultivation in some states or provinces. If you purchase plants from a good nursery, they will provide you with the information you need. If you collect your own plants, follow all state, provincial, and federal regulations. If you collect from private land, have the permission of the owner. If you harvest from public land, obtain the proper permits.

31

How to Grow a Garden in the Woods

The plans are complete, the garden area is prepared, the paths are installed, and you've selected the herbs, shrubs, and trees you want to plant. Now it is time to put the plants in the ground and learn how to keep them healthy and growing for many years to come.

Soil and Soil Fertility

If you haven't done so already, now is the time to collect soil samples and send them to a soil testing laboratory. The test results will only be as good as the samples you send in, so don't take this task lightly. Look at your garden area closely and evaluate where you might have different kinds of soils and soil conditions. Is your garden on a hill? If yes, you should collect soil samples from the top of the hill *and* the bottom of the hill. If you have a wet spot that you want to plant, take a sample there. A dry area that gets more sun than the rest of the garden deserves its own test. Sketch out a map of your garden and mark down where you are going to collect each sample and give it a number. Put that number on your soil test bag or box so you can match the results up to the right area of the garden. Select a soil testing laboratory; there are many public and private labs to choose from. Just be sure to select one within your general geographical region. This is important because laboratories use different methods depending on the soils in their region. If you send soils from Oregon to a testing lab in North Carolina, the recommendations might not be quite right for your conditions. First, find the home garden form, which is different from the commercial form. Then select what you are going to grow. Some forms will let you make multiple selections. The form might have a shade garden category that you can check. If not, look for perennials and ground covers, deciduous trees and shrubs, and azaleas and rhododendrons. Give yourself plenty of time to get your soil tests results back; it can take up to a month, particularly at public labs.

Alison Dressler and Margaret Bloomquist mixing soil amendments into the garden soil before planting. Photo by Kelly Gaskill.

When you get your soil tests results back, see if there are any recommendations for soil pH adjustments or fertilizer additions. In my part of the country, I expect to see recommendations to add lime and phosphorus. Purchase your recommended supplies and have them on hand when you start planting. I suggest that you purchase organic amendments. If this were a vegetable garden, you would evenly distribute your amendments over the whole garden and till them in. That doesn't work in the woods. You need to do some calculations to estimate how much you need to apply to the small areas you will be planting.

Let's do a little example. You want to plant wild ginger in an area that measures approximately 5' × 3'. That is 15 square feet. The soil test recommendation is for 20 pounds of fertilizer for 1,000 square feet. First convert pounds of fertilizer to ounces; remember this is all done by weight. 20 pounds × 16 ounces per pound = 320 ounces. The formula you will use is (recommended ounces × square feet of area to be planted) ÷ square feet the recommendation was for = ounces of fertilizer for your area. Using our example that is (320 × 15) = 4,800 ÷ 1,000 = 4.8 ounces of fertilizer for 15 square feet, which I will round off to five ounces. I have a little plastic scale that weighs up to 32 ounces that I use for this purpose; I take it right out into the garden with me.

Rake back the leaf litter and work up the soil to a depth of six inches or so. Sprinkle the recommended amendments over the area, and rake them in. Now is a good time to work in some additional organic matter in the form of compost or chopped up leaves. The area is now ready for planting.

Additional fertilization is rarely needed in the shade garden, with a few exceptions. One is if you planted in an area with poor soil to begin with. You will probably need to keep building the soil up for several years through annual additions of compost and a little balanced organic fertilizer. The other situation is having pH that is too high or too low. Soil pH is important for proper uptake of nutrients by the plant. If the soil pH of an area tested way off in the beginning, you should consider retesting it after a few years of adding amendments to see if you've got the pH into the proper range. For the rest of the garden, the annual deposit of leaves from the trees and plants and your own additions of mulch will add nutrition to the soil as worms, insects, and soil microorganisms break down the organic matter and release the nutrients into the soil.

Didn't Do a Soil Test?

If you didn't get around to getting your soil tested, you can still put plants and seeds in your garden. But it does help to know something about the soils in your area. For example, where I live in western North Carolina, I know the soils in the woods are very acidic. I also know that many of the medicinal herbs I want to grow benefit from a little additional calcium in the soil. So my five-gallon planting supplies bucket always has a small container of lime in it.

Margaret planting young goldenseal plants in beds in the woods gardens at the research station. It was a windy day, so she temporarily covered up the plants with the pots to prevent them from drying out. Photo by Kelly Gaskill.

When I am gifted a new plant, I carry it and my bucket out into the woods, scrape back the leaf litter with my shoe, dig a hole with a trowel, sprinkle in some lime, stir it up, put in my plant, firm the soil around it, spread the leaves around again, and that's it. It almost always grows just fine. To make this work for you, learn something about your soils. I suggest talking to someone in your area with a shade garden that you admire.

Put the Plants or Seeds in the Ground

Now that your soil is amended, you can start putting plants and seeds in the ground. In the paragraph above, I explained how easy it is to plant a single plant. My suggestion, though, is to plant most of your herbaceous perennial species in groups of three or more. Consider how large the plant will get and space multiple plants accordingly. Allow for good air circulation around your plants to help prevent fungal problems on the foliage. This is particularly important in humid and rainy climates. I have a tendency to plant things too close initially to get a full cluster effect quickly. This just means that I have to thin the plants out in a few years and move the plants I dug out to a new area or trade them for something else.

If you have a single large plant, when you take it out of its pot, look at the root system. Is it big enough to divide into two or three plants? Most of the herbaceous perennial herbs can be divided this way. The book chapters on goldenseal, black cohosh, and bloodroot provide more information on vegetative propagation, but in short, break or cut the big rhizome into several pieces, trying to have a bud and roots on each piece. Plant them under several inches of soil and cover with leaves or other mulch.

Seeds benefit from a little more bed preparation. Work the top few inches of soil into a finer seedbed, sprinkle your seeds on top, cover with about an inch of soil, tamp the soil down, and cover with mulch. If wild turkeys, other wildlife, or slugs tend to disturb your seeds or seedlings, you can protect them with window screening held down with landscape staples.

Mulches

Adequate and appropriate mulching makes your garden more attractive and less labor intensive to maintain. A good mulch layer holds in soil moisture, eliminating or reducing the need for watering. It should look attractive and provide a nice background for the foliage plants. A thick layer of mulch also provides a weed barrier.

Every region has commonly used shade garden mulches. They are usually readily available and inexpensive. In my area, hardwood bark is the mulch of choice. In other areas, pine bark, pine needles, and wood chips are common. The leaves that fall from your trees provide a natural mulch. These can be chopped or composted and reapplied.

Wood and bark mulches can harbor the artillery fungus, which can be a problem if they are close to your house or cars. Artillery fungus can infest wood-based mulch and shoot sticky, black spores onto nearby surfaces. These tiny dots are most evident on light colored siding and car finishes. They are difficult to remove and leave a stain. Dr. Donald Davis at Pennsylvania State University is an authority on artillery fungus. He has two suggestions at this time. The first one is not to use wood-based mulches near the foundation of the house or near where you park your cars. The second one is to mix 40 percent spent mushroom compost (available in bags at nurseries, home improvement centers, and discount stores) with 60 percent bark mulch. The mushroom compost apparently inhibits the fungus. Wood chips tend to be more infested than bark mulch, so he recommends you avoid using the wood chips near the house, even if you add the mushroom compost. If you live in an area where mushroom compost is readily available, it makes a good mulch used alone, too.

Irrigation

Most established woodland herb gardens don't need irrigation—unless there is a dry area that gets several hours of late afternoon sun. Newly planted areas and seedbeds might also need some water. This can be provided with hoses or with water carried to the site in carboys or buckets.

Weeds

A woodland herb garden is usually easier to care for than a vegetable or flower garden. Properly mulched and with adequate shade for the species, weed control should be minimal. Some areas have invasive weed species that are taking up residence in the

When Margaret was finished planting the goldenseal, she covered the beds with a thick layer of leaves. Photo by Margaret Bloomquist.

Some gardeners will let their medicinal plants share the space with the weeds, and others will remove them and add more mulch to suppress new ones. Photo by Jeanine Davis.

forests. Be aware of these and remove them when you first notice them. If something happens to reduce the shade in an area of your garden (a tree comes down or your neighbor takes a tree down), do something to replace the shade as soon as possible. In my experience, grasses move into these areas very quickly. You can plant a tree and then construct inexpensive, temporary shade with latticework or shade cloth to provide shade until the tree grows.

Slugs

Slugs can be the nemesis of the shade gardener. They will eat a bed of seedlings overnight, consume all your goldenseal fruit, and eat holes through the foliage on their way up and down the plant. The subject of slugs is covered in the commercial ginseng and goldenseal parts of this book, but I think it is important enough to cover again here.

Slugs live in the mulch and leaves below your plants. They come out at night to feed. They overwinter as eggs in the mulch and plant debris and hatch out in the spring to dine on your plants. You may never have a problem with them, or you might fight with them for years, and then suddenly they are gone. Or, you might need to fight them forever. They like moist, shady places, just like your woodland herb garden. Dealing with slugs can be a frustrating experience, and the only solution that seems to work everywhere is the use of chemical baits made of iron phosphate, metaldehyde, or Mesurol. These are toxic compounds that I will not use in my gardens for fear of dogs, cats, children, or wildlife consuming them. So they are not even on my list of options.

If you start noticing holes in the leaves of your plants, put some short pieces of board or heavy cardboard on the ground in that area. These should be attractive hiding places for the slugs. In the early morning, look under the boards. If there are slugs there, you have identified your pest. Some people use this method for control. They put little boards or something similar all over their gardens and every morning look underneath and knock the slugs into a bucket of soapy water. Personally, I don't have time for that! Another slightly more practical method that works well for some gardeners is to have beer traps set in the ground. These can be made out of almost anything—just some kind of shallow dish that you put level with the ground and fill with beer. You can actually buy plastic ones with lids called Slug Saloons. In theory, the slugs are attracted to the beer, fall in, and drown. I have not found these to be effective at all.

Slugs don't like to crawl over copper, so you can buy copper strips to surround your gardens. I tried this too. It was expensive and didn't work for me. A two-inch border of lime or diatomaceous earth sprinkled around your plants is a decent slug barrier, because they don't like to crawl across anything dry and scratchy. But these have to be replaced every time it rains; again, not practical in my life.

There are two methods that work for me. One involves metal window screen. If slugs are a problem in an area where I want to plant seeds, I cover the seeded area with window screen pinned down with landscape pins. I have also constructed little protective boxes out of window screen, pushing the edges of the open end several inches into the ground. You can also cut window screen into 4" lengths that you push into the ground on edge so a few inches stick up with a nice jagged edge that slugs won't cross. Window screen is readily available, inexpensive, easy to cut with a scissors, and malleable. Now, one thing to keep in mind, if the slug eggs are already present inside the area you want to protect, you might just trap them inside your border! If you suspect this is the case, remove all the mulch, right down to the bare soil. Then erect your barrier and put fresh mulch down.

My second and best method for slug control in the shade garden is to create good wildlife habitat. Slugs are a food source for toads, frogs, birds of all sorts, ground beetles, mice, and firefly larvae. Some people have a few guinea hens around just to keep their plants free of slugs and insects. But beware: guinea hens can be noisy. My husband hates guineas, so they are not an option in my garden.

Do not let chickens loose in your shade garden; they will eat the slugs but also consume the tender little plants and seedlings you are trying to grow.

Insects

Insects are rarely a problem in the shade garden. There are some cutworms and leaf-eating beetles that will cause damage on the foliar portions of black cohosh, but I have not found them to be very serious, especially in a mixed garden. If you can accept a little damage from time to time, a well-established shade garden with a wide diversity of plants, insects, and birds will usually maintain a good balance, and major damage to any particular species would be unusual. If you do find an insect that appears to be doing significant damage, get a specimen or a really good photograph of it, and get it identified. Learn everything you can about its lifecycle, preferred food sources, and natural predators. Then decide if control measures are warranted.

Larger Pests

Deer, groundhogs, wild turkeys, and dogs cause more damage to woodland herb gardens than diseases and insects. Deer will browse (eat) the tops of many plants, and in the winter they can strip the bark from the trees. What they eat, and how much they eat, varies from year to year. Some years you might spot several deer in your woods every morning, but they cause no noticeable damage. Another year, one doe might eat an unbelievable number of plants.

There are dozens of suggested solutions for deer control, from tall fences to keeping a dog, to spreading human hair around your property. If deer become a problem, you can try all these methods one by one. I suggest talking to your neighbors first to see if any method works particularly well in your area. Something that I noted in my area in 2013 was that it was a banner year for jewelweed (*Impatiens capensis*) and the

We built these cages to prevent deer, wild turkey, and groundhogs from browsing on our baby ginseng plants. Photo by Jeanine Davis.

deer just loved it. When I put out a query about this, others reported that where there was lots of jewelweed, the deer were not consuming other plants.

Groundhogs can also cause a great deal of damage and create large, ugly holes in the ground throughout the garden. There are as many suggested solutions for groundhogs as there are for deer, and they range from trapping them, to putting hot peppers, dirty cat litter, or garlic in their holes. Groundhogs used to eat all the leaves off my bloodroot plants. They did this for three or four years. Then, one year they quit, and they haven't done it again. I know the groundhogs are still there, but they aren't eating any of my medicinal herbs anymore—or, I should probably say, so far.

Wild turkeys are a problem in some areas, scratching up mulch and plants in gardens, and eating flowers and seeds along with the insects. Like the deer and the groundhogs, the online forums and blogs are full of suggestions for how to control them, but no one method stands out. Netting supported by metal t-posts can be effective to protect small areas.

The next time I have a deer, groundhog, or turkey problem, I want to try using motion-sensing sprinklers. I haven't spoken to anyone personally about them, but the articles and videos about them are impressive.

Voles, also referred to as meadow mice, can be a serious problem in some shade gardens. Unfortunately, many of the things we do to maintain a beautiful shade garden creates the perfect habitat for voles. Voles make above-ground runways under the cover of grass, mulch, dead leaves, and heavy vegetation. All that bark mulch you so carefully put around your plants can be the perfect cover for them. Voles will eat just about any plant material, including roots, foliage, flowers, seeds, fruit, bulbs, stems, and bark. Their populations can be quite low for many years and then suddenly explode. It is important to learn what their runways and the damage they cause look like so control measures can be started before populations get too high in your garden. Links to several detailed information pages on voles and their control are included in the Resource section at the end of the book.

Control methods include moving mulch at least three feet away from the base of trees and plants that are serving as the voles' food sources, putting protective collars around tender young trees, and installing wire barriers around planted areas using hardware cloth with holes ¼" or less in size. These barriers should be inserted at least 6" into the ground and be at least 12" tall. The last time I planted ginseng, we put ¼" hardware cloth 12" below ground all the way around the beds at the time we constructed the beds.

There are many baits, traps, poisons, and sprays that can be purchased for vole control. They may or may not be effective, and I don't like to use them in my garden. Try to maintain healthy populations of predators, including owls, hawks, shrews, and foxes. Cats and dogs will also catch and kill voles.

Cutting bloodroot into planting-size pieces. Photo by Alison Dressler.

Propagation

Refer to the chapters on specific plants to learn about rhizome division and seed collection, stratification, and sowing. All methods described are easily adapted to the home garden. For example, if seeds need to be stratified, they must be exposed to cold and moist conditions. This can take place outdoors or you can put fresh seeds in moist sand (clean sandbox sand) in a Ziploc bag and store them in the refrigerator for a few months. For plants other than the 14 covered in detail in this book, you will need to do a little research. Do an Internet search for "plant name" seed germination. Most plants will have some information available on how to germinate them, but not all. Some seeds need light to germinate, e.g., lobelia. For other plants, the seed needs to be scarified (breaking through the seed coat by nicking, cracking, scratching it, or exposing it to acid). Wild indigo and passionflower are examples of seeds that need to be scarified. Rhizome division is the easiest method for propagating plants (if they have rhizomes, of course). But probably the most enjoyable method of propagation available to the home gardener is swapping seeds and rhizomes with gardening friends.

Harvest

Since profit making is not an objective, home gardeners have much more flexibility when it comes to harvesting their plants to make medicine. Once a plant herb or root is big enough to provide adequate material for you to make what you want, you can harvest it. Follow the guidelines in the Forest Botanicals Bought and Sold in the United States and Canada appendix about what part of the plant to harvest and when. Herbs can be cut off with a hand pruners;

Sometimes the best way to harvest roots is on your hands and knees. Photo by Jeanine Davis.

the rhizomes of herbaceous perennials can be dug with a shovel or fork; and branches or roots for bark harvest can be cut with loppers.

Wash roots carefully using a tub of water and a hose with a nozzle that provides a strong stream of water. Roots can be dried in a kitchen tabletop dehydrator or just spread out in front of a fan—with the air conditioner running if it is hot and humid outside. Herbs (the foliage) can be dried in the tabletop dehydrator or hung from the ceiling or on a rack out of direct sunlight.

Thoroughly dried herbs should be stored in glass jars or other sturdy containers in a dark cabinet or closet to maintain quality and color.

32

Ginseng—A Horticultural Challenge

If it were easy or just moderately difficult to grow American ginseng (*Panax quinquefolius*) in the woods, the roots would not be nearly so valuable. Woodlands where ginseng thrives—growing vigorously, reproducing well, staying healthy—are uncommon. It is unlikely that your backyard woodlot is such an uncommon site. Ginseng does grow in less-than-ideal environments, but rarely is it big and impressive, producing balls of brilliant crimson berries, like the plants 'sang diggers seek and the home shade gardener can take pride in.

Outside its ideal habitat, ginseng is easily stressed and quite intolerant of stress. Sometimes under stress, such as summers with too much heat or too little rainfall, the top will senesce early, and the root will actually shrink during dormancy from consuming its own starches in order to grow a new bud and top for the following spring. If this happens repeatedly, eventually the root will be too small to create a new top, and the plant will die. I've seen this happen once to young roots throughout a thriving wild-simulated planting that was subjected to consecutive excessively hot and dry summers, and I've seen it happen with normal weather in several instances to rootlets transplanted to convenient, but not ideal, wooded areas, such as a home shade garden. Unfortunately, this kind of slow death is not the usual outcome of stress on ginseng. Usually, the plant is attacked and imperiled by fungi, which have an advantage against ginseng because it has no waxy coating on its foliage to form a physical barrier to fungal spores. Cultivated ginseng, even woods-cultivated ginseng, is particularly susceptible to fungi because faster-than-normal growth stresses most plants and because ginseng needs, or at least greatly benefits from, mycorrhizal fungi, micronutrients, and perhaps other constituents of the rich woodland soil that are typically lacking when it's grown outside its fertile native habitat.

Creating, in a home shade garden, the relatively stress-free conditions where ginseng will survive and thrive is difficult. The

Ginseng harvested from the Jeanine's woodland gardens at the research station. Photo by Margaret Bloomquist.

general advice that Jeanine has provided in the last three chapters will certainly be helpful, but ginseng is a special case—a horticultural challenge. Chapters 4 and 5 on commercial woods growing are essentially consistent with Jeanine's counsel, but, of course, are more specific to ginseng. They should provide you with much of what you will need to know, making you aware of what problems to look for, and at least keeping you from making the rookie mistakes typical of first-time 'sang growers. You will already appreciate, then, the critical importance of cooling shade, excellent drainage but constant moisture, high calcium levels and their maintenance, etc. Here, I will add a few suggestions that might be impractical and/or not cost effective for a commercial grower.

It is instructive to note that the Chinese do not try to amend their field soil for growing ginseng; rather, they expend considerable time and effort to trek up into the mountains, gather the richest woodland soil they can find, and then cart it back to their fields to create beds to grow ginseng in (under constructed shade). Such a trek is an option worth considering. If your woodlot does not have a loamy fertile soil, at least the equivalent of a good topsoil, you will need to get some rich loamy soil from somewhere. Whatever soil you end up with should have a pH between 5.5 and 6.5. Add sulphur to lower the pH, lime to raise it, but if it's already within that range but low in calcium, add gypsum, not lime, to raise (and maintain) the calcium level.

When I created my own minimalist

Ginseng growing in the woodland gardens at the research station where Jeanine works. Photo by Alison Dressler.

shade garden, I transplanted half a dozen local, wild, mature ginseng roots, four of which are still there after 28 years. If you live in wild 'sang country, this is your best option. A wild root will almost certainly be disease free and perfectly adapted to your climate. Ideally, you'd want to plant roots about eight years old, dug carefully so as to inflict minimal damage on the hair roots. (Roots less than five years old are illegal to dig, and old roots are more susceptible to transplant stress.) Dig them legally or buy them in the fall from a local digger or dealer, expecting to pay between $5 and $10 each for modest roots. Outside 'sang country, order from one of the seed/seedling dealers listed in chapter 9. Seeds are about 100 times cheaper, but slower and a bit more problematic.

Plant your rootlets at least two feet apart both ways to minimize the spread of fungal diseases. If you use seeds, you may want to plant more densely and thin to at least 2' × 2'. If you have tulip poplar, sugar maple, white ash, or black walnut in your woodlot, your ginseng will probably do best under them, but you must have good drainage as well as good soil moisture wherever you plant. A high canopy for good air circulation is also desirable. It's important to interspace other medicinals between your ginseng plants. Ginseng is adapted to compete with other woodland herbs. Goldenseal and bloodroot are particularly appropriate for interplanting because they have anti-fungal properties. I also interspaced maidenhair fern in my garden, because it's a natural and elegant companion to ginseng. Mulch your ginseng and maintain that mulch conscientiously, preferably with several inches of well-rotted leaf litter, which will add micronutrient and organic matter content. (Mushroom compost has proven not to make good mulch for ginseng.) Avoid nitrogen amendments.

Should one of your plants be attacked by a fungus, be willing and unhesitant to sacrifice it. If it is attacked by a foliar fungus, then cut off the top and remove it and the mulch around it quickly, then replace the mulch. A new top will probably emerge the next spring. If your foliage discolors, then carefully dig down and check the root for rot. (Discoloration, usually a reddish-bronze tint, can also indicate stress caused by heat or malnutrition.) If the root is rotting, remove it from the garden immediately and do not replant ginseng in that spot.

If you grow 'sang, you are likely to encounter some problems. Be patient, be persevering, and be vigilant. If you're successful in growing big, handsome ginseng plants, don't naively assume that no one would ever be so bold as to try to rob you of your green gold, even right out of your own backyard. It happens.

33

Making Some Simple Products from Your Woodland Medicinals

People who attend my presentations at conferences, workshops, and festivals often ask if there are simple things they can make out of the herbs they grow. Included are five of my own recipes that have evolved from ones I found in cookbooks and online. Ginseng tea is medicinal; if you are not accustomed to taking ginseng, consult with an herbalist or your physician before ingesting, especially if you take prescription medicines.

Elderberry Syrup

- 1 cup fresh or ⅔ cup dried elderberries, completely de-stemmed
- 3 cups water
- 1 cup raw honey or sugar

Optional: 2 or more tablespoons fresh grated ginger, 1 to 2 teaspoons ground cinnamon, and/or ½ teaspoon ground cloves (I got the ideas for these additions from some recipes I found online. They make for a nice change, although I still prefer just elderberries with a little sweetener).

Put the elderberries and water in a medium saucepan. One or more of the optional ingredients can be added at this time. Bring to boiling and then reduce to a simmer. Simmer, uncovered, for 30 minutes to one hour, stirring and mashing the berries occasionally, until the mixture is reduced by about half. Strain through a metal kitchen strainer or cheesecloth into a bowl. Let the liquid cool till warm (compost the berries). Add the honey or sugar to the warm liquid, stirring well. Pour into a clean pint mason jar or other sturdy glass bottle. Label and date. Store in the refrigerator and use within two to three months.

Fresh Ginseng Green Tea

- 1 tablespoon of fresh ginseng root cut into thin strips
- 1 green tea teabag
- 8 ounces of water
- Raw honey or sugar

Put the ginseng strips into a metal tea ball. Put in a mug with the green tea teabag. Boil the water. Let it cool for a couple minutes, then pour it over the teabag and tea ball. Steep for 5 minutes. Remove teabag and tea ball. Sweeten the tea with honey or sugar to taste. You can eat the ginseng strips.

Ramps Quiche

- 2–3 cups whole ramps
- 2 tablespoons butter
- 4 large eggs, beaten
- ¼ cup milk, plain yogurt, or half-and-half
- ½ pound cubed or grated Swiss cheese
- Salt and fresh ground pepper, to taste
- 1 (9-inch) unbaked pie shell, homemade or purchased

Clean and trim the ramps the same way you would scallions, leaving some of the green on. Cut into ½ inch pieces and sauté in butter until golden. Beat together the eggs, milk, yogurt, or half-and-half, Swiss cheese, salt, and pepper. Stir in the ramps. Pour into the pie shell. Bake at 375°F for 30 to 40 minutes or until a knife inserted near the center of the pie comes out clean and the crust is golden brown. Let cool for 10 minutes before cutting.

Simple Bloodroot Dye for Natural Fibers

- 2 gallons water
- 1 to 2 cups fresh, clean bloodroot root, cut up
- 1 cup salt

Wear rubber or latex gloves during this process. Put all ingredients together in a large pot that you will not use for cooking food. The bloodroot can be put into a mesh bag made from fine cheesecloth or pantyhose. Boil the mixture for 10 minutes. Add the natural fiber you want to dye, e.g., reed, cane, or oak splits for baskets or cotton. Boil for several more minutes. Remove from the heat and let set for several hours until fiber is desired color. Rinse item and dry. The dye solution can be refrigerated and used again.

Antiseptic Soap

- 8 ounces of melt-and-pour glycerin or goat milk soap base, cut into small cubes
- 2 teaspoons of goldenseal root powder (can use electric coffee grinder to make powder)
- 2 tablespoons olive oil
- Optional: 10 drops essential oil for fragrance and/or 10 drops soap coloring (remember the goldenseal will add a slight yellow color)
- Soap molds (3 or 5 ounce waxed disposable cups work fine)
- Rubbing alcohol in a spray bottle

Put oil in glass or ceramic cup or mug and heat for just a few seconds in a microwave just until warm. Stir in the goldenseal root powder. Put the soap cubes in a medium-sized glass mixing bowl. Heat on high in the microwave for 30 seconds. Stir gently with a silicone spatula. Reheat for 30 seconds at a time until melted. Do not overheat! Remove from microwave. Stir in goldenseal oil, and fragrance and color, if desired. Pour into molds. Spray the surface of the soap lightly with rubbing alcohol to get rid of bubbles. Let set for 30 minutes or longer. Pop out of mold (or peel off the paper cup).

34

Home Gardeners' Stories

The Bullington Botanical Bunch

When I started my search for a home gardener to interview, I called John Murphy, Director for the Bullington Gardens, a public horticultural education center and gardens located just outside of Hendersonville in the southern mountains of North Carolina. The beautiful gardens there are maintained by a group of dedicated volunteers and I thought that there might be some good candidates among them. Indeed there were and John arranged for me to visit with them.

The Native Gardens at the Bullington Gardens

I arrived at the Bullington Gardens early on a bright September morning in 2013. As soon as I climbed out of my car, I spotted large beds full of brightly colored dahlias. They looked so beautiful with the green mountains behind them and a Carolina blue sky above that I started snapping pictures before I had even shut the car door. I strolled up the pathway and found John taking pictures of a group of adults creating a tabletop mandala from flowers, berries, and other organic materials they had gathered from the property. He escorted me to the native plant gardens and introduced me to my hostess for the day, Bonnie Arbuckle. Bonnie was dressed for gardening. She had work gloves on her hands and big sunglasses and a warm smile on her face. After quick introductions, John headed back to his group, and Bonnie led me down a mulched path to meet two of her volunteer team members, Juanita Lambert and Frances Jones. She pointed out the fourth volunteer, Larason Lambert, who was walking down another path with a chainsaw in hand. Frances looked familiar to me and after a short discussion we realized that she had been a member of a group of Master Gardeners who several years ago had helped me plant ramps and native medicinals in my shade gardens at the research station. We soon discovered that Juanita

Sign at the entrance to the Native Woodland Garden at the Bullington Gardens in Hendersonville, North Carolina. Photo by Jeanine Davis.

had also been to the research station to collect mayapple plants after I had finished a project with them. Sharing those connections made for a great ice-breaker, and we were soon chatting like old friends.

These four volunteers call themselves the Bullington Botanical Bunch. Every Tuesday morning they spend three hours working in the Native Woodland Garden and the Rain Garden. They are members of the Western Carolina Botanical Club, which is responsible for maintaining these gardens. The woodland garden is situated on a gentle slope covered in tall trees and a variety of native shrubs, including hollies and azaleas. The gardens are about 13 years old and include many plants that Bob Bullington planted when the property was

Overlooking the Native Garden area at the Bullington Gardens. Photo by Jeanine Davis.

Running cedar (*Lycopodium digitatum*) covered large sections of the garden. Photo by Jeanine Davis.

his home and commercial nursery. One of his favorite plants, shortia (*Shortia galacifolia*), also known as Oconee bells, are well represented in the gardens.

As the three women escorted me along the path winding through the woods, Larason was busy felling a few small trees to increase light for some of the understory and herbaceous plants in the garden. As the trees shading the garden grow, the amount of light reaching the ground gradually changes. Without some intervention, areas of the garden will eventually become too dark for some of the plants growing there. Canopy thinning is a necessary practice to maintain optimum light levels for plants.

Bonnie, Juanita, and Frances led me down the path, enthusiastically pointing out and naming plants along the way. Many of the plants were familiar to me, but many more were not. I enjoyed learning about new plants from them, and they were interested in hearing about the medicinal values of many of them from me. For example, they asked why I was taking pictures of Joe-Pye weed (*Eupatorium purpureum*) because it is a common wild flower growing in meadows and along roadsides throughout our region. I explained that it is also called gravel root, and it is a medicinal herb used to treat kidney stones. Some of the other medicinal plants growing there were bloodroot, black cohosh, dewberry (*Rubus villosus*), false unicorn, ginseng, lobelia (*Lobelia cardinalis* and *L. siphilitica*), partridge berry (*Mitchella repens*), pinkroot, skullcap (*Scutelllaria lateriflora*), Solomon's seal (*Polygonatum biflorum*), wild ginger, and witch hazel (*Hamamelis virginiana*). There was also lots of galax, which

The amphitheater surrounded by blooming Joe-Pye weed, otherwise known as gravel root. Photo by Jeanine Davis.

is used more for decorative purposes than medicinal use.

They told me how they cared for the garden by removing weeds, invasive plants, and other exotics, and introducing more native plants. They plant each species in a group, with a weather-resistant label displaying the common and scientific names. Occasionally, a new plant is added that turns out not to be a native. It doesn't matter how pretty it is; once identified as "not a native," it is removed—to be replaced with a better, hopefully native, selection. Sometimes a plant has to be allowed to reach flower stage before a positive identification can be made. As we walked, the three women never stopped pulling weeds and picking up sticks.

Larason graciously shared the journal he has kept of his work in the Native Woodland Garden for the past 13 years. His journal chronicles the progression of activities that converted a nature trail into the beautiful gardens that exist today. The journal documents the building of the amphitheater, entrance gate, rail fence, and a rock feature; extensive grading and trail building; bringing in the water supply; adding water erosion control features and some benches. It also describes the many plants the botanical club transplanted and labeled. It is quite an impressive effort.

As noon drew near we left the gardens, for we had all been invited to have lunch at Bonnie's house. As much as I enjoyed the public garden, I really wanted to visit a

The Bullington Botanical Bunch, from left to right, Juanita Lambert, Larason Lambert, Bonnie Arbuckle, and Frances Jones. Photo by Jeanine Davis.

Medicinal herb garden at the Bullington Gardens with hop plants climbing the structure in the middle. Photo by Jeanine Davis.

Black cohosh and ferns line the driveway in front of Bonnie's house. Photo by Jeanine Davis.

shade garden at someone's home, because I intended to provide my readers with an example of what a gardener could do on their own property.

Bonnie's Gardens

Bonnie lives on a large lot in the historic village of Flat Rock, North Carolina, a community of quaint cottages, old summer estates, and neighborhoods nestled in the woods. After enjoying a delicious lunch on the deck of her home overlooking the back yard, she led us on a tour of her gardens.

When Bonnie and her husband purchased their property about 16 years ago it had the typical suburban landscaping of that time, with foundation plantings and lots of lawn. There were tall mature trees shading the front of the house and woods toward the back of the property. She arrived at her new home with a little alternate leaf dogwood tree (*Cornus alternifolia*) in a five-gallon pot and planted it in what would eventually become a garden near the back of her house. That dogwood is now a tall,

Bonnie walked us around her house and property pointing out special plants along the way. Photo by Jeanine Davis.

Stone stairs lead you down through the gardens to the back yard. Photo by Jeanine Davis.

beautiful tree and one of Bonnie's favorites. She has enjoyed sharing many of its seedlings with friends.

Gradually, Bonnie removed most of the grass in the yard and replaced it with a variety of native plants. She put stone pathways in the shady side yards and bordered them with many native herbaceous perennials. The narrow space in the front yard between the driveway and the main road was once covered with invasive English ivy. That was torn out and replaced with native shrubs and small trees.

The backyard makes up most of property. There is an open grassy area in the middle, with woods behind it. There is also a dry, sunny garden off to one side. The garden at the back of the house surrounding her favorite dogwood is now filled with small trees, shrubs, and vines. A coral honeysuckle vine (*Lonicera sempervirens*), a favorite of hummingbirds, winds up and over the railing of the second floor deck where we ate our lunch.

Over the years Bonnie has purchased native plants at plant sales, festivals, and nurseries, and traded many more with gardening friends. Some of the native plants in her gardens weren't planted, but emerged from seed that had been sitting dormant in the soil and germinated after she removed the invasive English Ivy. These plants add a variety of textures and color to her gardens and include crane fly orchid (*Tipularia discolor*), spotted wintergreen (*Chimaphila maculate*), Solomon's plume (*Maiamthemum racemosum*), and partridge berry. Interesting plants also randomly appeared

The gardens continue around the entire house. Photo by Jeanine Davis.

There is a sunny, grassy lawn area in the backyard. Photo by Jeanine Davis.

There is a smooth, log-lined path through the woodland garden. Photo by Jeanine Davis.

in the dry, sunny garden as she cleaned that up. Makes you wonder what the area looked like before houses were built there.

The whole time we walked around her gardens, Juanita and Frances asked about various plants they saw, and Bonnie explained what they were and where she got them. She also told them what they could dig—sharing plants as avid gardeners do.

She is not quite as strict about the "native plants only" rule in her own gardens as she is in the Bullington Gardens. Some of the exotics growing in amongst the native plants were from family gardens; others she just enjoys; and some serve as useful groundcover that is more interesting and requires less maintenance than grass. Mosses that appear are also allowed to grow and develop into a groundcover. Bonnie is constantly moving plants from here to there, adding new ones, and removing ones that no longer fit her vision for the gardens. It is clear that she gardens daily and gets great pleasure from doing so.

There were so many different kinds of plants in her gardens, I can't begin to remember them all, but here are a few that stand out in my memory: pinkroot, sassafras, native St. John's wort, hollies, dogwood, mosses, shortia, bloodroot, black cohosh, mountain mint, ferns, rhododendrons, and gravel root. There were some culinary herbs, too, but interestingly, I didn't see a single vegetable plant.

There was a cistern under the deck that she had installed during a drought about five years ago. She has water from some of her gutters diverted there to irrigate her gardens with. She has done such a marvelous job of creating a natural space for

Bonnie has a cistern to collect water from the roof so she will always have water for her gardens. Photo by Jeanine Davis.

plants, animals, and birds that her yard is certified as a Backyard Wildlife Habitat by the National Wildlife Federation.

Bonnie's approach to gardening is casual in appearance but deliberate in its execution. She can identify every plant in her garden and knows what conditions each prefers. She takes careful note of which plants grow together in the wild and tries to replicate that in her gardens; "companion planting" is what vegetable gardeners would call it; planting "in guilds" is what naturalists call it.

I greatly enjoyed spending a few hours with Bonnie, Frances, Juanita, and Larason. I accomplished what I set out to do—conduct an interview and take pictures for this

There is a birdbath in a small, sunny clearing with wildflowers. Photo by Jeanine Davis.

Bonnie's gardens are a tranquil place nestled in a neighborhood. Photo by Jeanine Davis.

Bonnie's property has many beautiful, mature trees. Photo by Jeanine Davis.

book—but I came away with more than I expected. All four of these people are at the stage in their lives when they could be doing whatever they want. Others might choose to spend their days shopping, playing bridge, swinging a golf club, or watching TV, but not these four. They volunteer in numerous organizations, spend time with people with similar interests, read non-fiction, cook good food, and garden daily. I was impressed with how physically fit they all were and how much they laughed and smiled. When my daughter came home late that evening, I showed her a picture of the four of them, and we agreed they should be role models for both of us. I didn't ask them for advice for new gardeners, but after spending a few hours with them, the lessons I took home were to have a passion for something and try to make a difference in the world, even if it is in a small way. I think they have all done a fine job on both accounts.

Ceara Foley: An Herbalist in Paradise

Ceara Foley is a very talented and popular herbalist in western North Carolina. I have known her for many years and enjoy taking plant walks with her or attending classes she teaches at conferences in our region. Ceara is the director/owner of the Appalachia School of Holistic Herbalism in Asheville, North Carolina. It is the oldest school

There are many special places on Ceara's property, like this beautiful root cellar she had built into the hillside. Photo by Jeanine Davis.

The plants Ceara likes to use the most she grows close to the house. Photo by Jeanine Davis.

of herbal studies in the Southeast and the one where she herself studied. I visited Ceara at her home in the Sandy Mush area of Madison County on a beautiful sunny afternoon in September 2013.

To reach Ceara's place, I headed north out of Asheville along a road that grew more rural, winding, and mountainous as the miles rolled on. Turning off the main road onto a narrow gravel road, I drove slowly, as directed, giving myself time to study the scenery as it passed by. Then I turned onto an even rougher dirt road for the final approach to her place. Soon I passed a faded sign indicating the land was a United Plant Savers Sanctuary. Finally, at the end of the road I entered a little paradise of greenery with a beautiful house and gardens situated in a small clearing butted up against the mountainside, with a small stream running along the side.

As I pulled up to the house, Ceara's young daughter came running out to greet me, immediately asking if I would like a glass of water or a cup of tea. What a great little hostess. She shut the gate behind my car and led me into the house. Ceara and her daughter share a charming log house comfortably nestled up against the hillside. Big windows let in lots of sunlight and offered wonderful views of greenery and colorful late-season flowers. The house was comfortable and welcoming—and totally off the grid. Ceara has lived that way most of her adult life. I'm impressed!

We sat at the kitchen table and spent some time talking about mutual friends, local activities, and food. Then we got down

Whimsical little touches like the glass toadstool and a Tinker Bell figurine were fun to find. Photo by Jeanine Davis.

Entering the vegetable and flower garden. Photo by Jeanine Davis.

to business. I wanted to learn about all the medicinal and edible plants that Ceara nurtured on her homestead.

Ceara moved to the Asheville area to study with our mutual friend, herbalist Peggy Ellis. Fifteen years ago, she bought her homestead (over 30 acres, far away from town) so she could surround herself with the plants she studies. She grows a wide array of herbs, vegetables, fruits, and flowers, for her practice, personal use, and enjoyment.

Ceara had a plant inventory done years ago, something I would recommend anyone with an expanse of wooded property do if they can afford it. That inventory told her what was on her land to begin with, where it was located, and helped her decide what she needed to bring in. Remember that a good plant inventory cannot be done all at one time. Woodland plants in particular are very seasonal. Someone needs to return several times during each season to catch the plants at their different stages of growth and bloom. Ceara's plant inventory showed that she didn't have goldenseal on the property, so that was one of the first plants she introduced. There was ginseng growing there already, but she seeded more so she could have several populations in different locations.

Plants and nature are a source of comfort to Ceara. She has had physical challenges her whole life, and plants have helped her cope with them. She feels she is giving back to the plants that have helped her when she educates others on how to grow and protect them.

A small stream runs near the house and gardens. Photo by Jeanine Davis.

We walked outdoors, through the gate, and around to a narrow little garden with a rock wall that runs along one side and the back of her house. This is at the base of the hill, so the garden rises right up in front of you. I was charmed by the beautiful moss-covered rock work and the heavy wooden door on a root cellar nestled into the hillside. This garden was shady, cool, and damp, filled with a wide variety of plants and trees, The Solomon's seal was dripping with big purple berries, and the jewelweed was ablaze in yellow, orange, and red flowers. Ceara's daughter enjoyed pointing out a little glass mushroom and a figurine of Tinker Bell tucked up under the ferns. This is the garden where Ceara keeps her woodland herbs close to her. There is black cohosh, blue cohosh, goldenseal, bloodroot, and many other herbs all growing contentedly together in this charming little area.

Next, we headed out to her big garden that is surrounded by a wooden fence with big wooden doors for the gate. Ceara was embarrassed by what she thought was a messy garden. She had been out of town for two weeks, and the day she had planned to clean it up, it had rained heavily. Ceara calls herself a lazy gardener. I don't agree with that assessment, but if she compares her gardens to the average American garden, I can understand why she might think that is what people would say. Ceara's gardens are not manicured, but they are big and full of life and color. Many of the weeds growing there are medicinal, and all of them add to the biodiversity of the garden, providing

Ceara has created a very special place to reside. Photo by Jeanine Davis.

Ceara and her daughter. Photo by Jeanine Davis.

homes for beneficial insects. Everywhere, there were birds, butterflies, insects of all kinds, chickens, cats, and probably a few snakes (we joked about them as we carefully watched where we placed our feet). Ceara makes medicine from plants that most of us consider weeds. She lets most weeds grow freely, then has a big weeding party and turns the plants she "weeded" from her gardens into tinctures. Jewelweed, for example, was lush and blooming everywhere, and that is a favorite plant for her to tincture. I enjoyed looking at all the vegetables and flowers she had growing in this garden, but she really took me out there to see the black cohosh plants that came up unexpectedly in her asparagus beds one year and have lived there quite happily ever since.

Ceara believes that we should use the plants that naturally grow in the region where we live. These are probably the plants we need the most. She plants the herbs she uses frequently in the gardens next to her home; bloodroot is an example of a plant that she likes to have close to the house where she can see it often and have it at hand for making medicine. Black cohosh is another plant she cultivates and keeps close by. Black cohosh is good for arthritis and other conditions she is coping with. Ceara grows most of her woodland plants from root divisions. With the exception of ginseng, she doesn't grow many of them from seeds.

Ceara's gardens serve more than her own family. She uses her homestead for teaching and often holds weekend classes there. She welcomes these times when many eager hands are available to weed, plant, and make medicine.

Advice to New Gardeners

Ceara would like to see everyone have a garden, even if it is very small. But if you are fortunate enough to own a little land, she strongly suggests that you take the time to get to know it before you start planting. Spend an entire year observing what grows naturally on your property. Observe the kinds of places and conditions each type of plant grows in. Get to know what the plants look like and what likes to grow together. Ceara refers to these as "guilds." Remember that many of the woodland plants are seasonal, and their bloom times are often short. So you need to walk your property frequently all year long to observe them all.

When you are ready to plant, start close to your house with small plantings of many kinds of plants. Consider the plants that you use or want to learn more about. For example, if you or a family member tends to have respiratory problems or infections, plant herbs that are good for those conditions. Boneset, mullein, goldenseal, and elderberry are just a few examples of respiratory-support herbs. Choose some plants that have multiple uses, for example, goldenseal, black cohosh, or ginseng. And you don't have to plan to use every herb you plant. Ceara, for example, grows wild ginger because it is pretty, native, and easy to grow from root divisions, but she doesn't use it in her practice.

Blue cohosh is a favorite plant of hers. It is super easy to grow and a very pretty plant in a mixed garden. She does make medicine from it, although she doesn't use it herself. I noted a lot of Solomon's seal growing close to her house. She likes the plant, but she only uses a little of it.

Ceara wants to remind all gardeners, but particularly new gardeners, to be patient with woodland plants. They are different from the vegetables and culinary herbs that you are probably used to growing in your gardens. Some of the woodland plants can take several years before they emerge, so mark the areas where you planted them and wait. Sometimes plants do disappear—moles or wild turkeys might get into them—but often they just need some time.

When you are in Wallace's woods, you would never know there is a house right nearby. Photo by Jeanine Davis.

Be patient, enjoy all the plants around you, and create your own herbal paradise.

Wallace Souther: A Hobby Gone Wild!

I knew Wallace because he had worked in my building for a number of years. We used to eat lunch together and talk about a wide variety of topics, including herbs. I knew that he had planted a few woodland herbs somewhere nearby, but I didn't know how extensive his interest and plantings were until I started looking for gardeners to interview for this book, and a mutual friend suggested I contact him.

The Land

One sunny, warm August morning in 2013, I drove out to meet Wallace at his property in the Cane Creek area in Henderson County, North Carolina. The land, which had once been a mink and hog farm, has been in his family for four generations. He showed me the remains of an outside furnace where his ancestors cooked large kettles of potatoes to feed to the hogs. Originally the farm was about 50 acres, but it has been divided up several times over the years and now the family owns two parcels totaling about 20 acres. The family wants to keep the remaining land for future generations and is in the process of creating a trust or some other legal arrangement to do so. Two of Wallace's aunts live on the land. One lives in the old family home where she grew up, and the other in a much newer house in the center of the property. Eventually, Wallace would like to build a vacation home there, too. He does have a small trailer on the property that he spends some time in, but usually he just visits and works the land during the day.

Wallace serves as a caretaker for the land and has a few money-making projects going on in the open areas. There, he is growing vegetables, a row of hops, and a big bed of flowers. He recently constructed a big hoop house so he can grow vegetables in the cooler months. He received a USDA Conservation grant to pay for it. He received another USDA Conservation grant to put in a well for irrigation purposes. That, combined with a Pasture Management Plan, will provide water for watering troughs for livestock and fencing to control the grazing of horses that are being raised by a local trail riding business leasing some of his bottom lands. He is also growing some field crops in the bottom land. He raises worms and, until recently, he kept a small flock of chickens. His newest interest is in raising insects for food. He hasn't started that venture yet, but he's sure it will be popular in the future, and he wants to be one of the early adopters.

Wallace enjoys growing many kinds of mushrooms. These are shiitakes. Photo by Jeanine Davis.

Wallace checking on his little vermicomposting bin. Photo by Jeanine Davis.

The Woods Garden

But the reason I was there was to see his woods gardens. Wallace loved showing off his acre or so of woods, and it was beautiful with morning sun beams shining through the bright green foliage. Although there is a house right next to the driveway along the forest edge and another right across

Wallace with the mushrooms he planned to eat for dinner. Photo by Jeanine Davis.

the street, it is a quiet area, and when you stand in the middle of his woods gardens you could just as well be in the middle of a big forest. Wallace first started caring for the woods and planting herbs there six years ago. Now he is writing a forest management plan and intends to incorporate herbs, mushrooms, and other non-timber forest products into it.

Unfortunately, the major weed Wallace must contend with is poison ivy. Since he has a bad reaction to its oil, getting it under control is a top priority. He tries to grow as organically as possible, but when preparing a new area for planting he often uses

Wallace looking for his young ginseng. Photo by Jeanine Davis.

Wallace's woods are mostly partial shade, which results in more plant growth on the forest floor. Photo by Jeanine Davis.

the herbicide Roundup to knock back the poison ivy. He did apply for a USDA Conservation grant to pay for goats to control the poison ivy instead, but the USDA was out of funds by the time he applied. He will try again next year.

Ginseng was one of the first plants he established in his woods garden. He did a lot of reading about growing it and then ordered stratified seed online from a company in Pennsylvania. Then he used Scott's wild-simulated planting method to plant several areas. He now has small patches of ginseng scattered throughout the woods, and he wants to plant some every year or two so there will be always be ginseng on the property for future generations of his family to use and enjoy. He wouldn't mind making a little money from it eventually, but mostly he is growing it to conserve it and for posterity. He is sure there used to be ginseng in those woods years ago.

He has also planted ramps, goldenseal, black cohosh, and St. John's wort. These are scattered in small plots across the woods. Some were planted as seeds and some as transplants, and they are all different ages. There are many native plants there, too, including ferns and Solomon's seal. Wallace does not have his plants in manicured beds.

This is not a display garden. It is a naturalized woodland garden with faint paths that you can wander along. You have to look closely to find the botanicals, but it's like a treasure hunt. Wallace got so excited leading me through the woods and pointing out various plants and telling me stories of how they got there or what he uses them for. Because of the wide diversity of plants, both native and planted, it is an ever-changing landscape of colors, textures, and smells. Once young plants are established, Wallace pretty much leaves them alone, letting nature create a wonderful polyculture of plant species that probably helps keep disease and insect pressure low. It is a good example of a low-input method of maintaining a woods garden. It is not for everyone, but it's well-suited to Wallace's relaxed, fun, permaculture, self-sufficient approach to life.

Wallace also has a passion for mushrooms and was experimenting with different methods of growing them—some more successfully than others. He has blue oyster, angel wings, hericium, and shiitake mushrooms. He pointed out many logs and stumps that he had inoculated, and some were fruiting while we were there, including the shiitake logs.

Future Plans

Wallace's long range plan for the woods is to make it a beautiful, relaxing forest haven for his family to enjoy. He wants to put in solar power and a rainwater collection system and build an outdoor canning and tincture making-kitchen. He might even build some tree houses to rent out so others can enjoy the beauty of what he is creating there.

Wallace grows these plants and fungi because he enjoys them, uses them, and wants them there for his daughter and her children to enjoy. He makes tinctures from the herbs and eats all the mushrooms. He might sell a few mushrooms and herbs from time to time but mostly this is a labor of love. He is a passionate forest gardener.

PART SIX

Sustainable Wild-harvesting

A book about ginseng and other woodland botanicals would not be complete without at least a short discussion about wild-harvesting. The history and culture of these plants are based on collecting these plants from wild populations. This is still a vibrant industry in much of Appalachia and on the west coast of the United States. The markets for some plants are such that wild-harvesting or managing of wild populations might be the best option for making a profit, but conservation of native populations must always be an important consideration.

35

What Is Wild-harvesting?

Wild-harvesting, also referred to as wildcrafting, is the harvesting of plants, ferns, and fungi from wild populations growing naturally in the environment. In most cases, a person did not plant them there. I say "in most cases" because wild-harvesters, or gatherers, have been sowing ginseng seeds throughout the Appalachian mountains for over 100 years. But for everything else there, has been little or no human involvement with the plants, ferns, and fungi that wild-harvesters gather.

I think most people would be surprised to learn just how big an industry wild-harvesting is. This is not just a historical activity. Many people who hunt, fish, and trap also wild-harvest herbs, roots, mushrooms, and ferns. This has recently been brought to the attention of the public through some reality television shows highlighting rather extreme examples of wild-harvester lifestyles.

Who are the Wild-harvesters?

Where I live in western North Carolina, wild-harvesting is part of the culture. There are traditional and non-traditional gatherers. The traditional gatherers are people who have wild-harvested ginseng, ramps, and other herbs for generations. For years, the plants were gathered for use at home as food and medicine and to sell. For some families, the money they made from wild-harvesting was a substantial part of their annual income. It is still common, for example, for grandfathers to take their grandchildren out into the woods in the fall to teach them how to find and harvest ginseng to make a little extra income. Hundreds of people harvest roots and herbs to sell each year. There is also a non-traditional group of wild-harvesters composed of herbalists, survivalists, and permaculturists who also gather these plants and fungi for making herbal medicine and for food. And we can't leave out the notorious poachers. Unfortunately, these are people who steal these plants and fungi from private and public lands. Poaching is a serious problem in some areas, especially for ginseng.

Wild-harvesters gather plants of value from forests, meadows, and fields. These may be on public lands or private lands. It is illegal to collect plants from most state and national parks, but state and national forests have different rules concerning wild-harvesting. Some require permits and limit harvest periods. Some charge fees and limit harvest amounts. Wild-harvesting can also take place on private land. Permission from the landowner to harvest is required in all states and provinces.

Wild-harvest Methods

There are different "methods" of wild-harvesting. You might see the term "ethically wild-harvested" on a herbal product label. This means that the wild-harvester was concerned about the long-term sustainability of the plants he or she harvested, being careful to harvest in such a manner that the plant population will continue to grow and prosper. For example, many ethical wild-harvesters plant the seeds of a ginseng plant before digging the root, only harvest 10 to 15 percent of a patch of ramps, or refuse to harvest from a tiny population of goldenseal. These harvesters want there to be plants for future generations to enjoy and collect from. There is another group of harvesters who are just in it for the money, and when they find a patch of ginseng or ramps, they dig it all. They want to make as much money as they can right now, and there is no concern for future harvests or the conservation of the plant. Poachers use this method, too, but they also dig out of season, without permits, and take from private land, and state and national parks.

The American Herbal Products Association created leaflets for 19 states on sustainable harvesting of ginseng. They call this "good stewardship" and are encouraging harvesters to follow their guidelines so wild ginseng will be here for generations to come.

Who Do Wild-harvesters Sell To?

Most wild-harvesters sell the roots they dig and herbs they gather to raw-botanicals buyers, also referred to as dealers. Many ginseng dealers buy other herbs, also. These buyers are usually not very visible to those outside the industry. They rarely advertise and few have webpages. Wild-harvesters find buyers by talking to other wild-harvesters. The exception is ginseng dealers. Most states post a list of registered ginseng dealers, complete with addresses and phone numbers. I do not know who poachers sell their stolen herbs and roots to. Most of the buyers I communicate with tell me that they don't knowingly buy illegally harvested material, so where that material enters the supply chain is a mystery to me.

Wild-harvesting on Your Own Land

If you are fortunate enough to own property that has populations of valuable herbs, ferns, and fungi, you can harvest from these populations for sale or use. You can also use these populations as a source of planting stock, rhizomes, and seeds for starting production areas. Many people manage the wild populations on their property. They harvest some of the plants to offer for sale and also expand existing populations by planting seeds and making some root divisions. The majority of seeds that are sown naturally in the forest never germinate because they don't ever make contact with the soil. The simple practice of gathering seeds, scratching back the leaf litter, pressing the seeds into the soil, and covering them again will increase germination percentages dramatically.

36

Why There Will Always Be a Place for Wild-harvesting

Some of my botanist friends who see the damage caused by wild-harvesting would like to have the practice banned forever. As a conservationist I understand this view, but as someone who works in the medicinal herb industry, I know this will never work. We can make a strong case to the consumer that cultivated herbs can be of higher quality, are more consistent in their biochemical profiles, are less likely to contain unknown species, and are better for protecting wild populations. The reality is that it costs money to farm herbs, and for some herbs the demand and prices don't make them a viable economic option for farming. This could be because the market for a particular herb is very erratic, with buyers only looking to purchase it every few years. Or the volume of an herb required may be very small. Or the prices are just too low to cover the costs of production. Some growers have enough land and other resources available to plant some plants and "just let them grow" or to carefully manage wild populations. But for many herbs, the number of people doing this is not sufficient to satisfy the demand when it is there. For those reasons, it is my opinion that there will always be a place for wild-harvesting in this industry.

Wild-harvesting for "minor herbs," as I refer to them, can be sustainable if we adopt ethical practices that protect wild populations. This could involve training harvesters in ethical harvesting practices, requiring them to get licenses, and monitoring what and how much they collect. I am not a hunter, but this sounds like what I'm told is required of hunters in many states. Efforts to start programs such as these are being discussed in many areas.

There is also the tradition and activity of wild-harvesting that many people want to maintain. Many Native Americans, herbalists, hunters, and wild-food enthusiasts enjoy spending time in the woods gathering herbs to make medicine or share with friends. We should always maintain places where people can personally wild-harvest just as we maintain places for people to fish, hunt, and hike.

37

Rules and Regulations for Wild-harvesters

There are rules and regulations that wild-harvesters must follow, and they vary from place to place. It would be impossible for me to list all of them for all the states and provinces in North America, but by providing some examples you should know what to look for in your own area.

Protected Plant Status

Find out what the protected plant status of the plant you want to harvest is. Every state and province has a list of protected plant species. An Internet search on "protected plant species" along with the state or province should reveal it. Some of these lists are quite extensive, e.g., there are 419 plants on the North Carolina list as of September 2013. There are also federal lists. Read what the rules are pertaining to plants on these lists. The plants will be grouped in categories, such as endangered, threatened, rare, exploitably vulnerable, and special concern.

In North Carolina, for example, to propagate or sell any plant on the protected plant species list, a person has to get a permit, identify the source of the initial planting stock, demonstrate how they are going to grow the plants, allow for an annual inspection, and maintain records on the plants. And a new rule in North Carolina is that galax can only be harvested from July 1 through March 31.

State, Federal, and International Regulations

CITES regulations on the international movement of protected plants have been explained in the ginseng and goldenseal sections of this book. If you have any plans to move a protected plant across state or national lines, check the CITES listings to see if any of those rules apply.

The ginseng rules are ones most wild-harvesters need to understand. If you wild-harvest ginseng and want to sell it to someone in another state or country, you must sell it to a state-licensed ginseng dealer or be a state-licensed ginseng dealer. Many states, such as Wisconsin, have leaflets available that clearly spell out all the rules pertaining to wild-harvesting ginseng. West Virginia has a webpage devoted to the topic, including a useful Frequently Asked Questions section. If you live in Canada, you cannot legally wild-harvest or sell wild ginseng. Also, check the Custom and Border Protection websites before trying to move any plant material across country borders.

PART SEVEN

Supplemental Information

Don't let the title "Supplemental Information" fool you. This section contains some of the most valuable information in the book! You'll find descriptions of 60 botanical plants currently being traded in the industry, a historical pricing table, information on the new regulations for dietary supplements that affect growers and manufacturers, Joe-Ann's disease list, and the comprehensive resources directory so you don't always have to flip through the book to find an email and URL. I think you will find that you return to this section over and over again.

Joe-Ann McCoy's Disease List is a very special chapter in this book. I don't think anything like it exists anywhere else. Not only does this list let you know what problems you might encounter when growing these plants, it can help you when your plants are in trouble. When you have a disease on your plants, getting a fast and accurate diagnosis is critically important. As plant disease clinics at universities and plant pathology departments experience more budget cuts, faculty have increasing responsibilities. They have less time than they ever did to spend diagnosing diseases on minor crops such as these botanicals. By supplying them with the list of known diseases when you submit a diseased plant sample for diagnosis, you give them an important tool that can speed up their job, increasing the chances that you will find out what is wrong with your plants in time to save them.

Dr. Joe-Ann McCoy provides many valuable services to growers, wild-harvesters, researchers, and botanists in North America. She is the director of the North Carolina Arboretum Germplasm Repository in Asheville, North Carolina. The mission of that facility is to maintain a diverse collection of native medicinal plant germplasm to support research and conservation. She educates growers and wild-harvesters on proper plant identification and teaches them how prepare plant vouchers. She also conducts research on many aspects of medicinal plant conservation, ecology, reproduction, and chemistry.

APPENDIX 1

Forest Botanicals Bought and Sold in the United States and Canada

There are hundreds of different plants and plant parts bought and sold in the botanicals industry. The list below contains many of the forest botanicals commonly traded in the United States. It is not a complete list by any means, but it does include the plants that many companies are currently buying and selling as raw botanicals. The majority of the plants listed are only wild-harvested at this time, but you might be able to make money growing some of them. Then again, it may be that some of these plants cannot be cultivated economically, but could be sustainably (and profitably) wild-harvested from your property. That is for you to explore. Historical and current prices for these plants are included in Appendix 2.

Be sure to check on any state, provincial, or federal legislation concerning these plants. Some of them are protected plants, and that puts restrictions on collection from public lands and sometimes results in the need for permits to cultivate certain plants. It definitely affects where you can collect planting stock from. I always recommend that you obtain planting stock from your own land or a reputable nursery. Looking over some of the protected plant lists, I noted that some plants showed up on a large number of those lists. They are blue cohosh, butternut, ginseng, goldenseal, and stargrass.

The list below includes the common names and scientific name (*in italics*) for each plant, along with a brief description of the plant, where it grows naturally, the plant part or parts that are harvested to sell (**in bold**), and a few of the uses for the plant material gathered. Pay close attention to the scientific names for each plant. The first part of the name is the genus and the second part is the species. Be sure that you are collecting the right plant for your market. The same common name can refer to several different plants. For example, the name "yellowroot" is used for both *Hydrastis canadensis* and *Xanthorhiza simplicissima*. The species name is also important. For example, there are many species of skullcap, including *Scutellaria lateriflora* and *Scutellaria baicalensis*. Both are important medicinal herbs and both can be cultivated. But the first one, *S. lateriflora*, is the native skullcap that can be wild-harvested, whereas *S. baicalensis* is the Chinese skullcap, which must be cultivated.

A few definitions:

- **annual:** a plant that completes its life cycle in one year
- **biennial:** a plant that grows two years, blooming in the second year
- **deciduous:** a plant with leaves that fall off each year
- **evergreen:** a plant with leaves that stay on all year

- **herbaceous perennial:** the above-ground part of the plant dies each year, but the underground part lives on
- **perennial:** at least part of the plant lives on from year to year

Balm of Gilead (*Populus balsamifera*): Also known as poplar buds and balsam poplar. A medium-sized, perennial tree that grows in wet woods, swamps, and river bottoms throughout the northern United States and most of Canada, south to Virginia. In southern states, **buds** from other poplars are collected, including cottonwood trees (*Populus deltoides*). The sticky, fragrant buds are harvested in the winter or early spring before they open. They are steeped in oil or made into a salve used to treat muscle soreness, cuts, abrasions, and sunburn.

Balmony (*Chelone glabra*): Also known as turtlehead. A two- to three-foot tall, herbaceous perennial with leafy stems and white to pinkish flowers that grows in wet woodlands in eastern North America from Manitoba, south to Alabama, and west to Minnesota. **Leaves** are harvested in the spring. They are steeped into a tea used as a laxative, for indigestion, as a tonic, and to treat worms and jaundice.

Bayberry (*Myrica cerifera*): Also known as wax myrtle. A large, perennial, evergreen shrub with leathery leaves that grows in coastal areas, swampy areas, and pine barrens from New Jersey, south to Florida, and west to about Arkansas. **Root bark** is harvested in the fall. It is used to treat uterine hemorrhages, jaundice, diarrhea, sore throats, and canker sores.

Bethroot (*Trillium erectum*): Also known as red trillium, birthroot, purple trillium, squaw root, wake-robin, and ill-scented trillium. A herbaceous, perennial, spring wildflower with three big leaves and a brown, maroon, or purplish flower that grows in rich moist forests in eastern North America from Quebec, south to Georgia, and west to Michigan. **Roots** are harvested in late summer and early fall. It is used as a labor stimulant and to treat a wide range of disorders including internal bleeding, lung disorders, skin infections, and problems with menstruation and menopause.

Black cohosh (*Actaea racemosa* or *Cimicifuga racemosa*): Also known as rattlesnake root, squaw root, papoose root, and bugbane. A herbaceous perennial, about four feet tall, with feathery leaves and tall spikes of white flowers that grows in rich woods in the eastern half of North America from Quebec, south to Georgia, and west to Missouri. **Roots** are harvested in the late summer or fall. It is used to treat menopausal symptoms and uterine pain, it also has anti-inflammatory and sedative properties.

Black haw (*Viburnum prunifolium*): Also known as arrow-wood, crampbark, and Southern black haw. A small, perennial, deciduous tree or shrub that grows along forest edges, creeks, fence rows, and roadsides in eastern North America from Quebec, south to Florida, and west to Iowa. **Root bark** and **tree bark** are harvested in the early spring or fall. It is used as a tonic and to treat diarrhea, uterine cramps, and spasms.

Black Indian hemp (*Apocynum cannabinum*): Also known as dogbane and Indian hemp. A shrub-like, herbaceous perennial that grows on forest edges and old fields throughout most of North America. **Roots** are harvested in the fall. It is used for respiratory health and as a laxative and to induce vomiting. All parts of this plant are poisonous.

Black root (*Veronicastrum virginicum*): Also known as Culver's root. A two- to five-foot-tall herbaceous perennial with spikes of tiny white to purple flowers that grows in rich moist woods, uplands, meadows, and prairies throughout most of the eastern half of North America. **Roots** are harvested during the summer. It is used as a mild laxative, to stimulate the liver, and to induce vomiting. The fresh root is toxic.

Black walnut (*Juglans nigra*): Also known as walnut. A tall, perennial tree that grows in rich woods throughout most of the eastern half of North America. **Hulls** are gathered in the fall. **Leaves** are harvested any time. A decoction is made from the hulls to treat ringworm and other fungal infections, and a strong tea is made of the leaves to eliminate internal parasites.

Blackberry (*Rubus villosus*): Also known as dewberry and brambleberry. A perennial trailing bramble that grows on forest edges, in old fields, and pastures throughout the eastern half of North America. **Roots** and **root bark** are harvested in the spring and fall. **Leaves** may also be used. It is used in a tea to treat cold symptoms, diarrhea, and a variety of female problems.

Bloodroot (*Sanguinaria canadensis*): Also known as red Indian paint, redroot, and puccoon. A low-growing, herbaceous perennial, with a showy white spring flower, that lives in moist forests and forest edges throughout most of the eastern half of North America. **Roots** are harvested in late summer and fall. It is used to treat skin lesions of all sorts, sore throats, and gingivitis. It has antimicrobial and antifungal properties. It is also a dye plant. The root can be toxic, and some people have a skin reaction to the root sap.

Blue cohosh (*Caulophyllum thalictroides* and *C. giganteum*): Also known as papoose root, squaw root, and blue ginseng. A three-foot-tall, herbaceous perennial with bluish-purple foliage in the spring, that grows in forests in the eastern half of North America south to Georgia. **Roots** are harvested in the early fall. It is used to treat many gynecological problems, asthma, and other respiratory issues. It has antimicrobial and antifungal properties.

Blue flag (*Iris versicolor*): Also known as wild iris and flag lily. A two- to three-foot tall perennial with attractive blue flowers that grows in swampy areas and along pond edges throughout northeastern North America, south to Virginia. **Roots** are harvested in the fall. It is used to treat various skin diseases.

Boneset (*Eupatorium perfoliatum*): Also known as feverwort and sweat plant. A single-stemmed, herbaceous perennial, about five feet tall, topped with a cluster of white flowers, that grows in woods, swamps, pastures, and prairies throughout eastern North America, west to the Dakotas. The **entire top** of the plant is harvested when the plant is blooming in late summer and early fall. It is used as a laxative and to treat coughs, fevers, and chest illnesses.

Bugleweed (*Lycopus virginicus*): Also known as gypsywort. A perennial mint that grows in rich woods, swamps, and other shady, damp areas throughout eastern North America, west to Texas. The **entire top** of the plant is harvested in summer and fall. It is used as a sedative and treatment for diabetes, digestive problems, and lung disorders.

Butternut (*Juglans cinerea*): Also known as white walnut. A perennial tree that grows in rich woods in eastern North America, south to Alabama, and west to Iowa. **Bark** is harvested in the summer. It is used as a dye in traditional crafts.

Cleavers (*Galium aparine*): Also known as bedstraw. A low-growing annual with whorls of leaves surrounding the stem that can be found in woods and pastures throughout North America. The **entire top** of the plant is harvested in the summer. It is used to increase appetite, reduce fever, and cure vitamin C deficiency.

Cramp bark (*Viburnum opulus*): Also known as cranberry tree and black haw. A tall, deciduous, perennial shrub that grows in low areas throughout the northern United States and Canada, south to Tennessee. **Bark** is harvested in the fall. It is used to treat uterine cramps and as an astringent.

Cranesbill (*Geranium maculatum*): Also known as wild geranium. A herbaceous perennial, one to two feet tall, with deeply cut leaves and pink to lavender flowers, that can be found growing in moist woods, meadows, and along fence rows throughout eastern North America, south to Georgia, and west to Kansas. **Roots** are harvested from late spring through summer. It is used in a tea to treat sore throats and diarrhea. It is also used as an antiseptic and to reduce swelling.

Elderberry (*Sambucus nigra* L. ssp. *canadensis*): Also known as elder. A large, perennial shrub that produces fragrant white flowers and bunches of small purple berries. It can be found growing along forest edges, streams, and in open areas throughout most of North America. **Flowers** are collected in mid-summer and **berries** in late summer. It is commonly used to treat colds, flu, and coughs.

False unicorn (*Chamaelirium luteum*): Also known as star root, star grubroot, grubroot, fairywand, devil's bit, helonias, and starwort. A herbaceous perennial with a rosette of leaves and a tall, curved flower-stalk that grows in forests, bogs, and meadows from Ontario, south to Florida, and west to Illinois. **Roots** are harvested in the fall. It is used to treat many gynecological problems, including menopause. It is also used to treat pain, depression, and coughs.

Figwort (*Scrophularia marilandica* and *S. nodosa*): A tall, shrub-like, herbaceous perennial that grows in forests, forest edges, fields, and roadsides in eastern North America, west to Oklahoma. The **entire top** of the plant or **entire plant** is harvested in the summer. Figwort is used internally and externally to treat skin disorders.

Fringe tree (*Chionanthus virginicus*): Also known as white ash. A small, perennial tree that grows on stream banks and in other damp places in woods from Pennsylvania, south to Florida, and west to Texas. **Root bark** and **tree bark** are harvested in spring and fall. It is used as a liver tonic and to improve digestion.

Galax (*Galax urceolata*): Also known as beetleweed, coltsfoot, and wandflower. A low growing perennial with glossy leaves that grows in the southeastern United States. **Leaves** are harvested year round. It is used in the florist trade.

Ginseng (*Panax quinquifolius*): Also known as sang, manroot, and American ginseng. A herbaceous perennial that grows in rich forests in eastern North America, south to Georgia, and west to Oklahoma. **Roots** are harvested in the fall. It is used as a tonic, aphrodisiac, and to treat a wide range of disorders.

Goldenseal (*Hydrastis canadensis*): Also known as yellowroot, ground raspberry, yellow puccoon, wild circuma, eye-balm, yellow paint, wild turmeric, and yellow-eye. A herbaceous perennial that grows in rich woods in eastern North America from Ontario, south to Georgia, and west to Kansas. It is a protected plant in some states and provinces. **Roots** are harvested in the fall. **Leaves** are also gathered anytime. It is used as an eyewash and a treatment for sore throats, nasal congestion, and stomach disorders. It is also a good surface antiseptic and has anti-inflammatory properties.

Indian turnip (*Arisaema triphyllum*): Also known as jack in the pulpit. A perennial that grows in damp forests and bogs in eastern North America, west to Kentucky. **Roots** are harvested in spring and fall. It is used to treat coughs, and lung and digestive disorders.

Life root (*Senecio aureus*): Also known as squaw

weed, cough weed, and false valerian. A perennial that grows in moist woods and swamps from Maryland, south to Florida, and west to Minnesota. The **top of the plant** is harvested in the summer. It is used as a uterine tonic, for kidney problems, and to treat coughs and poor circulation.

Lobelia (*Lobelia inflata*): Also known as Indian tobacco. An annual about three feet tall with pale blue to violet flowers that grows in open woods and pastures in eastern North America, south to Georgia, and west to Nebraska. **The entire top of the plant** is harvested when the seed pods have matured. It is used for respiratory problems and in homeopathic preparations and smoking blends.

Maidenhair fern (*Adiantum pedatum*): Also known as finger fern and ginseng pointer. A perennial fern that grows in moist woods and on creek banks in eastern North America, west to Oklahoma. **Fronds** are harvested throughout the growing season. It is used to treat coughs and colds.

Mayapple (*Podophyllum peltatum*): Also known as mandrake, umbrella plant, Indian apple, and devil's apple. An early blooming, herbaceous perennial with a single umbrella-like leaf that grows in rich forests and on forest edges in eastern North America, west to Nebraska. **Roots** are harvested in late summer. It is used to treat warts, cancer, hepatitis, constipation, and liver problems.

Missouri snakeroot (*Parthenium integrifolium*): Also known as wild quinine and American feverfew. A three- to five-foot-tall herbaceous perennial that grows in open woods and prairies throughout the eastern United States. **Roots** are harvested in the fall. **Leaves** are also sometimes harvested. It is used as a diuretic and as a treatment for fevers and kidney problems.

Mountain maple (*Acer spicatum*): A small, perennial tree or shrub that grows in moist woods in Canada from Saskatchewan to Newfoundland and in the eastern United States, south to Georgia, and west to Iowa. **Bark** is harvested any time. It is used to increase the appetite and as a treatment for eye infections.

Mullein (*Verbascum thapsus*): Also known as Indian tobacco, wooly mullein, and common mullein. A fuzzy-leafed biennial with a tall flower-stalk with small yellow flowers that grows in open woods, forest edges, and along roadsides throughout North America. **Leaves** are harvested before the plant blooms. **Flowers** and **roots** are also sometimes harvested. The leaves are used to treat respiratory problems. Flowers are steeped in oil and used for ear aches.

New Jersey tea (*Ceanothus americanus*): Also known as red root and Jersey tree. A perennial shrub that grows in open woods and on rocky areas in eastern North America, west to Texas. **Roots** are harvested in the fall and **leaves** in the summer. Leaves are used as a tea, for skin problems, skin cancer, and for sore throats. Roots are used as a sedative.

Oregon grape (*Mahonia* [or *Berberis*] *aquifolium*): Also known as holly-leaved barberry and mountain grape. A small, evergreen, perennial shrub with shiny leaves and blue-black berries. Different *Mahonia* species grow throughout most of North America, south to North Carolina. **Roots** are dug in the fall. It is used as a goldenseal substitute. It has antimicrobial properties and is a liver stimulant. It is also used for constipation, skin diseases, acne, and rheumatism.

Partridge berry (*Mitchella repens*): Also known as squaw vine and squaw berry. A low-growing perennial vine with tiny red berries that is found in moist woods of western North America, west to Oklahoma. The **top of the plant** is harvested in the late summer to fall. It is used to treat many gynecological problems.

Passionflower (*Passiflora incarnata*): Also known as maypop. A perennial, climbing vine with striking purple and white flowers that grows on forest edges, in open woods, and on fence lines from Pennsylvania, south to Florida, and west to Oklahoma. The **top of the plant** is harvested in the summer, just as the plant begins to bloom. It is used as a mild sedative, and to treat high blood pressure and anxiety.

Pinkroot (*Spigelia marilandica*): Also known as Carolina pinkroot, Carolina pink, Indian pink, American wormroot, and wormgrass. A herbaceous perennial with red-and-yellow flowers that grows in woods and thickets throughout the southeastern United States, west to Texas. It is a protected plant in some states. **Roots** are harvested in the fall. It is used to treat indigestion.

Pipsissewa (*Chimaphila umbellata*): Also known as dragon's tongue and spotted wintergreen. A very short, evergreen perennial with shiny, leathery leaves that grows in forests throughout northern North America, south to about North Carolina. **Leaves** are harvested during the growing season. It is used as a tonic and as a treatment for urinary tract and kidney problems.

Pokeweed (*Phytolacca americana*): Also known as poke, pokeroot, and poke sallet. A large, shrub-like perennial with long clusters of shiny purple berries that grows on forest edges, fence rows, and anywhere the soil has been disturbed in eastern North America, west to Minnesota, and the southwestern and west coast states. **Roots** are harvested in the fall. It may have anti-tumor properties and is used to treat arthritis and infections.

Prickly Ash (*Zanthoxylum americanum* and *Z. clava-herculis*): Also known as tooth ache tree. A perennial shrub or small tree that grows in forests and on riverbanks in the eastern United States. *Z. clava-herculis* is the more southern species. **Bark of stems and roots** is harvested in spring and fall. It is used for joint pain and inflammation.

Queen of the meadow (*Eupatorium purpureum*): Also known as gravel root and Joe-Pye weed. A tall, herbaceous perennial with a mound of purplish flowers on top that grows in sunny spots in moist woods, old fields, forest edges, and along stream banks throughout eastern North America. **Roots and leaves** are harvested in the fall. It is used for urinary tract infections, interstitial cystitis, and joint pain.

Queen's delight (*Stillingia sylvatica*): Also called Stillingia root. A two- to four-foot-tall perennial that grows in woods, pine barrens, forest edges, and old fields from Virginia, south to Florida, and west to Texas. **Roots** are harvested in the fall. It is used to treat constipation, skin diseases, and liver ailments.

Ramps (*Allium tricoccum*): Also known as wild leeks and wood leeks. A perennial bulb that grows in moist forests and bottoms, mostly in the mountains from Canada, south to North Carolina, and west to Missouri. It is a protected plant in some states and provinces. **Bulbs with tops** are harvested in the early spring. It is used as a food, flavoring, and tonic.

Sarsaparilla (*Aralia nudicaulis*): Also known as American sarsaparilla and spikenard. A two-foot-tall perennial with red-purple berries that grows in moist woods from Canada, south to North Carolina, and west to Missouri. **Roots** are harvested in the fall. It is used as a general tonic and for treating stress.

Sassafras (*Sassafras albidum*): Also known as cinnamon wood and smelling stick. A perennial tree with three different-shaped leaves that grows in woods, fence rows, and old fields from New England, south to Florida, and west to Texas. **Leaves** are harvested in the summer, **root bark** is harvested in late winter and early spring, and **tree bark** is harvested in spring and early summer. It is used to treat skin conditions and stomach distress. It is also used to make gumbo filé powder and tea.

Senega snakeroot (*Polygala senega*): Also known as Seneca snakeroot. A perennial that grows in dry woods and rocky areas across southern Canada and the eastern half of the United States, south to Georgia, and west to the Dakotas. **Roots** are harvested in the fall. It is used as a laxative and as a treatment for respiratory problems and uterine disorders.

Skullcap (*Scutellaria lateriflora*): Also known as scullcap. A two-foot-tall, herbaceous perennial with tiny flowers that grows in moist woods, swamps, and meadows throughout North America. The **entire top of the plant** is harvested during full bloom in the summer. It is used as a sedative, and as a treatment for headaches and back pain.

Skunk cabbage (*Symplocarpus foetidus* or *Dracontium foetidum*): Also known as polecat weed and swamp cabbage. A skunk-scented, herbaceous perennial that grows in moist woods, along stream banks, and in swamps from Quebec, south to Georgia, and west to Iowa. **Roots and leaves** are harvested throughout the growing season. It is used to treat asthma and skin problems.

Slippery elm (*Ulmus rubra*): A tall, deciduous perennial tree that grows in woods and open areas throughout most of eastern North America, west to North Dakota. The **inner bark** is harvested in spring and fall. It is used as a laxative and as a treatment for skin conditions, sore throats, coughs, stomach ulcers, diarrhea, and for wound healing.

Solomon's seal (*Polygonatum biflorum*): A herbaceous perennial with bell-shaped flowers and, later, purple berries that hang under the leaves. It grows in woods throughout most of eastern North America and west to Wyoming. **Roots** are harvested in the fall. It is used as a mild laxative and to treat skin conditions, stomach problems, coughs, insomnia, and arthritis.

Spikenard (*Aralia racemosa*): Also known as spignet, American spikenard, spiceberry, fleabane, Indian root, and old-man's-root. A tall, herbaceous perennial that grows in woods and on riverbanks in eastern North America, south to Mississippi, and west to Utah. **Roots** are harvested in summer and fall. It is used to treat backaches, colds, asthma, and skin problems.

Stargrass (*Aletris farinosa*): Also known as true unicorn root, star root, blazing star, and colic root. A herbaceous perennial with a rosette of leaves and tall white flower spikes that grows in moist woods and meadows from Ontario, south to Florida, and west to Wisconsin. **Roots** are harvested in the fall. It is used as a tonic and as a sedative.

Stone root (*Collinsonia canadensis*): Also known as horsemint, horseweed, and horsebalm. A two- to four-foot-tall perennial with tall spikes of yellowish flowers that grows in moist woods from Quebec, south to Florida, and west to Missouri. **Roots** are harvested in the fall. It is used to treat respiratory problems and inflammation.

Sumac (*Rhus glabra*): Also known as scarlet sumac and smooth sumac. A perennial shrub with hard red berries in the fall that grows on forest edges, old fields, and meadows throughout most of North America. **Stem bark and root bark** are harvested in spring and fall. It is used to treat asthma, skin problems, diarrhea, colds, stomach problems, fevers, and infections.

Virginia snakeroot (*Aristolochia serpentaria*): A perennial vine-like plant with brown, S-shaped flowers that grows in moist woods from New England, south to Florida, and west to Kansas. **Roots** are harvested in the fall. It is used to stimulate digestion.

Wahoo (*Euonymus atropurpureus*): Also known as burning bush, Indian arrow, and bitter ash. A perennial shrub or small tree that grows in moist forests, swamps, and on stream banks from Quebec, south to Florida, and west to Montana. **Root and stem bark** is harvested any time. It is used for uterine problems, as a liver stimulant, laxative, and diuretic, and to treat fevers.

White pine (*Pinus strobus*): Also known as eastern white pine. A tall, perennial tree that grows in forests throughout much of North America The **inner bark** is harvested during the growing season. It is used to treat coughs and wounds.

White willow (*Salix alba*): Also known as willow and willow bark. A large, perennial tree that grows in rich woods in the eastern United States. **Bark** is harvested any time. It is used to reduce fever, treat diarrhea and arthritis, and as a tonic.

Wild cherry (*Prunus serotina*): Also known as cherry and black cherry. A perennial tree that grows in woods, pastures, and fence rows throughout eastern North America. **Bark** is harvested at any time but best in early spring. **Fruit** is also sometimes harvested in late summer. It is used to treat coughs, colds, diarrhea, and pain.

Wild ginger (*Asarum canadense*): Also known as Canadian snakeroot, Indian ginger, Southern snakeroot, black snakeroot, heart snakeroot, snakeroot, coltsfoot, false coltsfoot, catfoot, colic root, and little brown jug. A low-growing perennial with glossy leaves and little brown jug-like flowers that grows in moist woods in eastern North America from Canada, south to Georgia, and west to Kansas. **Roots** are harvested in the fall. It is used to relieve flatulence, as a stimulant, and as a treatment for fevers, sore throats, colds, and indigestion.

Wild hydrangea (*Hydrangea arborescens*): Also known as hydrangea and smooth hydrangea. A five- to six-foot-tall perennial shrub that grows in woods and on stream banks throughout much of the eastern United States, west to Kansas. **Roots** are harvested in the fall. It is used as a tonic and laxative. It also helps expel bladder and kidney stones.

Wild indigo (*Baptisia tinctoria*): Also known as false indigo, yellow indigo, dyer's baptisia, and rattle bush. A shrubby perennial with little pea-like leaves and yellow flowers that grows in drier woods and clearings from Ontario, south to Florida, and west to Minnesota. **Roots** are harvested in the fall. It is used as a laxative, an astringent, and as an antiseptic for the skin and mouth.

Wild yam (*Dioscorea villosa*): Also known as colic root. A perennial vine that grows in forests, swamps, and along roadsides in Ontario and the eastern United States, west to Nebraska. **Roots** are harvested in the fall. It is used to treat asthma, gastrointestinal problems, menopause, and pain.

Witch hazel (*Hamamelis virginiana*): A small, perennial tree with yellow fall flowers that grows in forests and on forest edges throughout most of eastern North America. **Leaves** are harvested in the summer or fall, and **bark** is harvested in the spring and fall. It is used to treat bruises, wounds, varicose veins, sore muscles, bug bites, and hemorrhoids.

Yellowroot (*Xanthorhiza simplicissima*): A shrubby, woody herbaceous perennial that grows in moist woods along streams in the eastern United States west to Texas. **Roots** are harvested in the spring or fall. It is used as a topical antiseptic, mouthwash, and to treat stomach ulcers.

APPENDIX 2

Prices Paid for Forest Botanicals from 2004 through 2013

The table below includes prices paid and charged by various sectors of the industry. The prices were obtained from a wide variety of sources, including dealers' price lists (what they pay wild-harvesters and growers), personal quotes from dealers, wholesale and retail catalogs, websites, growers, processors, and manufacturers. In Table 20, the first set of columns (Raw Material Prices Dealers Paid to Growers and Wild-harvesters) represents the prices paid by local buyers of raw botanicals to harvesters and growers for raw materials. These are the raw-material suppliers and importers/exporters to whom most wild-harvesters and growers sell their botanicals. These buyers source a wide range of products from all over the world, consolidate them, and sell directly to manufacturers. They are usually the easiest ones for growers and wild-harvesters to sell their products to and may be the best way to move large volumes; however, the grower will usually receive the lowest price per pound through this market. Many of the ginseng buyers also buy other medicinal herbs.

The second set of columns (Wholesale Bulk Herb Prices) lists the wholesale prices charged by farmers/harvesters and other companies selling raw material direct to commercial users. In order to get a higher price per pound, growers sometimes cut out the middleman—that is, the raw material supplier listed in the first set of columns. You can often make more money per pound if you sell your raw product in bulk directly to a manufacturer, herbalist, homeopathic doctor, or health food store. However, many people find it difficult to locate these markets; develop and maintain the necessary relationships; or meet the quality, quantity, or delivery requirements. Sometimes, it is simply that the quantities required per delivery are too small to justify the time and expense to make the delivery. That said, there are many small herbal product manufacturers, herbal schools, and individuals making products who buy on a wholesale level now. The Internet makes it easy for these people to find you.

The third set of columns (Retail Bulk Herb Prices) lists retail prices charged for raw material sold by various kinds of companies direct to the consumer. This category includes just about every conceivable kind of company, including mail-order seed companies, Internet and catalog sales, herbal soap companies, organic farms, herb shops, large herbal products companies, and many alternative health practitioners. They sell small amounts of herbs, usually dried and sifted, for extremely high prices. If you have a way to sell retail and are willing to do the packaging and marketing, this is the way to make the most money per unit volume.

The fourth set of columns (Wholesale Nursery Prices) contain the wholesale prices growers charge retail nurseries and other growers for plants and seeds. It includes wholesale prices for large volumes of medicinal herbs for planting stock to commercial growers who want to produce these plants or for large landscaping jobs or habitat remediation projects. A market is also growing for using these native plants in repopulation projects on public land.

The fifth set of columns (Retail Nursery Prices) contains the retail nursery prices charged for plants and seeds. There is a good market for medicinal herb plants and seeds for home gardens and landscapes, community gardens, display gardens at tourist attractions, and herb-related business landscapes. Managing a good retail nursery, however, is not for everyone. You have to be willing to deal with the public on a daily basis, maintain a diverse inventory, and do your own marketing. For the right grower, this is certainly a good option, and I know several people making good money this way. There are also many online retail nurseries.

Because I thought it would be of interest to put a historical perspective on the pricing of these botanicals, the 2004 and 2007 prices from the first two versions of this book are included. I obtained some dealer prices for raw materials for 2010 and 2011. Then as complete a listing for 2013 as possible was provided.

The 2013 prices were current as of September of that year. They should be considered a snapshot of prices for that specific time only. Do not make any business decisions based on these prices, because they can—and will—change very rapidly. Most raw material suppliers publish a price list two to four times a year. Others do not publish a price list but will fax you one if you request it and will only guarantee those prices for a very short time. Ginseng prices are usually indicated as "market," and you have to call the buyer and learn the price being paid that day. The intent of this list is to introduce the wide range of forest botanicals bought and sold and show the huge price differences between products and markets and how they have changed over a ten-year period.

Another caution: Do not make the mistake of deciding to grow a particular plant just because it is selling for the highest price. What this price list does not contain are the quantities of each product bought and sold in each year. For example, a 2005 price list included the estimated volumes of herbs needed and prices paid by one particular company for one year. The price paid for black cohosh root was listed as $2.50 per pound, and the volume needed was 250 dried pounds. In contrast, false unicorn was listed at $35.00 per dried pound, but the volume needed was only 20 pounds. The point is, some of these herbs are used in such small volumes by the whole industry that, if just a few growers produced them commercially, they would quickly saturate the market. In fact, some of these herbs are used in such small volumes that it may never be economically feasible for them to be grown on a commercial scale. Developing a sustainable wild-harvesting method for some forest plants will probably be the only answer to satisfying the market demand without depleting the native populations.

Organic prices in bold (Prices are per dried pound unless otherwise noted)	Raw Material Prices Dealers Paid to Growers and Wild-harvesters				
	2004	**2007**	**2011**	**2012**	**2013**
Balm of Gilead, buds (poplar buds, balsam poplar) (*Populus balsamifera*)	3.00–5.00	3.00–5.00	4.00–8.00	14.00	12.00–16.00
Balmony, leaves (turtlehead) (*Chelone glabra*)	on contract	2.00–3.00			0.50
Bayberry, root bark (wax myrtle) (*Myrica cerifera*)	2.00–3.00	2.50–3.50		2.00	2.00–3.00
Bethroot, root (red trillium, birthroot, squaw root, purple trillium, wake-robin, ill-scented trillium) (*Trillium erectum*)	1.00–2.50	1.00–3.00	2.50	2.00	1.00–3.00
Black cohosh, root (rattlesnake root, bugbane, squaw root, papoose root) (*Actaea racemosa*)	2.50–3.00	2.50–4.00	3.00–5.00	1.50–4.00	3.00–4.00
Black haw, root bark and tree bark (arrow-wood, crampbark, Southern black haw) (*Viburnum prunifolium*)	0.50	1.00–4.00		1.00–2.00	1.00–3.00
Black Indian hemp, root (dogbane, Indian hemp) (*Apocynum cannabinum*)	3.00	3.00–4.00			2.00
Black root, root (Culver's root) (*Veronicastrum virginicum*)	4.00	5.00–10.00			10.00
Black walnut, hulls (walnut) (*Juglans nigra*)	on contract	on contract			2.00
Black walnut, leaves	0.25				1.00
Blackberry, root and root bark (Dewberry, brambleberry) (*Rubus villosus*)	on contract	on contract			
Bloodroot, root (red Indian paint, red root, puccoon) (*Sanguinaria canadensis*)	7.00–9.00	5.00–6.00	8.00	8.00	8.00–16.00; 8.00 green
Blue cohosh, root (squaw root, blue ginseng, papoose root) (*Caulophyllum thalictroides, C. giganteum*)	1.00	1.00–1.50	2.25	0.75	1.00
Blue flag, root (wild iris, flag lily) (*Iris versicolor*)				8.00	3.00–12.00
Boneset, herb (feverwort, sweat plant) (*Eupatorium perfoliatum*)	0.50	0.75–1.00			1.00
Bugleweed, herb (Gypsywort) (*Lycopus virginicus*)	1.00	1.50–2.00			1.00
Butternut, tree bark (white walnut) (*Juglans cinerea*)	0.20	0.25–0.40			0.25
Cleavers, herb (bedstraw) (*Galium aparine*)	0.75	1.00–1.25			
Cramp bark, bark (cranberry tree, black haw) (*Viburnum opulus*)				10.00	3.00–12.00

Organic prices in bold (Prices are per dried pound unless otherwise noted)	Raw Material Prices Dealers Paid to Growers and Wild-harvesters				
	2004	2007	2011	2012	2013
Cranesbill, root (wild geranium) (*Geranium maculatum*)	1.50–2.00	2.00–2.50	2.50	2.00	1.00–3.00
Elderberry, flowers and fruit (Elder) (*Sambucus nigra* L. ssp. *canadensis*)					1.00 fruit
False unicorn, root (star root, star grub root, grub root, fairywand, devil's bit, helonias, starwort) (*Chamaelirium luteum*)	27.50–45.00	30.00–85.00	40.00		50.00–65.00
Figwort, herb (*Scrophularia marilandica* and *S. nodosa*)	1.00	1.00			0.75
Fringe tree, tree bark and root bark (white ash) (*Chionanthus virginicus*)	2.50–3.50	3.00–7.00		5.00–6.00	5.00–7.00
Galax, leaves, fresh, (beetleweed, coltsfoot, wandflower) (*Galax urceolata*)	0.02/leaf; 40.00–80.00/ box of 5,000 leaves	0.02/leaf; 40.00–80.00/ box of 5,000 leaves	0.01/leaf; 20.00/box of 5,000 leaves	0.02/leaf; 100.00/box of 5,000 leaves	0.02/leaf; 100.00/box of 5,000 leaves
Ginseng, root (sang, manroot, American ginseng) (*Panax quinquifolius*)	30.00 cultivated; 362.50 wild	12.00–30.00 cultivated; 800.00–1,000.00 wild	300.00–340.00	600.00–650.00	800.00–850.00 wild; 200.00–250.00 wild, green
Goldenseal, herb (yellowroot, ground raspberry, yellow puccoon, wild circuma, eye-balm, yellow paint, wild turmeric, yellow-eye) (*Hydrastis canadensis*)	1.00–6.00	1.00–3.00	2.00–4.00	2.00–4.00	3.00–5.00
Goldenseal, root, dried	12.00–22.00	11.00–25.00	20.00–22.00	12.00–22.00; 10.00 green	**67.00–70.00 organic;** 16.00–35.00; 5.00–12.00 green
Indian turnip, root (jack in the pulpit) (*Arisaema triphyllum*)	3.00–4.00	7.00–10.00	6.00		1.00
Life root, herb (squaw weed, cough weed, false valerian) (*Senecio aureus*)	2.00	1.00–1.50			0.50
Lobelia, herb (Indian tobacco) (*Lobelia inflata*)			2.00		2.00
Maidenhair fern, herb (finger fern, ginseng pointer) (*Adiantum pedatum*)	1.50	1.00–1.50			
Mayapple, root (mandrake, umbrella plant, Indian apple, devil's apple) (*Podophyllum peltatum*)	0.10	on contract	2.50	2.75–3.00	2.00–6.00

Organic prices in bold (Prices are per dried pound unless otherwise noted)	Raw Material Prices Dealers Paid to Growers and Wild-harvesters				
	2004	2007	2011	2012	2013
Missouri snakeroot, herb (wild quinine, American feverfew) (*Parthenium integrifolium*)		0.50–0.75			1.00
Missouri snakeroot, root	on contract	1.25–1.75			
Mountain maple, bark (*Acer spicatum*)	0.25	0.25			
Mullein, herb (Indian tobacco, wooly mullein, common mullein) (*Verbascum thapsus*)					0.75
New Jersey tea, herb (red root, Jersey tree) (*Ceanothus americanus*)	4.00	4.00			1.00
New Jersey tea, root	0.75	0.75			1.50
Oregon grape, root (holly-leaved barberry, mountain grape) (*Mahonia* or *Berberis aquifolium*)	2.00–2.50	on contract			
Partridge berry, herb (squaw vine, squaw berry) (*Mitchella repens*)	2.50–3.50	5.00–7.00	8.00 green	8.00 green	8.00; 7.00 green
Passionflower, herb (Maypop) (*Passiflora incarnata*)	0.70	0.50–6.50			1.00
Pinkroot, root (Carolina pinkroot, Carolina pink, Indian pink, American wormroot, wormgrass) (*Spigelia marilandica*)	3.00–12.00	5.00–7.00	8.00		5.00
Pipsissewa, herb (dragon's tongue, spotted wintergreen) (*Chimaphila umbellata*)	2.00	2.00			
Pokeweed, root (poke, pokeroot, poke sallet) (*Phytolacca americana*)			1.00	0.50–0.75	0.50
Prickly ash, northern, tree bark and root bark (toothache tree) (*Zanthoxylum americanum*)	1.25–2.00	1.25–2.25			1.00
Prickly ash, southern, tree bark and root bark (toothache tree) (*Zanthoxylum clava-herculis*)	1.50–2.00	1.20–3.50			1.50
Queen of the meadow, herb (Joe-Pye weed, gravel root) (*Eupatorium purpureum*)	0.30	0.30–0.75	2.25		0.50–1.00
Queen of the meadow, root	1.00	1.00–1.75		1.50	1.00
Queen's delight, root (Stillingia root) (*Stillingia sylvatica*)	2.00–3.00	4.00–5.00			5.00
Ramps, fresh bulbs (wild leeks, wood leeks) (*Allium tricoccum*)	1.50/bunch	1.50–1.75/ bunch	2.00/bunch	15.00–25.00/ pound	17.00–25.00/ pound

Organic prices in bold (Prices are per dried pound unless otherwise noted)	Raw Material Prices Dealers Paid to Growers and Wild-harvesters				
	2004	2007	2011	2012	2013
Sarsaparilla, root (American sarsaparilla, spikenard) (*Aralia nudicaulis*)	1.00	1.50–2.50			1.00
Sassafras, herb (cinnamon wood, smelling stick) (*Sassafras albidum*)	1.00	1.00–1.50	1.50	1.00–1.40	1.00–1.50
Sassafras, root bark	2.50–6.00	3.00–10.00			3.00
Sassafras, tree bark	0.10	0.10–0.20			
Senega snakeroot, root (Seneca snakeroot) (*Polygala senega*)	6.00–8.00	7.00–10.00		16.00	7.00–16.00
Skullcap, herb (scullcap) (*Scutellaria lateriflora*)	1.50–2.50	1.50–2.00			2.00
Skunk cabbage, herb (polecat weed, swamp cabbage) (*Symplocarpus foetidus, Dracontium foetidum*)	1.00	1.00			1.00
Skunk cabbage, root	2.00	1.00–3.00	4.00–6.00	4.00–6.00	
Slippery elm, inner bark (*Ulmus rubra*)	2.00–3.00	2.00–3.00		3.00	2.00–3.00
Solomon's seal, root (*Polygonatum biflorum*)	1.50–2.00	2.50–3.00			4.00
Spikenard, root (spignet, American spikenard, spiceberry, fleabane, Indian root, old-man's-root) (*Aralia racemosa*)	1.00–2.00	1.50–2.50	4.00		1.00
Star grass, root (true unicorn root, star root, blazing star, colic root) (*Aletris farinosa*)	12.00	25.00–30.00	18.00	20.00	20.00–25.00
Stone root, root (horsemint, horseweed, horsebalm) (*Collinsonia canadensis*)	1.00–1.50	1.00–1.75	2.25	2.00	1.00–2.00
Sumac, root bark (scarlet sumac, smooth sumac) (*Rhus glabra*)	0.45–0.50	0.40–0.50			0.50–1.00
Sumac, tree bark	0.45–0.50	0.40–0.50		0.50	0.50
Virginia snakeroot, root (*Aristolochia serpentaria*)	20.00–30.00	50.00–60.00	70.00	60.00	60.00–80.00
Wahoo, tree bark and root bark (burning bush, Indian arrow, bitter ash) (*Euonymus atropurpureus*)	3.00	5.00–6.00			5.00
White pine, inner tree bark (Eastern white pine) (*Pinus strobus*)	on contract	on contract			
White willow, inner tree bark (willow, willow bark) (*Salix alba*)	0.25	0.25–0.30			0.50
Wild cherry, tree bark (cherry, black cherry) (*Prunus serotina*)	0.25–1.00	0.70–1.00		0.25–1.00	0.10–1.40

Organic prices in bold (Prices are per dried pound unless otherwise noted)	Raw Material Prices Dealers Paid to Growers and Wild-harvesters				
	2004	2007	2011	2012	2013
Wild ginger, root (Canadian snakeroot, Indian ginger, Southern snakeroot, black snakeroot, heart snakeroot, snakeroot, coltsfoot, false coltsfoot, catfoot, colic root, little brown jug) (*Asarum canadense*)	0.50	on contract		1.00	1.00; 5.00 green
Wild hydrangea, root (hydrangea, smooth hydrangea) (*Hydrangea arborescens*)	1.25–1.50	1.25–1.50	2.25	2.00–2.5	2.00
Wild indigo, root (false indigo, yellow indigo, dyer's baptisia, rattle bush) (*Baptisia tinctoria*)	2.00–4.00	7.00–9.00	16.00	12.00	8.00–12.00
Wild yam, root (colic root) (*Dioscorea villosa*)	0.75–1.50	0.50–4.25	2.25	2.00	2.00–2.50
Witch hazel, leaves (*Hamamelis virginiana*)	1.20–1.50	1.25–6.00	1.25	1.25–1.50	1.00–2.00
Witch hazel, tree bark	1.20–1.50	0.75–1.00	1.75	1.25	1.00–4.00
Yellowroot, root (*Xanthorhiza simplicissima*)		on contract			

	Wholesale Bulk Herb Prices (per dried pound unless noted otherwise)			Retail Bulk Herb Prices (per dried pound [per dried oz])		
Organic prices in bold	2004	2007	2013	2004	2007	2013
Balm of Gilead, buds (poplar buds, balsam poplar) (*Populus balsamifera*)	10.00–35.00	12.00–38.00	12.00–50.00	29.00 [5.00]	29.00–160.00 [2.00–24.00]	38.50–150.00 [8.75]
Balmony, leaves (turtlehead) (*Chelone glabra*)	8.25–8.83	8.25–9.00		18.84–40.00 [4.00]	18.00–40.00 [4.50]	79.20
Bayberry, root bark (wax myrtle) (*Myrica cerifera*)	8.01–8.50	10.00–30.50	9.60–12.00	10.20–34.50 [7.00]	10.50–80.00 [0.78–24.00]	30.00–47.50 [3.71]
Bethroot, root (red trillium, birthroot, squaw root, purple trillium, wake-robin, ill-scented trillium) (*Trillium erectum*)	5.65–6.00	5.50–21.50	5.00–27.00	7.20–25.50 [3.50]	7.25–40.00 [1.75–40.00]	29.00 [2.60]
Black cohosh, root (rattlesnake root, bugbane, squaw root, papoose root) (*Actaea racemosa*)	4.48–4.75	11.40–16.25	**19.50 organic;** 13.20–17.50	5.70–33.30 [4.00]	14.75–30.70 [1.10–13.00]	**18.50–37.50 [1.56–3.06] organic;** 21.00–42.00 [2.00]

	Wholesale Bulk Herb Prices (per dried pound unless noted otherwise)			Retail Bulk Herb Prices (per dried pound [per dried oz])		
Organic prices in bold	**2004**	**2007**	**2013**	**2004**	**2007**	**2013**
Black haw, root bark and tree bark (arrow-wood, crampbark, Southern black haw) (*Viburnum prunifolium*)		6.60–7.75	12.20–16.00	[3.75]	11.00–23.79 [0.78–2.40]	17.00–34.00 [1.50]
Black Indian hemp, root (dogbane, Indian hemp) (*Apocynum cannabinum*)		60.00	9.20–11.70	[4.00]	72.00 [4.00]	12.80
Black root, root (Culver's root) (*Veronicastrum virginicum*)				24.95 [12.00]	25.00–35.00 [14.00]	71.93
Black walnut, hulls (walnut) (*Juglans nigra*)	4.48–4.85	3.45–6.25	7.50–10.20	5.82–7.76 [3.00]	5.75–12.50 [0.75–1.50]	**23.58 [4.40] organic;** 12.00–17.00 [1.15]
Black walnut, leaves		3.25–3.85		9.95 [3.00]	5.50–11.90 [0.63–1.40]	**22.08 [1.81] organic;** 12.00–21.00 [1.15]
Blackberry, root and root bark (Dewberry, bramble-berry) (*Rubus villosus*)	4.95–5.25	7.00–8.40	10.50	6.30–20.50 [3.50]	14.00 [1.00–1.19]	14.00–27.12 [1.13]
Bloodroot, root (red Indian paint, red root, puccoon) (*Sanguinaria canadensis*)	9.66–10.25	10.00–40.00	27.00–40.50	12.30–30.00 [7.50]	32.00–75.00 [2.25–35.00]	36.00–111.89 [3.00–3.50]
Blue cohosh, root (squaw root, blue ginseng, papoose root) (*Caulophyllum thalictroides, C. giganteum*)	5.61–5.95	7.80–14.50	7.90–10.30	6.95–21.00 [3.00]	7.50–33.20 [3.20]	11.00–40.58 [0.88–3.06]
Blue flag, root (wild iris, flag lily) (*Iris versicolor*)			**17.00–26.00 organic**			**27.00 [2.25] organic;** 59.33–69.50
Boneset, herb (feverwort, sweat plant) (*Eupatorium perfoliatum*)	4.01–4.25	4.00–8.75	**9.38–13.35 organic;** 13.35–14.48	5.10–23.00 [2.75]	8.50–25.00 [0.90–1.20]	**12.50 [1.06] organic;** 16.50–25.19 [1.50]
Bugleweed, herb (Gypsywort) (*Lycopus virginicus*)	7.30–7.75	7.50–10.00	**8.00–11.30 organic**	9.30–12.40 [3.50]	8.00–16.70 [0.60–1.80]	**12.00 [1.00] organic;** 42.38
Butternut, tree bark (white walnut) (*Juglans cinerea*)		4.80–6.83	6.00	(2.50)	9.72–9.75 [0.69–0.81]	8.00 [0.69]

	Wholesale Bulk Herb Prices (per dried pound unless noted otherwise)			Retail Bulk Herb Prices (per dried pound [per dried oz])		
Organic prices in bold	**2004**	**2007**	**2013**	**2004**	**2007**	**2013**
Cleavers, herb (bedstraw) (*Galium aparine*)		3.65–34.00	**7.50–13.90 organic**	10.95 [4.00]	7.25–44.00 [0.53–3.07]	**10.00–20.83 [0.81] organic;** 16.67–19.37 [1.69]
Cramp bark, bark (cranberry tree, black haw) (*Viburnum opulus*)			25.13–30.00			33.50–45.58 [2.75–3.96]
Cranesbill, root (wild geranium) (*Geranium maculatum*)	6.83–7.25	9.00–20.50	13.13	7.95–22.00 [4.00]	15.00–22.00 [1.01–1.90]	17.50–36.81 [1.44–2.20]
Elderberry, flowers and fruit (Elder) (*Sambucus nigra* L. ssp. *canadensis*)			**7.50 organic fruit**			**26.00 [2.04] organic flowers;** 10.00–60.00 [1.25] flowers; **10.00–19.67 [1.06] organic fruit;** 19.00–21.00 [1.69–1.70] fruit
False unicorn, root (star root, star grub root, grub root, fairywand, devil's bit, helonias, starwort) (*Chamaelirium luteum*)	40.50–43.00	95.00–140.00	78.00	51.60–146.00 [22.00]	135.00–150.00 [13.00]	165.00–254.28 [14.27–21.00]
Figwort, herb (*Scrophularia marilandica* and *S. nodosa*)		12.00	**8.00 organic**	[4.00]	14.00–25.00 [1.03–3.99]	**12.00 [1.00] organic;** 40.69
Fringe tree, tree bark and root bark (white ash) (*Chionanthus virginicus*)	14.60–15.50	16.00–19.25	20.00	12.95–24.80 [9.00]	30.00–32.00 [2.31–2.56]	30.00 [2.50]
Galax, leaves, fresh, (beetleweed, coltsfoot, wandflower) (*Galax urceolata*)	7.40/100 leaves	18.00–23.00/100 leaves	14.50–169.00/100 leaves	1.85/25 leaves	8.33/25 leaves	36.00–42.50/25 leaves
Ginseng, root (sang, man-root, American ginseng) (*Panax quinquifolius*)		39.00–60.00 cultivated; 850.00 wild	**66.75–91.00 organic**	29.90–68.00 [35.00] cultivated	39.00–89.00 [4.75–6.25] cultivated; 1,200.00 [80.00–120.00] wild	**89.00–205.33 [8.50–13.13] organic;** 67.13

	Wholesale Bulk Herb Prices (per dried pound unless noted otherwise)			Retail Bulk Herb Prices (per dried pound [per dried oz])		
Organic prices in bold	**2004**	**2007**	**2013**	**2004**	**2007**	**2013**
Goldenseal, herb (yellowroot, ground raspberry, yellow puccoon, wild circuma, eye-balm, yellow paint, wild turmeric, yellow-eye) (*Hydrastis canadensis*)	10.60–11.25	10.60–38.50	15.00–40.00	7.50–81.00 [10.00]	18.00–44.00 [1.25–4.00]	**21.00–78.17 [1.75] organic;** 43.00–92.00
Goldenseal, root, dried	49.91–53.00	42.00–100.00	**58.00–166.00 organic;** 51.00–119.00	42.95–127.00 [18.00]	60.00–130.00 [5.50–12.00]	**70.00–200.00 [5.00–11.61] organic;** 71.00–175.00 [6.25–12.00]; **70.00 organic, green;** 50.00 [3.75] green
Indian turnip, root (jack in the pulpit) (*Arisaema triphyllum*)		20.50		[3.00]	22.00–25.00 [1.90–3.00]	
Life root, herb (squaw weed, cough weed, false valerian) (*Senecio aureus*)				[4.00]	[4.00]	50.86
Lobelia, herb (Indian tobacco) (*Lobelia inflata*)			**15.90–23.25 organic**			**31.00 [2.56] organic;** 10.75–38.75 [1.21]
Maidenhair fern, herb (finger fern, ginseng pointer) (*Adiantum pedatum*)	6.13–6.50	5.40–9.45		7.80–13.95 [3.00]	12.00–45.00 [0.97–34.90]	42.38
Mayapple, root (mandrake, umbrella plant, Indian apple, devil's apple) (*Podophyllum peltatum*)	6.83–7.25	9.50–11.25	8.50–10.50	6.45–21.00 [5.00]	12.00–16.00 [0.90–1.10]	11.50–16.50 [1.50]
Missouri snakeroot, herb (wild quinine, American feverfew) (*Parthenium integrifolium*)						
Missouri snakeroot, root						
Mountain maple, bark (*Acer spicatum*)				7.95	7.95	

	Wholesale Bulk Herb Prices (per dried pound unless noted otherwise)			Retail Bulk Herb Prices (per dried pound [per dried oz])		
Organic prices in bold	**2004**	**2007**	**2013**	**2004**	**2007**	**2013**
Mullein, herb (Indian tobacco, wooly mullein, common mullein) (*Verbascum thapsus*)			**7.88–11.10 organic**			**10.50–20.67 [0.88–1.73] organic;** 12.70–23.25
New Jersey tea, herb (red root, Jersey tree) (*Ceanothus americanus*)				24.00	25.00	
New Jersey tea, root				24.00	25.00	30.17
Oregon grape, root (holly-leaved barberry, mountain grape) (*Mahonia* or *Berberis aquifolium*)	7.30–7.75	5.00–25.00	7.88–10.40	9.30–19.50 [5.00]	11.00–36.00 [1.16–1.37]	10.50–26.33 [0.88–2.23]
Partridge berry, herb (squaw vine, squaw berry) (*Mitchella repens*)	9.66–12.00	16.00–35.00	27.75	12.30–32.00 [5.00]	21.00–43.00 [1.88–3.00]	37.00–676.42 [3.06–7.00]; 45.00–59.33 [7.00] green
Passionflower, herb (Maypop) (*Passiflora incarnata*)	4.01–4.25	6.50–8.25	**10.00–14.00 organic;** 10.80–14.00	5.10–20.50 [4.00]	12.00–17.29 [0.91–2.30]	**15.00–31.25 [1.25–2.65] organic;** 14.75–21.31 [1.41]
Pinkroot, root (Carolina pinkroot, Carolina pink, Indian pink, American wormroot, wormgrass) (*Spigelia marilandica*)	36.73–39.00	43.00	57.00	46.80–62.40 [30.00]	65.50 [32.00]	115.00 [9.95–32.00]
Pipsissewa, herb (dragon's tongue, spotted wintergreen) (*Chimaphila umbellata*)	6.83–7.25	5.00–28.00	8.00–15.90	8.70–27.00 [4.00]	10.25–44.00 [0.72–1.70]	12.00–32.94 [1.00]
Pokeweed, root (poke, pokeroot, poke sallet) (*Phytolacca americana*)			**7.00–9.70 organic**			27.12–30.17
Prickly ash, northern, tree bark and root bark (toothache tree) (*Zanthoxylum americanum*)	9.42–10.00	8.50–9.00		11.95–25.50 [6.00]	16.00–20.59 [1.25–1.55]	39.42–42.38 [2.92]
Prickly ash, southern, tree bark and root bark (toothache tree) (*Zanthoxylum clava-herculis*)	9.42–10.00	10.00	12.25–13.75	11.95–25.50 [6.00]	12.00–25.00 [2.00]	18.00–40.69 [1.50]

	Wholesale Bulk Herb Prices (per dried pound unless noted otherwise)			Retail Bulk Herb Prices (per dried pound [per dried oz])		
Organic prices in bold	**2004**	**2007**	**2013**	**2004**	**2007**	**2013**
Queen of the meadow, herb (Joe-Pye weed, gravel root) (*Eupatorium purpureum*)			12.50–15.50	6.95–25.50		**28.50 organic;** 15.00–32.42 [1.35]
Queen of the meadow, root	5.89–6.25	6.50–11.25	12.75	6.95–25.50 [5.00]	10.00–15.10 [0.90–1.10]	17.00–39.75 [1.38–3.33]
Queen's delight, root (Stillingia root) (*Stillingia sylvatica*)	11.54–12.25	12.00		14.70–19.60 [6.00]	21.00 [6.50]	59.33–83.33
Ramps, fresh bulbs (wild leeks, wood leeks) (*Allium tricoccum*)	9.50	10.00	25.00	15.00 [1.50–5.00/ bunch]	15.00–20.00 [1.50–7.00/ bunch]	12.50–30.00
Sarsaparilla, root (American sarsaparilla, spikenard) (*Aralia nudicaulis*)	3.77–4.00	4.00		4.80–44.60 [5.00]	8.50–14.00 [2.30–5.00]	17.44
Sassafras, herb (cinnamon wood, smelling stick) (*Sassafras albidum*)	4.95–5.25	5.50–10.00	**8.25 organic**	6.30–35.00	6.25–13.00 [0.78–1.00]	**11.00 [0.88] organic;** 27.12
Sassafras, root bark		18.00	13.13	35.00–47.00 [4.00]	20.00–28.00 [1.87–2.20]	17.50–46.33 [1.44–4.02]
Sassafras, tree bark			13.13			17.50–46.33 [1.44–4.02]
Senega snakeroot, root (Seneca snakeroot) (*Polygala senega*)				[5.00]	[5.00]	36.81
Skullcap, herb (scullcap) (*Scutellaria lateriflora*)	9.66–10.25	9.00–12.75	**13.88–18.30 organic**	12.30–33.50 [5.00]	18.00–32.00 [1.20–1.50]	**18.50–57.17 [1.56–3.13] organic;** 44.08
Skunk cabbage, herb (polecat weed, swamp cabbage) (*Symplocarpus foetidus*, *Dracontium foetidum*)						
Skunk cabbage, root	4.95–5.25	5.25		6.30–22.30 [6.00]	8.95–22.00 [6.00]	73.52
Slippery elm, inner bark (*Ulmus rubra*)	6.83–7.25	10.50–20.00	**14.25 organic;** 14.40–16.50	7.95–28.50 [5.00]	18.50–25.00 [1.38–1.95]	**19.00 [14.25] organic;** 28.00–50.00 [2.00–3.17]
Solomon's seal, root (*Polygonatum biflorum*)	7.30–7.75	9.30–23.00	17.25	9.30–12.95 [6.00]	11.50–29.00 [1.50–2.00]	23.00–37.30 [1.94–2.65]

	Wholesale Bulk Herb Prices (per dried pound unless noted otherwise)			Retail Bulk Herb Prices (per dried pound [per dried oz])		
Organic prices in bold	**2004**	**2007**	**2013**	**2004**	**2007**	**2013**
Spikenard, root (spignet, American spikenard, spiceberry, fleabane, Indian root, old-man's-root) (*Aralia racemosa*)	7.77–8.25	8.25–19.00	14.30–18.30	9.90–30.50 [5.00]	11.50–24.00 [1.70–1.90]	24.00–47.47 [2.00]
Star grass, root (true unicorn root, star root, blazing star, colic root) (*Aletris farinosa*)				[15.00]	[15.00]	
Stone root, root (horsemint, horseweed, horsebalm) (*Collinsonia canadensis*)	4.95–5.25	10.00–11.75	9.00–11.50	6.30–20.50 [6.00]	13.00–18.00 [0.95–1.38]	15.00–38.99 [1.25–2.33]
Sumac, root bark (scarlet sumac, smooth sumac) (*Rhus glabra*)	55.50	55.00		7.95 [5.00]	7.45 [5.00]	23.25
Sumac, tree bark		55.00		7.95 [5.00]	7.45 [5.00]	23.25
Virginia snakeroot, root (*Aristolochia serpentaria*)	49.91–53.00	65.00		63.60–84.80 [10.00]	87.00 [15.00]	
Wahoo, tree bark and root bark (burning bush, Indian arrow, bitter ash) (*Euonymus atropurpureus*)	11.54–12.25	14.50		14.70–44.50 [6.00–8.00]	23.00–42.00 [8.00]	57.64
White pine, inner tree bark (Eastern white pine) (*Pinus strobus*)		5.25		7.95–19.50 [3.00]	8.25–11.00 [0.69–0.82]	
White willow, inner tree bark (willow, willow bark) (*Salix alba*)	3.77–4.00	4.00	**7.50 organic;** 7.90–11.50	4.80–8.00 [4.00]	6.50–8.75 [0.53–1.00]	**10.00 [0.81] organic;** 11.83–36.81 [1.25]
Wild cherry, tree bark (cherry, black cherry) (*Prunus serotina*)	2.12–3.00	4.05	**6.00 organic**	2.70–17.00 [3.00]	5.06–10.00 [0.47–1.00]	**8.00–22.00 [0.69–1.96] organic;** 17.44–21.58
Wild ginger, root (Canadian snakeroot, Indian ginger, Southern snakeroot, black snakeroot, heart snakeroot, snakeroot, coltsfoot, false coltsfoot, catfoot, colic root, little brown jug) (*Asarum canadense*)	4.71–5.00	5.75–12.25		6.00–25.00 [5.00]	8.00–25.00 [0.95–5.00]	17.00–19.00 [1.70]

Organic prices in bold	Wholesale Bulk Herb Prices (per dried pound unless noted otherwise)			Retail Bulk Herb Prices (per dried pound [per dried oz])		
	2004	2007	2013	2004	2007	2013
Wild hydrangea, root (hydrangea, smooth hydrangea) (*Hydrangea arborescens*)	4.71–5.00	5.50–10.50	8.63–13.00	6.00–23.50	6.75–19.50 [0.85]	11.50–36.33 [0.94–2.61]
Wild indigo, root (false indigo, yellow indigo, dyer's baptisia, rattle bush) (*Baptisia tinctoria*)	5.00–12.75	27.00–43.00	42.75	15.30–35.00 [7.00]	35.00–52.00 [7.25]	49.00–71.93 [4.50–4.69]
Wild yam, root (colic root) (*Dioscorea villosa*)	5.89–6.25	6.00–15.00	8.20–10.70	7.50–24.00 [6.00]	8.50–19.50 [0.70–1.50]	**26.50 [2.06] organic;** 13.00–32.42 [1.06–2.50]
Witch hazel, leaves (*Hamamelis virginiana*)	5.89–6.25	6.00–11.75	11.00–14.00	7.50–21.00 [4.00]	7.50–13.00 [0.60–0.95]	19.00–34.87 [1.70]
Witch hazel, tree bark	5.89–6.25	14.00	8.63	9.95–20.50 [3.00]	15.00 [0.95–1.25]	11.50–19.37 [0.94–1.70]
Yellowroot, root (*Xanthorhiza simplicissima*)				[17.00]	[17.00]	

Organic prices in bold	Wholesale Nursery Prices (one pound of seed [100 plants/roots])			Retail Nursery Prices (small pkg seed [single plant])		
	2004	2007	2013	2004	2007	2013
Balm of Gilead, buds (poplar buds, balsam poplar) (*Populus balsamifera*)		[250.00]	[110.00–270.00]	[6.95]	3.95 [4.00–5.00]	5.00 [5.00–8.00]
Balmony, leaves (turtlehead) (*Chelone glabra*)	9.75/1,500 seeds [150.00–500.00]	10.50/1,500 seeds [150.00]	[420.00]	2.65–3.50 [6.00–7.00]	2.50–3.50 [6.00–8.00]	2.00; 50.00/oz seeds [4.20–7.00]
Bayberry, root bark (wax myrtle) (*Myrica cerifera*)	68.00–91.00	540.00 [200.00]		1.75–2.65 [4.00]	2.25–2.85 [4.00–5.00]	
Bethroot, root (red trillium, birthroot, squaw root, purple trillium, wake-robin, ill-scented trillium) (*Trillium erectum*)	50.00 [135.00 roots]	50.75 [140.00–200.00]	65.00 [155.00–220.00]	9.95 [5.00–15.00]	9.95 [7.00–15.00]	**9.95 organic;** [2.85–8.00]
Black cohosh, root (rattlesnake root, bugbane, squaw root, papoose root) (*Actaea racemosa*)	308.00–434.50 [200.00 plants; 175.00–300.00 roots]	545.00 [150.00–300.00]	[100.00–125.00]	2.70–3.95 [4.00–10.00]	3.30–3.95 [4.00–8.00]	**3.95/50 seeds organic;** [1.85–6.00]

	Wholesale Nursery Prices (one pound of seed [100 plants/roots])			Retail Nursery Prices (small pkg seed [single plant])		
Organic prices in bold	**2004**	**2007**	**2013**	**2004**	**2007**	**2013**
Black haw, root bark and tree bark (arrow-wood, crampbark, Southern black haw) (*Viburnum prunifolium*)	[250.00–340.00]	[88.00–340.00]	700.00		[12.00]	[7.00–8.00]
Black Indian hemp, root (dogbane, Indian hemp) (*Apocynum cannabinum*)					3.00 [5.00]	**2.95/100 seeds [4.50–5.00 root] organic;** 2.00; 100.00/oz seeds
Black root, root (Culver's root) (*Veronicastrum virginicum*)	296.00 [500.00]	54.00/oz seeds [500.00]	[300.00]	2.65 [3.95–7.00]	2.65–3.00 [5.00–7.00]	**2.95/300 seeds organic;** 2.00–2.50; 50.00/oz seeds [2.42–6.00]
Black walnut, hulls (walnut) (*Juglans nigra*)					[20.00]	
Black walnut, leaves						
Blackberry, root and root bark (Dewberry, brambleberry) (*Rubus villosus*)		[275.00]			[3.50]	[2.16]
Bloodroot, root (red Indian paint, red root, puccoon) (*Sanguinaria canadensis*)	70.00/1,000 seeds [60.00 plants; 100.00–150.00 roots]	75.00/1,000 seeds [140.00–175.00]	**[70.00] organic;** [40.00];	7.00–9.95 [4.00–7.00]	10.00 [2.00–8.00]	**9.95/100 seeds [1.25–5.00 root] organic;** 2.00; 80.00/oz seed [2.50–10.00]
Blue cohosh, root (squaw root, blue ginseng, papoose root) (*Caulophyllum thalictroides*, *C. giganteum*)	33.00/100 seeds [750.00 plants; 100.00 roots]	33.00/100 seeds [750.00 plants; 150.00 roots]	225.00 [90.00–420.00]	7.00–7.95 [5.00–10.00]	7.95 [4.00–8.00]	**9.95/10 seeds [3.75–4.50 root; 5.00–7.95 plant] organic;** 2.00; 15.00/oz seeds [1.65–7.00]

	Wholesale Nursery Prices (one pound of seed [100 plants/roots])			Retail Nursery Prices (small pkg seed [single plant])		
Organic prices in bold	**2004**	**2007**	**2013**	**2004**	**2007**	**2013**
Blue flag, root (wild iris, flag lily) (*Iris versicolor*)			225.00 [240.00]			2.00; 15.00/ oz seed [2.80–7.00]
Boneset, herb (feverwort, sweat plant) (*Eupatorium perfoliatum*)	7.75/7,000 seeds	7.75/7,000 seeds; 36.00/oz seeds	450.00 [360.00]	1.75–2.65 [3.25–7.00]	2.85–3.00 [7.00]	**2.95/100 seeds organic;** 2.00; 30.00/ oz seed [4.20–7.00]
Bugleweed, herb (Gypsywort) (*Lycopus virginicus*)		[525.00]		2.65 [7.00]	2.85 [7.00]	
Butternut, tree bark (white walnut) (*Juglans cinerea*)		[95.00]				
Cleavers, herb (bedstraw) (*Galium aparine*)				2.85 [7.00]	2.85 [7.00]	**2.95/50 seeds organic**
Cramp bark, bark (cranberry tree, black haw) (*Viburnum opulus*)						**4.95/20 seeds organic**
Cranesbill, root (wild geranium) (*Geranium maculatum*)	[350.00]	[150.00–350.00]	1,200.00 [120.00–160.00]	1.75–2.00 [3.95]	1.75–2.50 [3.95–4.50]	**3.95/20 seeds [9.00] organic;** 2.00; 80.00/ oz seed [1.40–4.00]
Elderberry, flowers and fruit (Elder) (*Sambucus nigra* L. ssp. *canadensis*)						**4.95/50 seeds [8.00–9.00] organic**
False unicorn, root (star root, star grub root, grub root, fairywand, devil's bit, helonias, starwort) (*Chamaelirium luteum*)	200.00/oz seeds	200.00/oz seeds		4.95–7.00 [7.00]	4.95–7.00 [7.00]	4.95/30 seeds
Figwort, herb (*Scrophularia marilandica* and *S. nodosa*)	[150.00]	[150.00]		3.95 [7.00]	3.95 [7.00]	2.00; 10.00/ oz seeds
Fringe tree, tree bark and root bark (white ash) (*Chionanthus virginicus*)	[22.00/7 gal pot]	[22.00/7 gal pot]			[10.00]	
Galax, leaves, fresh, (beetleweed, coltsfoot, wandflower) (*Galax urceolata*)	[60.00]	[60.00]	[72.00]	[6.49]	[6.49]	[5.50–6.99]

	Wholesale Nursery Prices (one pound of seed [100 plants/roots])			Retail Nursery Prices (small pkg seed [single plant])		
Organic prices in bold	2004	2007	2013	2004	2007	2013
Ginseng, root (sang, manroot, American ginseng) (*Panax quinquifolius*)	104.50–180.00 [105.50–300.00 roots]	131.50–220.00 [75.00–225.00 roots]	30.00–400.00 [50.00–85.00]	5.90–22.50 [4.00–7.95]	3.40–25.00 [3.00–5.00]	**[3.00–10.00] organic;** 3.75; 22.50–25.00/oz [0.62–6.00 root; 3.13–4.00 plant]
Goldenseal, herb (yellowroot, ground raspberry, yellow puccoon, wild circuma, eye-balm, yellow paint, wild turmeric, yellow-eye) (*Hydrastis canadensis*)	70.00/500 seeds [500.00 plants; 140.00–299.00 roots]	70.00/500 seeds [150.00 plants; 240.00 roots]	**[200.00–426.90 roots] organic;** 60.00; 50.00/500 seeds [40.00–427.00 plants]	9.95–11.00 [5.00–7.00]	9.95 [1.40–7.00]	**[2.65–6.47 root] organic;** 2.00; 3.57–10.00/50 seeds [0.75–7.48]
Goldenseal, root, dried						
Indian turnip, root (jack in the pulpit) (*Arisaema triphyllum*)		[150.00]	90.00–375.00 [560.00]	[15.00]	[1.60–15.00]	2.00; 25.00/oz seeds [1.20–9.00]
Life root, herb (squaw weed, cough weed, false valerian) (*Senecio aureus*)						
Lobelia, herb (Indian tobacco) (*Lobelia inflata*)						**2.95/1,000 seeds organic;** 2.00; 80.00/oz seeds
Maidenhair fern, herb (finger fern, ginseng pointer) (*Adiantum pedatum*)		[375.00]	[90.00–800.00]			[1.65–9.00]
Mayapple, root (mandrake, umbrella plant, Indian apple, devil's apple) (*Podophyllum peltatum*)	[60.00–500.00 plants; 150.00 roots]	[60.00–500.00 plants; 150.00 roots]	[360.00]	[2.50–7.00]	[7.00–15.00]	**[7.95 plant; 4.50 roots] organic;** 2.00 [3.60–7.00 roots]
Missouri snakeroot, herb (wild quinine, American feverfew) (*Parthenium integrifolium*)	[150.00]	[30.00/oz seed [150.00]	225.00 [240.00]	2.50–3.50 [3.25–7.00]	2.50–3.50 [3.25–7.00]	**2.95/20 seeds organic;** 2.00–2.50; 15.00/oz seeds
Missouri snakeroot, root						

	Wholesale Nursery Prices (one pound of seed [100 plants/roots])			Retail Nursery Prices (small pkg seed [single plant])		
Organic prices in bold	**2004**	**2007**	**2013**	**2004**	**2007**	**2013**
Mountain maple, bark (*Acer spicatum*)						
Mullein, herb (Indian tobacco, wooly mullein, common mullein) (*Verbascum thapsus*)						**2.95/100 seeds organic;** 3.50/20 seeds
New Jersey tea, herb (red root, Jersey tree) (*Ceanothus americanus*)	[350.00]	[350.00]		3.50	3.50 [5.00]	2.00; 2.95/30 seeds; 40.00/oz seeds [2.42–7.00]
New Jersey tea, root						
Oregon grape, root (holly-leaved barberry, mountain grape) (*Mahonia* or *Berberis aquifolium*)		90.00 [400.00]		2.65 [4.00–10.00]	2.85 [1.50–10.00]	
Partridge berry, herb (squaw vine, squaw berry) (*Mitchella repens*)	[60.00–75.00]	[60.00–75.00]		[5.00–5.95]	[8.00]	
Passionflower, herb (Maypop) (*Passiflora incarnata*)	1,732.00	1,975.00	300.00	3.50–4.60 [3.25–10.00]	3.95–5.25 [7.00–10.00]	**3.95/20 seeds [10.00] organic;** 2.00; 20.00–30.00/oz seeds [5.00]
Pinkroot, root (Carolina pinkroot, Carolina pink, Indian pink, American wormroot, wormgrass) (*Spigelia marilandica*)	[550.00]	[550.00]	[175.00–325.00]	[7.75]	[7.50]	[7.50–16.00]
Pipsissewa, herb (dragon's tongue, spotted wintergreen) (*Chimaphila umbellata*)					3.95	3.95/500 seeds
Pokeweed, root (poke, pokeroot, poke sallet) (*Phytolacca americana*)						**2.95/100 seeds organic**
Prickly ash, northern, tree bark and root bark (toothache tree) (*Zanthoxylum americanum*)					4.95	
Prickly ash, southern, tree bark and root bark (toothache tree) (*Zanthoxylum clava-herculis*)						

	Wholesale Nursery Prices (one pound of seed [100 plants/roots])			Retail Nursery Prices (small pkg seed [single plant])		
Organic prices in bold	**2004**	**2007**	**2013**	**2004**	**2007**	**2013**
Queen of the meadow, herb (Joe-Pye weed, gravel root) (*Eupatorium purpureum*)	296.00; 67.00/ 100,000 seeds [200.00–500.00]	296.00; 65.00/ 100,000 seeds [200.00–500.00]	225.00 [420.00]	1.75–2.90 [3.25–7.00]	2.00–3.00 [3.00–7.00]	**2.95/300 seeds organic;** 2.00; 15.00–25.00/oz seed [2.42–7.00]
Queen of the meadow, root						
Queen's delight, root (Stillingia root) (*Stillingia sylvatica*)						
Ramps, fresh bulbs (wild leeks, wood leeks) (*Allium tricoccum*)	[15.00 bulbs]	[15.00 bulbs]	600.00 [19.40–600.00 bulbs]	[0.19–0.33 bulbs]	[0.10–0.33 bulbs]	2.00; 5.50/50 seeds; 40.00–45.00/oz seeds [0.69–7.00 bulb]
Sarsaparilla, root (American sarsaparilla, spikenard) (*Aralia nudicaulis*)	9.75/300 seeds	9.75/300 seeds		2.65 [7.00]	2.85 [7.00]	
Sassafras, herb (cinnamon wood, smelling stick) (*Sassafras albidum*)				3.50	3.50 [8.00]	
Sassafras, root bark						
Sassafras, tree bark						
Senega snakeroot, root (Seneca snakeroot) (*Polygala senega*)	19.66/1,000 seeds	21.00/1,000 seeds		3.95–5.00	3.95–5.00	3.95/30 seeds
Skullcap, herb (scullcap) (*Scutellaria lateriflora*)	200.00–573.00 [35.00–500.00]	191.00 [35.00–250.00]		1.75–4.20 [2.25–7.00]	2.85–3.40 [6.00–7.00]	**2.95/100 seeds [6.50–7.95] organic;** 2.00–2.35; 30.00 oz/ seeds
Skunk cabbage, herb (polecat weed, swamp cabbage) (*Symplocarpus foetidus, Dracontium foetidum*)	40.00/100 seeds	40.00/100 seeds		5.95	5.95	
Skunk cabbage, root						
Slippery elm, inner bark (*Ulmus rubra*)				4.96–7.00 [7.95]	7.00 [9.00]	**[15.00] organic;** 4.95/100 seeds

	Wholesale Nursery Prices (one pound of seed [100 plants/roots])			Retail Nursery Prices (small pkg seed [single plant])		
Organic prices in bold	**2004**	**2007**	**2013**	**2004**	**2007**	**2013**
Solomon's seal, root (*Polygonatum biflorum*)			[78.00]	7.00 [5.00]	7.00 [5.00]	**[6.50–7.95] organic;** [1.50–5.00]
Spikenard, root (spignet, American spikenard, spiceberry, fleabane, Indian root, old-man's-root) (*Aralia racemosa*)				2.50 [3.95]	3.95 [6.00]	2.00–2.50; 2.95/30 seeds; 100.00 oz/seed [6.00–10.00]
Star grass, root (true unicorn root, star root, blazing star, colic root) (*Aletris farinosa*)				5.00 [6.95]	3.95 [5.00]	3.95/500 seeds
Stone root, root (horsemint, horseweed, horsebalm) (*Collinsonia canadensis*)	[500.00 plants; 350.00 roots]	[500.00 plants; 330.00 roots]		5.00 [6.95]	5.00 [5.00–10.00]	[4.50–5.00]
Sumac, root bark (scarlet sumac, smooth sumac) (*Rhus glabra*)				2.50 [4.00–5.00]	2.85 [10.00]	
Sumac, tree bark						
Virginia snakeroot, root (*Aristolochia serpentaria*)					9.95 [12.00–15.00]	**[8.10–9.00] organic**
Wahoo, tree bark and root bark (burning bush, Indian arrow, bitter ash) (*Euonymus atropurpureus*)						2.50; 30.00/oz seeds [7.00]
White pine, inner tree bark (Eastern white pine) (*Pinus strobus*)						
White willow, inner tree bark (willow, willow bark) (*Salix alba*)						
Wild cherry, tree bark (cherry, black cherry) (*Prunus serotina*)					[7.00–9.00]	
Wild ginger, root (Canadian snakeroot, Indian ginger, Southern snakeroot, black snakeroot, heart snakeroot, snakeroot, coltsfoot, false coltsfoot, catfoot, colic root, little brown jug) (*Asarum canadense*)	[60.00]	[60.00–225.00]	[36.00]	11.00 [3.25–3.95]	11.00 [5.00–7.00]	[0.90–4.00]
Wild hydrangea, root (hydrangea, smooth hydrangea) (*Hydrangea arborescens*)				[3.25]	[3.00]	2.50; 30.00/oz seeds [6.00]

	Wholesale Nursery Prices (one pound of seed [100 plants/roots])			Retail Nursery Prices (small pkg seed [single plant])		
Organic prices in bold	**2004**	**2007**	**2013**	**2004**	**2007**	**2013**
Wild indigo, root (false indigo, yellow indigo, dyer's baptisia, rattle bush) (*Baptisia tinctoria*)		[490.00]		2.50–5.95 [5.00]	5.95 [7.00]	5.95/50 seeds [8.00–9.00]
Wild yam, root (colic root) (*Dioscorea villosa*)	112.00/ 1,000 seeds [18.00 roots]	112.00/ 1,000 seeds [19.00]	[180.00]	4.95 [4.00–11.95]	4.95 [5.00–10.00]	[1.80–4.00]
Witch hazel, leaves (*Hamamelis virginiana*)	33.50/1,000 seeds [17.00/7 gal pot]	35.00/1,000 seeds [18.00/7 gal pot]	[800.00]	2.65 [6.95]	3.95 [8.00–9.00]	**3.95/20 seeds organic;** 9.00 [8.00]
Witch hazel, tree bark						
Yellowroot, root (*Xanthorhiza simplicissima*)	[60.00–150.00]	[60.00–150.00]		3.95	3.95	

APPENDIX 3

Interesting and Helpful Calculations, Tables, and Miscellaneous Information

Table 21 provides rough guidelines for the spacing of plants and the approximate number of plants per acre. Every situation will be different because of placement of posts in a shade structure, trees in the woods, equipment used, etc. These calculations were made assuming a solitary acre with approximately 100-foot-long beds. I can't stress enough how varied these numbers can be, especially in the woods, but this table gives you a starting point for estimating plant needs. If you have an older version of this book, you will see that I have reduced the number of plants per acre in the woods-cultivated and wild-simulated settings. This is based on observations of real-life situations over the past ten years.

Converting Pounds per Acre to Kilograms per Hectare

Most of the fertilizer rates and yields reported in this book are presented as pounds per acre. If you are more familiar with metric terminology, here is an easy conversion. It is not precise, but it's close:

1 pound/acre = 1.12 kilograms/hectare

To make the conversion, multiply the pounds by 1.12. For example, if a fertilizer rate is 2,500 pounds/acre. Multiple 2,500 × 1.12 to get 2,800 kilograms/hectare.

Estimating Yields (From Chapter 12)

If you know exactly how much actual bed space your total harvest will include, then you can easily estimate your final yield. As an example, if you harvested four pounds of roots from 20 square feet of bed space, and the total bed space to be harvested is 100 square feet, then multiply the pounds of root from the measured area by the total amount of bed space you are going to harvest. Then divide by the square feet of the measured test area. So, for my example, 4 × 100 = 400; 400 ÷ 20 = 20. I would expect to harvest 20 pounds of fresh root from the 100 square feet of actual bed space.

A way to get a rough estimate of yields per acre is to estimate the total bed space in your production area. For example, if you have four-foot-wide beds, on five-foot centers, and mark off a five-foot-long section of bed, you will be harvesting roots from 25 square feet of your production area. That includes the one-foot-wide walkway between your beds. To estimate your yield per acre, multiply the pounds of root from the measured section by 43,560 (the number of square feet in an acre). Then divide that number by the square footage of the measured area. So, if you harvested 4 pounds of fresh root from the 25 square feet, 4 × 43,560 = 174,240; 174,240 ÷ 25 = 6,970 pounds of fresh root per acre. Figuring on 30% dry weight, 6,970 × .30 = 2,091 pounds dried root per acre.

Table 21. Estimated Plants per Acre

	Intensive production under shade cloth		Woods-cultivated (⅔ of area is planted)		Wild-simulated (½ of area is planted)	
Spacing between plants	4 foot wide raised beds		4 foot wide raised beds		Spacing area per plant	Scattered pattern; no beds
	5' centers	6' centers	6' centers	8' centers		
6 inches (8 rows/bed)	139,200	115,200	76,800	57,600	12" x 12"	21,780
12 inches (4 rows/bed)	34,800	28,800	19,200	14,400	18" x 18"	9,680
18 inches (2 rows/bed)	11,600	9,600	6,400	4,800	24" x 24"	5,445

Table 22. Commonly Used Fahrenheit-to-Celsius Conversions

°F	°C
35	1-2
40	4
70	21
80	27
85	29
90	32
95	35
100	38
130	54
375	191

Table 23. Other Handy Conversions

1 gallon = 3.79 liters	1 liter = 0.26 gallons
1 tablespoon = 14.8 milliliters	1 milliliter = 0.07 tablespoons
1 ton (short) = 907 kilograms	1 kilogram = 0.001 ton
1 pound = 0.45 kilograms	1 kilogram = 2.2 pounds
1 ounce = 28.3 grams	1 gram = 0.04 ounces
1 inch=2.54 centimeters	1 centimeter=0.39 inches
1 foot=0.30 meters	1 meter=3.28 feet

APPENDIX 4

Good Agricultural, Collection, and Manufacturing Practices

As the natural products industry matures, attention to quality, purity, and safety has increased. There are guidelines and regulations affecting all aspects of the industry, from grower and wild-harvester to the final manufacturer. The United States has federal current Good Manufacturing Practices for Dietary Supplements. Although these regulations are for the manufacturer, many of them involve raw materials. As a result, manufacturers want to purchase raw materials that will keep them in compliance. Growers and wild-harvesters who sell material that helps manufacturers stay compliant with these regulations have a market advantage. Some of the product requirements that affect growers and harvesters are discussed below.

Botanical Identity

For the grower and wild-harvester, proper botanical identity is arguably the most important issue. You need to be absolutely sure that the plant material you sell is exactly what you say it is. This isn't just a problem for new growers and harvesters. Even experienced botanists are sometimes mistaken about plant identity; some species are easily confused. Wild-harvested yellow cohosh (*Actaea podocarpa*), for example, is easy to mistake for black cohosh. To prevent mistakes, which can cost you a sale or actually injure someone, learn your plants. Go on plant walks with experienced herbalists and naturalists, and consider taking a class on plant identification.

Take care when obtaining planting stock and seeds. Purchase from reputable companies with many years of experience in the industry, and get personal recommendations from other growers. Mistakes can still happen, so check your material while it is growing. Compare the flowers to those shown in plant identification books. Also, take advantage of any opportunity to have an expert visit your farm and see your plants. Last year a botanist friend of mine questioned a plant that was growing in one of my medicinal herb demonstration gardens. She took a sample plant with her and did some tests on it. Apparently it was a hybrid of two related species; definitely not what I wanted it to be. So we ripped all those plants out of the ground and started over.

Many buyers now require their suppliers to provide proof of identity. You can do this by providing a *voucher specimen* of the plant material. A voucher is a representative sample of a plant that can be used for identification. It is easy to make a pressed plant sample as a voucher. Collect a complete plant, including the stems, leaves, roots, and flowers or seeds.

Spread the plant out on paper, arranging it so you will be able to see all the parts clearly after it is pressed. Then put it between layers of newspaper and press it. (You can use a plant press, or you can make one with sheets of cardboard and some canvas belts with buckles or bungee cords.) There are dozens of instructions online on how to make a simple plant press. Once your sample is dry, number it and photograph it. Send the photograph along with the plant material you are selling. Store pressed specimens in a dry, dark cabinet. Maintain records so that you could match up the voucher with the material you sold if there were ever a question.

Adulteration

Contamination of plant material with species or compound is another serious problem within the industry. It is the high level of adulteration in imported material that is driving the increased demand for domestic raw material. Using the black cohosh example again, even one yellow cohosh root mixed into a large black cohosh order can cause the order to be rejected. To the best of my knowledge, yellow cohosh is not toxic, but it could cause a company's product to fail a laboratory test for purity of product.

Quality Testing

Some buyers require laboratory testing for species identification, bioactive constituents, microbial contamination, heavy metals, and pesticides. It varies from company to company as to who pays for this testing. If you have a very large quantity of a valuable herb to sell, you might want to have your product tested yourself. This could help you sell your crop. Testing, however, is not inexpensive. The several laboratories I am familiar with are listed in the References and Resources section below. Others can be found by searching online for "natural products testing laboratories."

Other Potential Regulations that Might Affect This Industry

There are many new regulations that will affect the food industry, most notably the US FDA's Food Safety Modernization Act. At the time I wrote this section in October 2013, what exactly this Act will mean to growers, particularly herb growers, was still unknown. However, there is discussion about making the processes related to washing and drying medicinal herbs fall under the same regulations as food products. If you are a herb grower or wild-harvester, it is in your best interest to pay attention to what is happening with these regulations. I also suggest that you start treating your medicinal herbs as if they were fresh produce. Consider this as you plan your washing, drying, packaging, and storage facilities and processes. Ensure that everything you use is food grade and completely washable. I am currently exploring how to build new herb dryers and convert existing herb dryers so that the interiors are completely washable and the screens or shelves are food grade. Any materials for the interior walls must be heat stable and not outgas any harmful fumes when heated. These are important issues that everyone in the industry needs to stay current with.

APPENDIX 5

Joe-Ann McCoy's Disease List

Dr. Joe-Ann McCoy, director of the North Carolina Arboretum Germplasm Repository in Asheville, has extensive training in plant pathology and much experience in identifying diseases on native plants. She has graciously compiled the following list of all the known diseases which affect the plants covered in this book. To the best of our knowledge, this disease list was complete and up to date as of October 2013.

One of the challenges of growing a plant that few people have cultivated is trying to identify and control diseases. If, for example, you are growing spikenard and it develops a leaf spot, and you take it in to your knowledgeable and helpful county extension agent, it is unlikely that the agent will know what the disease is because he or she will not have had any experience with spikenard. Spikenard probably will not be listed in any of the reference books in the extension office library, and a search of the Internet probably will not turn up anything either.

Even if you are fortunate enough to have a public plant disease clinic in your state, you probably will not get a diagnosis from them either, unless your unusual plant is infected with a fairly common disease. These laboratories process large numbers of samples every day. Most of these samples are commonly cultivated plants, such as tomatoes, soybeans, and rhododendrons, that the clinic pathologists are familiar with and can process quickly. When an unusual plant sample is received, especially during the busy growing season, the pathologists often do not have the time to do the literature searches and testing that may be necessary to give an accurate diagnosis. If, however, you provide a list of diseases to which that plant is known to be susceptible, the pathologists can immediately cross some off the list and then concentrate on those few remaining diseases which look probable. This greatly improves your chances of getting a correct diagnosis.

This important and unique disease list will be invaluable if you plan to produce any of these plants. Identifying the pathogen is the most important step in controlling or preventing disease. If you do not know what disease you are fighting, you do not know how to control it. However, if you can identify a particular disease, you can look it up in a plant disease book, which will describe the disease life cycle and environmental conditions necessary for that disease to grow and reproduce. Then, you can try to develop a strategy for prevention and control.

Diseases, Plant by Plant

Under the name of each plant is a listing of all the confirmed, positively identified diseases reported for that plant. In most cases, the state and country where that disease has been reported is listed in parentheses after the disease name.

Bethroot (*Trillium erectum* L.). (This is the list of diseases known to infect trilliums in general, not specifically *T. erectum.*)
Leaf blight:
Botrytis sp.
Ciborinia trillii (AL, GA, NY)
Cladosporium trillii (WA)
Leaf smut:
Urocystis trillii (ID, OR, Canada)
Leaf spot:
Colletotrichum lineola (United Kingdom)
Colletotrichum peckii (NY to NC, IL, MN, WI)
Colletotrichum trillii (WI)
Cylindrogloeum trillii (OR, WA)
Didymella sp. (ID)
Gloeosporium brunneo-maculatum (WI)
Gloeosporium trillii (CA, OR, WA)
Heterosporium trillii (ID, MT, WA)
Neomarssoniella trillii (ID)
Phylosticta trillii (NY, PA, WA, WI)
Ramularia trillii (ID)
Sclerotinia sp. (NC)
Sclerotinia sclerotiorum (FL, LA)
Septoria trillii (New England to SC, OK, WI)
Stem rot:
Sclerotium delphinii (NH, PA)
Sclerotium rolfsii (NH, PA)
Septoria recurvata (IN)
Septoria trillii (MS, OK, WI)
Rust:
Uromyces halstedii (IL, NY, WA, WI)
Smut:
Urocystis trillii (United Kingdom)

Black Cohosh (*Actaea racemosa* L. or *Cimicifuga racemosa* L. Nuttal)
Leaf spot:
Alternaria sp. (Canada)
Ascochyta actaeae (CT, NC, NY, VA)
Ascochyta sp. (Canada)
Ectostroma afflatum (CT, NY, VA)
Phyllosticta sp. (ID, MT, NC)
Nematode:
Ditylenchus destructor
Meloidogyne sp. (rootknot) (NJ)
Root and stem rot:
Armillaria mellea (United Kingdom)
Leptosphaeria clavigera (GA)
Ophiobolus nigro-clypeata (GA)
Phytophthora citricola, P. megasprema, P. kelmania (MO)
Pythium spp. (MO) :
P. dissotocum, P. irregular, P. sylvaticum, P. ultimum, P. vexans, P. chamaehiphon
Rhizoctonia solani (damping-off of seedlings) (Canada)
Rust:
Puccinea recondita (NC, TN, Canada)
Puccinea rubigo-vera (OH, MD, PA, VA, West Germany)
Puccinea rubigo-vera var. *agropyrina* (MD, NC, OH, VA)
Smut:
Urocystis carcinodes (NC, NY, OH, PA, TN, VA, Germany)

Bloodroot (*Sanguinaria canadensis* L.)
Gray mold, blight:
Botrytis cinerea (NY, WI)
Leaf spot:
Alternaria sp. (NC)
Ascochyta sp. (WI)
Cercospora sanguinariae (DE, MD, MO, NY, PA, TX, WI)
Colletotrichum sp. (WI)
Cylindrosporium circinans (MD, MO, WI)
Gloeosporium sanguinariae (OH, TX, WV)
Phyllosticta sanguinariae (MO, TX, WV)
Phyllosticta sp. (WI)

Root rot:
Pythium paroecandrum (brown rot) (VA)
Crown rot:
Fusarium oxysporum (CT)

Blue Cohosh (*Caulophyllum thalictroides* L. Michaux)
Leaf blight:
Botrytis sp. (NY, NJ)
Leaf spot:
Cercospora caulophylli (VT to VA, IA, MO, MS, WI, Canada)
Colletotrichum dematium (IL)
Vermicularia hysteriformis (synonym *Colletotrichum*) (WI)
Drechslera gigantea (IL, WI)
Phoma caulophylli (Canada)
Phyllosticta sp. (WI)
Powdery mildew:
Erysiphe cichoracearum (Canada)
Root blight:
Cylindrocephalum sp. (WI)
Stem blight:
Calloria caulophylli (NY)
Streptotinia caulophylli (Canada)
Orbilia caulophylli (Canada)

False unicorn (*Chamaelirium luteum* L.)
no pathogens currently reported

Galax (*Galax urceolata*)
Leaf spot:
Calonectria morganii (US)
Clypeolella leemingii (black spot) (MD to GA, FL, MS)
Cylindrocladium scoparium (FL)
Discohainesia oenotherae (NC)
Laestadia galactina (NC) (synonym *Laestadia*)
Glenospora melioloides (NY, OH) (synonym *Septobasidium*)
Phyllosticta galactis (NC, VA, WV)

Ginseng (*Panax quinquefolius* L.)
Blight:
Alternaria alternata (Canada)
Botrytis cinerea (gray mold) (eastern states, MI, WA, Canada)
Damping-off:
Alternaria panax (eastern states, IA, MN, MO, WI, OR, WA, Canada)
Fusarium sp. (NC,WI, Canada)
Pythium debaryanum (NC, NY, WI, Canada)
Rhizoctonia solani (AR, IN, MI, NJ, NY, WA, WI, Canada)
Leaf spot:
Septoria sp. (MI, MN)
Septoria araliae (WI)
Colletotrichum coccodes (WI)
Colletotrichum dematium (secondary, NY to NC, MO, MN, VT)
Nematode:
Meloidogyne sp. (CT, MI, NY, OH, PA, WI)
Powdery mildew:
Erysiphe sp. (Canada)
Root rot:
Armillaria mellea (WA)
Calonectria sp. (NC)
Cylindrocarpon destructans
Cylindrocarpon destructans var. *destructans* (Canada)
Fusarium avenaceum (Canada)
Fusarium equiseti (NC, Canada)
Fusarium oxysporum (Canada)
Fusarium scirpi (NY to AL, MO, WA, WI)
Fusarium scirpi var. *acuminatum* (eastern states, MO, WI)
Fusarium solani (Canada)
Phytophthora cactorum (eastern states, IA, MI, WA, WI, Canada, Bulgaria)
Phytophthora citricola
Phytophthora colocasiae (NC)
Pythium debaryanum (NY)
Pythium irregular (Canada)
Pythium ultimum (Canada)

Ramularia destructans (MI, NY, OR, WA, WI)
Ramularia panacicola (NY, WI)
Rhexocercosporidium panacis sp. *nov.* (WI, MI)
Rhizoctonia solani (AR, IN, MI, NJ, NY, WA, Canada)
Sclerotinia panacis (MI, MN, NY, WI)
Sclerotinia sclerotiorum (white rot) (MI, NY, OH, PA, WA, WI)
Sclerotinia smilacina (black rot) (MI, MN, NY, WI)
Septonema sp. (WI)
Stromatinia panacis (MI)
Thanatephorus cucumeris (WA)
Thielaviopsis basicola (black rot) (IL, MI, NJ, NY, OH)
Rust:
Puccinia araliae (MA, PA)
Wilt:
Fusarium avenaceum (Canada)
Verticillium albo-atrum (IN, KY, MI, NJ, NY, OH, OR, PA, TN, WI)
Verticillium dahlia (rare) (MI, WI)

Goldenseal (*Hydrastis canadensis* L.)
Leaf blight:
Alternaria sp. (MI, NY, OH)
Alternaria panax
Botrytis streptothrix (DE)
Botrytis sp. (CT to NC, IN, WA, WI)
Nematode:
Meloidogyne sp. (rootknot) (MI, OH, TX, WA)
Root rot:
Cylindrocladium colhounii (Canada)
Phymototrichum omnivorum (TX)
Rhizoctonia solani (root rot and damping-off) (NC)
Stem rot:
Pellicularia filamentosa
Virus:
Mosaic, unidentified (CT)
Wilt:
Fusarium sp. (IL, NY, OH, WA)
Fusarium oxysporum (Canada)

Mayapple (*Podophyllum peltatum* L.)
Blight, Gray Mold:
Botrytis cinerea (NJ)
Leaf blight:
Septotinia podophyllina (DE, MD, MO, NJ, NY, VA, WV)
Leaf spot:
Cercospora podophylli (IL)
Discohainesia oenotherae (VA)
Gloeosporium podophyllinum (MO)
Glomerella cingulata (DE, TX)
Pezizula oenotherae (VA)
Phyllosticta podophylli (NY to AL, AR, IA, MS, OH, OK, WI)
Phyllosticta podophyllina (IL)
Phyllostictina podophylli (MD)
Septoria podophyllina (NY to MS, MO, WI, WV)
Septotinia podophyllina (DE, MD, MO, NJ, NY, VA, WV)
Septosis podophyllina (MO, NJ, NY, PA)
Vermicularia podophylli (synonym *Colletotrichum*) (TX, VA, Canada)
Stem rot:
Rhizoctonia sp. (MO)
Rust:
Puccinia podophylli (GA, MS, NC, OK, WI, VA)

Pinkroot (*Spigelia marilandica* L. Linnaeus)
Powdery mildew:
Oidium sp. (IL)
Leaf spot:
Cercospora shoreae

Ramps (*Allium tricoccum* Aiton or *Allium tricoccum* var. *burdickii* (Hanes) A.G. Jones)

Leaf spot:
Colletotrichum circinans (smudge) (WI)
Septoria sp. (NC)
Septoria viridi-tingens (MD, MN, ND, NY, TN, WI)

Spikenard (*Aralia racemosa* L.)
Leaf spot:
Alternaria araliae (WI)
Alternaria sp. (NY)
Ascochyta marginata (WI)
Cercospora clavigera (Canada)
Cercospora leptosperma (IA, MI, NY, WI)
Mycosphaerelia sp. (AL)
Phyllosticta decidua (WI)
Ramularia repens (WI, Canada)
Sclerotium deciduum (WI)
Septoria sp. (WI)
Powdery mildew:
Phyllactinia corylea (MI, NE)
Rust:
Nyssopsora clavellosa (ME to PA, AZ, CA, MN, NY, OR, TX)
Stem blight:
Acrospermum album (NY)
Belonidium minimum (DE)
Leptosphaeria collinsoniae (NY)
Leptosphaeria conoidea (NY)
Leptosphaeria doliolum (NY)
Ophiobolus niesslii (NY)
Wilt:
Verticillium albo-atrum (NY)

Wild Ginger (*Asarum canadense* L.)
Leaf gall:
Synchytrium asari (CA, MI, MN, WI)
Leaf spot:
Ascochyta sp. (WI)
Ascochyta versicolor (ID)
Colletotrichum sp. (WI)
Leptothyrium symploci (SC)
Mycosphaerelia sp. (AL)
Plagiostoma asarifola (SC)
Powdery mildew:
Erysiphe communis (Canada)
Root rot:
Sclerotinia sclerotiorum (NY)
Rust:
Puccinea asarina (CA, ID, OR, NY, WA)

Wild Indigo (*Baptisia tinctoria* (L.) Ventenat)
Leaf blight:
Trichometasphaeria gloeospora (NJ)
Leaf spot:
Cercospora velutina (KS, IL, WI)
Marssonina baptisiae (IA, NM, VT, WI)
Mycosphaerella granulata (NJ)
Septoria baptisiae (SC, TX)
Stagonospora baptisiae (SC, WI)
Powdery mildew:
Erysiphe polygoni (MA)
Erysiphe alni (WI)
Erysiphe sp. (Austria)
Mycosphaerella baptisiicola (NJ, SC)
Mycosphaerella granulata (NJ)
Root rot:
Armillaria mellea (CA)
Phymatotrichum omnivorum (synonym *Botrytis*) (TX)
Rust:
Puccinia andropogonis (KS, NE, NC, OK)
Puccinia andropogonis var. *onobrychidis* (SC)
Stem blight:
Diaporthe arctii (NJ)
Physalospora obtusa (GA)

References and Resources for Parts 3 through 6

General References Used Throughout These Sections

Cech, R. 2002. *Growing At-risk Medicinal Herbs: Cultivation, Conservation and Ecology.* Horizon Herbs, Williams, OR.

Foster, S. 1998. *101 Medicinal Herbs: An Illustrated Guide.* Interweave Press, Loveland, CO.

Foster, S. 1993. *Herbal Renaissance: Growing, Using and Understanding Herbs in the Modern World.* Gibbs-Smith, Publisher, Salt Lake City, UT.

Hamel, P. B., and M. U. Chiltoskey. 1975. *Cherokee Plants and Their Uses—A 400-Year History.* Herald, Sylva, NC.

Harding, A. R. 1972. *Ginseng and Other Medicinal Plants,* rev. ed. A. R. Harding, Columbus, OH.

Krochmal, A., and C. Krochmal. 1984. *A Field Guide to Medicinal Plants.* Times Books, New York, NY

Millspaugh, C. F. 1974. *American Medicinal Plants.* Dover Publications., New York, NY. (Reprint of an 1892 book published under the title Medicinal Plants.)

Plant Hardiness Zone Maps. Agriculture and Agri-Food Canada. www.sis.agr.gc.ca/cansis/nsdb/climate/hardiness/index.html

Plant Hardiness Zone Map. United States Department of Agriculture, planthardiness.ars.usda.gov/PHZMWeb

Sturdivant, L., and T. Blakley. 1999. *Medicinal Herbs in the Garden, Field, and Marketplace.* San Juan Naturals, Friday Harbor, WA.

Tilford, G. L. 1998. *From Earth to Herbalist: An Earth Conscious Guide to Medicinal Plants.* Mountain Press, Missoula, MT.

United Plant Savers. 2004. unitedplantsavers.org

References Relevant to Specific Species, Chapters, and Appendices

Goldenseal References

Apsley, D., and C. Carroll. 2004. Growing American ginseng in Ohio: Selecting a site. Ohio State University Extension Factsheet, F-58-04, Ohio State University, Columbus, OH.

Cecil, C. E., J. M. Davis, N. B. Cech, and S. M. Laster. 2011. Inhibition of H1N1 influenza A virus growth and induction of inflammatory mediators by the isoquinoline alkaloid berberine and extracts of goldenseal (*Hydrastis canadensis*). *International Immunopharmacology* 11(11):1706–1714.

Convention on Trade in Endangered Species of Wild Fauna and Flora. (CITES). cites.org.

Davis, J. M., and A. Hamilton. 2008. Dryers for Commercial Herb Growers: A Construction Guide. North Carolina State University. ces.ncsu.edu/fletcher/programs/herbs/pdf/herb_dryer_leaflet.pdf

Dawes, M. L., and T. Brettell. 2012. Analysis of goldenseal, *Hydrastis canadensis* L., and related alkaloids in urine using HPLC with UV detection. *Journal of Chromatography* B 880(1):114–118.

Follett, J. M., J. A. Douglas, and R. A. Littler. 2005. The effect of gibberellic acid, potassium nitrate, and cold stratification on the germination of goldenseal (*Hydrastis canadensis*) seed. International Plant Propagators' Society, Combined Proceedings 5:165–170.

Foster, S. 1991. Goldenseal *Hydrastis canadensis*. Botanical Series No. 309. American Botanical Council, Austin, TX.

Foster, S. 1989. Goldenseal — Masking of drug tests from fiction to fallacy: An historical anomaly. *HerbalGram* 21:7, 35.

Greenbook Data Solutions. greenbook.net.

Haage, L. J., and L. J. Ballard. 1989. *A Grower's Guide to Goldenseal*. Nature's Cathedral, Norway, IA.

Hardacre, J. V., G. Henderson, F. B. Collins, E. L. Andersen, V. M. Harris, B. Fewster, R. Beck, D. Bowman, and E. L. Donzelot. 1962. *The Wildcrafters Goldenseal Manual*. Wildcrafters Publications, Rockville, IN.

Hus, H. 1907. The germination of *Hydrastis canadensis*. Missouri Botanical Garden Annual Report 1907:85–94.

Lloyd, J. U. 1912. The cultivation of hydrastis. *Journal of the American Pharmaceutical Association* 1:5–12.

Murray, M. T. 1995. *The Healing Power of Herbs: The Enlightened Person's Guide to the Wonders of Medicinal Plants*, 2nd ed. Prima Publishing, Rocklin, CA.

Reeleder, R. D. 2002. Diseases of shade-grown medicinal herbs. Poster presented at the American Phytopathological Society Annual Meeting, Milwaukee, WI.

Reeleder, R. D. 2003. The ginseng root pathogens *Cylindrocarpon destructans* and *Phytophthora cactorum* are not pathogenic to the medicinal herbs *Hydrastis canadensis* and *Actaea racemosa*. *Canadian Journal of Plant Pathology* 25(2): 218–221.

Reeleder, R. D. 2004. A new root disease of goldenseal (*Hydrastis canadensis*) caused by *Cylindrocladium colhouni*. *Canadian Journal of Plant Pathology* 26(4):596–600.

Rehman, J., J. M. Dillow, S. M. Carter, J. Chou, B. Le, and A. S. Maisel. 1999. Increased production of antigen-specific immunoglobulins G and M following in vivo treatment with the medicinal plants *Echinacea angustifolia* and *Hydrastis canadensis*. *Immunology Letters* 68(2-3):391–395.

Scazzocchio, F., M. F. Cometa, L. Tomassini, and M. Palmery. 2001. Antibacterial activity of *Hydrastis canadensis* extract and its major isolated alkaloids. *Planta Medica* 67(6):561–564.

Small, E., and P. M. Catling. 1999. Canadian medicinal crops. NRC Research Press, Ottawa, Canada.

Soil Food Web. soils.usda.gov/sqi/concepts/soil_biology/soil_food_web.html

Squier, T. B. B. 1998. *Herbal Folk Medicine: An A–Z Guide*. Henry Holt, New York, NY.

Van Fleet, W. 1914. Goldenseal under cultivation. USDA Farmers' Bulletin #613.

Veninga, L., and B. R. Zaricor. 1976. *Goldenseal/etc.: A pharmacognosy of wild herbs*. Ruka Publications, Santa Cruz, CA.

Ramps References

Allison, J., W. Jenner, N. Cappuccino, and P. G. Mason. 2007. Oviposition and feeding preference of *Acrolepiopsis assectella* Zell. (Lep., Acrolepiidae). *Journal of Applied Entomology* 131(9-10): 690–697.

Bernatchez, A., J. Bussières, and L. Lapointe. 2013. Testing fertilizer, gypsum, planting season and varieties of wild leek (*Allium tricoccum*) in forest farming system. *Agroforestry Systems* 87(5):977–991.

Davis, J. M., and J. Greenfield. 2002. "Cultivating Ramps: Wild Leeks of Appalachia." *Trends in New Crops and New Uses*, J. Janick and A. Whipkey, eds. ASHS Press, Alexandria, VA.

DeValue. J. 2013. Improving germination of *Allium tricoccum*: a plant threatened by foraging trends. Thesis, M.S., Alfred University, Alfred, NY.

Facemire, G., Jr. 1996. *Growing Your Own Ramps*. Richwood, WV: Self-published.

Facemire, G., Jr. 1997. *Ramps: From the Seed to the Weed*. Richwood, WV: Self-published.

Kushad, M. M., J. Masiunas, K. Eastman, W. Kalt, and M. A. L. Smith. 2010. Health promoting phytochemicals in vegetables. *Horticultural Reviews* 28:125–186.

Nault, A., and D. Gagnon. 1987. Some aspects of the pollination ecology of wild leek, *Allium tricoccum* Ait. *Plant Species Biology* 2:127–32.

Ritchey. K. D., and C. M. Schumann. 2005. Response of woodland-planted ramps to surface-applied calcium, planting density, and bulb preparation. *HortScience* 40(5): 1516–1520.

Rock, J. H., B. Beckageb, and L. J. Gross. 2004. Population recovery following differential harvesting of *Allium tricoccum* Ait. in the Southern Appalachians. *Biological Conservation* 116(2):227–234.

Vasseur, L., and D. Gagnon. 1994. Survival and growth of *Allium tricoccum* Ait. transplants in different habitats. *Biological Conservation* 68 (2): 107–14.

Whanger, P. D. 2004. Selenium and its relationship to cancer: an update. *British Journal of Nutrition* 91:11–28.

Whanger, P. D., C. Ip, C. E. Polan, P. C. Uden, and G. Welbaum. 2000. Tumorigenesis, metabolism, speciation, bioavailability, and tissue deposition of selenium in selenium-enriched ramps (*Allium tricoccum*). *Journal of Agriculture and Food Chemistry* 48 (11): 5723–5730.

Bethroot References

Miranda, E. 1980. *The Power of Nature's Medicine*. Deep South Publishing, Topton, NC.

Moerman, D. E. 1998. *Native American Ethnobotany*. Timber Press, Portland, OR. (This document is also available as an online database established in 1999 at herb.umd.umich .edu)

Pence, V. C., and V. G. Soukup. 1993. Factors affecting the initiation of mini-rhizomes from *Trillium erectum* and *Trillium grandiflorum* tissues in vitro. *Plant Cell, Tissue, and Organ Culture* 35:229–235.

Rousseau, J. 1947. Ethnobotanique Abenakise. *Archives de Folklore* 11:145–182.

Rousseau, J. 1945. Le Folklore Botanique De Caughnawaga. *Contributions de l'Institut botanique l'Universite de Montreal* 55:7–72.

Routhier, M. C., and Lapointe, L. 2002. Impact of tree leaf phenology on growth rates and reproduction in the spring flowering species *Trillium erectum* (Liliaceae). *American Journal of Botany* 89(3):500–505.

Black Cohosh References

Bhavneet, K., J. McCoy, and E. Eisenstein. 2013. Efficient, season-independent seed germination in black cohosh (*Actaea racemosa* L.). *American Journal of Plant Sciences* 4:77–83.

Greenfield, J., and J. M. Davis. 2003. Collection to commerce: Western North Carolina non-timber forest products and their markets. A report for the U.S. Forest Service. N.C. State University. ces.ncsu.edu/fletcher/programs/herbs/pdf/ntfpfinal17.pdf

Leach, M. J., and V. Moore. 2012. Black cohosh (*Cimicifuga spp.*) for menopausal symptoms. Database of Systematic Reviews. Issue 9. Art. No: CD007244.

New Hope 360. 2013. ABC Tackles black cohosh adulteration. newhope360.com/breaking-news/abc-tackles-black-cohosh-adulteration.

Strategic Reports. 2003. Analysis of the economic viability of cultivating selected botanicals in North Carolina. A report commissioned by J. M. Davis of North Carolina State University and the North Carolina Consortium on Natural Medicinal Products. Reading, Pennsylvania: Strategic Reports.

United States Food and Drug Administration, Current Good Manufacturing Practices for Dietary Supplements. fda.gov/Food/GuidanceRegulation/CGMP/

Bloodroot References

Bennett, B. C., C. R. Bell, and R. T. Boulware. 1990. Geographic variation in alkaloid content of *Sanguinaria canadensis* (Papaveraceae). *Rhodora* 92 (870): 57–69.

Dzink, J. L., and S. S. Socransky. 1985. Comparative in vitro activity of sanguinarine against oral microbial isolates. *Antimicrobial Agents and Chemotherapy* 27 (4): 663–665.

Frankos, V. H., D. J. Brusick, E. M. Johnson, H. I. Maibach, I. Munro, R. A. Squire, and C. S. Weil. 1990. Safety of sanguinaria extract as used in commercial toothpaste and oral rinse products. *Journal of the Canadian Dental Association* 56 (suppl 7): 41–47.

Godowski, K. C. 1989. Antimicrobial action of sanguinarine. *Journal of Clinical Dentistry* 1 (4): 96–101.

Graf, T. N., K. E. Levine, M. E. Andrews, J. M. Perlmutter, S. J. Nielsen, J. M. Davis, M. C. Wani, and N. H. Oberlies. 2007. Variability in the yield of benzophenanthridine alkaloids in wildcrafted vs cultivated bloodroot (*Sanguinaria canadensis* L.). *Journal of Agricultural Food Chemistry* 55(4):1205–1211.

Hannah, J. J., J. D. Johnson, and M. M. Kuftinec. 1989. Long-term clinical evaluation of toothpaste and oral rinse containing sanguinaria extract in controlling plaque, gingival inflammation, and sulcular bleeding during orthodontic treatment. *American Journal of Orthodontics and Dentofacial Orthopedics* 96 (3): 199–207.

Heithaus, E. R. 1981. Seed predation by rodents on three ant-dispersed plants. *Ecology* 62(1): 136–145.

Lyon, D. L. 1992. Bee pollination of facultatively xenogamous *Sanguinaria canadensis* L. *Bulletin of the Torrey Botanical Club* 119(4):368–375.

Mahady, G. B., S. L. Pendland, A. Stoia, and L. R. Chadwick. 2003. In vitro susceptibility of *Helicobacter pylori* to isoquinoline alkaloids from *Sanguinaria canadensis* and *Hydrastis canadensis*. *Phytotherapy Research* 17(3):217–221.

Marino, P. C., R. M. Eisenberg, and H. V. Cornell. 1997. Influence of sunlight and soil nutrients on clonal growth and sexual reproduction of the understory perennial herb *Sanguinaria canadensis* L. *Journal of the Torrey Botanical Society*. 124 (3): 219–227.

Ness, J. H. 2004. Forest edges and fire ants alter the seed shadow of an ant-dispersed plant. *Oecologia* 138(3):448–454.

Quiros, A. 2001. Domestication of bloodroot (*Sanguinaria canadensis* L.) for sanguinarine production: Improving cultivation and propagation techniques. Master of Science thesis, University of Georgia, Athens, GA.

Rockwood, L. L., and M. B. Lobstein. The effects of experimental defoliation on reproduction in four species of herbaceous perennials from northern Virginia. *Castanea* 59(1):41–50.

Salmore, A. K., and M. D. Hunter. 2001. Elevational trends in defense chemistry, vegetation, and reproduction in *Sanguinaria canadensis. Journal of Chemical Ecology* 27 (9): 1713–1727.

Salmore, A. K., and M. D. Hunter. 2001. Environmental and genotypic influences on isoquinoline alkaloid content in *Sanguinaria canadensis. Journal of Chemical Ecology* 27 (9): 1729–1747.

Schemske, D. W. 1978. Sexual reproduction in an Illinois population of *Sanguinaria canadensis. American Midland Naturalist* 100(2):261–268.

Southard, G. L., R. T. Boulware, D. R. Walborn, W. J. Groznik, E. E. Thorne, and S. L. Yankell. 1984. Sanguinarine, a new antiplaque agent: Retention and plaque specificity. *Journal of the American Dental Association* 108 (3): 338–341.

Blue Cohosh References

Alternative Nature Online Herbal. altnature .com/gallery/blue_cohosh.htm

Betz, J. M., D. Andrzejewski, A. Troy, R. E. Casey, W. R. Obermeyer, S. W. Page, and T. Z. Woldemariam. 1998. Gas chromatographic determination of toxic quinolizidine alkaloids in blue cohosh, *Caulophyllum thalictroides* (L.) Michx. *Phytochemical Analysis* 9 (5): 232–236.

Datta, S., F. Mahdi, A. Ali, I. A. Khan, M. B. Jekabsons, D. G. Nagle, and Y. D. Zhou. 2013. Toxins from botanical dietary supplements: Blue cohosh components disrupt mitochondrial membrane integrity and cellular respiration. *Planta Medica* 79.10:PH23.

Densmore, F. 1928. Uses of plants by the Chippewa Indians. SI-BAE Annual Report 44: 273–379.

Flom, M. S. 1971. The isolation and characterization of alkaloids of *Caulophyllum thalictroides* (L.) Michx. Ph.D. diss., Ohio State University, Columbus, OH.

Friedman, J. M. 2000. Tetratology society: Presentation to the FDA public meeting on safety issues associated with the use of dietary supplements during pregnancy. *Tetratology* 62 (2): 134–137.

Jhoo, J. W., S. Sang, and K. He. 2001. Characterization of the triterpene saponins of the roots and rhizomes of blue cohosh (*Caulophyllum thalictroides*). *Journal of Agricultural and Food Chemistry* 49 (12): 5969–5974.

Kight, J. 2004. Herbs for health. healthyherbs .about.com/library/weekly/aa060598.htm (no longer available).

Kowalchik, C., and W. H. Hylton, eds. 1987. *Rodale's Illustrated Encyclopedia of Herbs.* Rodale Press, Emmaus, PA.

Leung, A. Y., and S. Foster. 1996. *Encyclopedia of Common Natural Ingredients Used in Food, Drugs, and Cosmetics*, 2nd ed. John Wiley and Sons, New York, NY.

Lockard, A., and A. Q. Swanson. 1998. *A Digger's Guide to Medicinal Plants.* American Botanicals, Eolia, MO.

NatureServe. 2004. NatureServe explorer: An online encyclopedia of life. ver. 1.8. Arlington, Virginia: NatureServe. nature serve.org/explorer.

Purdue University, New Crops Center. 2004. From a USDA Publication from 1930 called The herb hunter's guide: American medicinal plants of commercial importance. hort .purdue.edu/newcrop/HerbHunters/blue cohosh.html

Small, E., and P. M. Catling. 1999. Canadian medicinal crops. NRC Research Press, Ottawa, Canada.

False Unicorn References

Baskin, C. C., J. M. Baskin, and E. W. Chester. 2001. Morphophysiological dormancy in seeds of *Chamaelirium luteum*, a long-lived dioecious lily. *Journal of the Torrey Botanical Society*. 128 (1): 7–15.

Matovic, N. J., J. M. U. Stuthe, V. L. Challinor, P. V. Bernhardt, R. P. Lehnmann, W. Kitching, and J. J. DeVoss. 2011. The truth about false unicorn (*Chamaelirium luteum*): Total synthesis of 23R,24S-chiograsterol B defines the structure and stereochemistry of the major saponins from this medicinal herb. *Chemistry—A European Journal* 17(27):7578–7591.

Mills, S. Y. 1991. *Out of the earth: The essential book of herbal medicine*. Viking Arkana, Middlesex, UK.

Galax References

Bir, R. E., and J. L. Conner. 2003. Galax: A crop for the future? *Commercial Horticulture News*. Buncombe County Extension, North Carolina Cooperative Extension, Asheville, NC, (December), 3–4.

Bir, R., and C. Deyton. 2003. Domesticating galax. Report to the North Carolina Specialty Crops Program, N. C. State University. cals.ncsu.edu/specialty_crops/publications/reports/bir.html

Creelin, J. R., and J. Philpott. 1990. *Herbal Medicine Past and Present, Vol. 2*. Duke University Press, Durham, NC.

Dressler, A. M., Bloomquist, and J. M. Davis. 2013. 2013 Report on special forest botanicals of Southern Appalachia. N.C. State University (in press; there will be a link to this report on ncherb.org).

Greenfield, J., and J. M. Davis. 2003. Collection to commerce: Western North Carolina non-timber forest products and their markets. A report for the U.S. Forest Service. N.C. State University. ces.ncsu.edu/fletcher/programs/herbs/pdf/ntfpfinal17.pdf

USDA Forest Service. 2001. Spring harvesting of galax to be restricted. National Forests in North Carolina News Release. (28 February). Asheville, NC: USDA Forest Service.

USDA Forest Service. 2003. Galax. Encyclopedia of southern Appalachian forest ecosystem. forestryencyclopedia.net

Mayapple References

Becker, H. 2000. Mayapple's cancer-fighting precursor. *Agricultural Research* 48 (7): 9, USDA.

Canel, C., R. M. Moraes, F. E. Dayan, and D. Ferreira. 2000. Molecules of interest: Podophyllotoxin. *Phytochemistry* 54 (2): 115–120.

Cushman, K., M. Maqbool, E. Bedir, I. Khan, and R. Moraes. 2003. Growth and podophyllotoxin content of mayapple under sun or shade. Abstract. *HortScience* 38 (5): 854.

Densmore, F. 1928. Uses of plants by the Chippewa Indians. SI-BAE Annual Report 44: 273–379.

Lee, C. T., V. C. K. Lin, S. X. Zhang, X. K. Zhu, D. VanVliet, H. Hu, S. A Beers, Z. Q. Wang, L. M. Cosentino, S. L. Morris-Natschke, and K. H. Lee. 1997. Anti-HIV activity of modified podophyllotoxin derivatives. *Bioorganic and Medicinal Chemistry Letters* 7 (22):2897–2902.

Lockard, A., and A. Q. Swanson. 1998. *A Digger's Guide to Medicinal Plants*. American Botanicals, Eolia, MO.

Maqbool, M., and K. Cushman. 2003. Mulch type, mulch depth, and rhizome planting depth for field-grown mayapple. Abstract. *HortScience* 38 (5): 855.

Moraes, R. M., C. Burandt, Jr., M. Ganzera, X.

Li, I. Khan, and C. Canel. 2000. The American mayapple revisited—*Podophyllum peltatum*—still a potential cash crop? *Economic Botany* 54(4): 471–476.

Small, E., and P. M. Catling. 1999. Canadian medicinal crops. NRC Research Press, Ottawa, Canada.

Smith, H. H. 1923. Ethnobotany of the Menomini Indians. *Bulletin of the Public Museum of the City of Milwaukee* 4: 1–174.

Pinkroot References

Pill, W. G., and B. Goldberger. 2010. Effect of IBA treatments, bottom heat, stock plant location, and cutting type on the rooting of *Spigelia marilandica* cuttings. *Journal of Environmental Horticulture* 28(1):53–57.

Taylor, L. A. 1940. Plants used as curatives by certain Southeastern tribes. Botanical Museum of Harvard University, Cambridge, MA.

Spikenard References

Clement, J. A., T. J. Willis, R. M. Kelly, J. A. McCoy, and J. D. Schmitt. 2009. Antitumor activity of *Aralia racemosa. Planta Medica* 75:75–97.

Manpreet, K., and K. Harinder. 2011. Analgesic activity of roots of *Aralia racemosa. Research Journal of Pharmacy and Technology* 4(12):1896–1897.

Wild Ginger References

Gilmore, M. R. 1933. *Some Chippewa Uses of Plants*. University of Michigan Press, Ann Arbor, MI.

Herrick, J. W. 1977. Iroquois medical botany. PhD thesis. State University of New York, Albany, NY.

Smith, H. H. 1923. Ethnobotany of the Menomini Indians. *Bulletin of the Public Museum of the City of Milwaukee* 4: 1–174.

Squier, T. B. B. 1998. *Herbal Folk Medicine: An A to Z Guide*. Henry Holt, New York, NY.

Taylor, L. A. 1940. Plants used as curatives by certain Southeastern tribes. Botanical Museum of Harvard University, Cambridge, MA.

Wild Indigo References

Beuscher, N., and L. Kopanski. 1985. Stimulation of immunity by the contents of *Baptisia tinctoria. Planta Medica* 5:381–384.

Chandler, R. F., L. Freeman, and S. N. Hooper. 1979. Herbal remedies of the Maritime Indians. *Journal of Ethnopharmacology* 1:49–68.

Miranda, E. 1980. *The Power of Nature's Medicine*. Deep South Publishing, Topton, NC.

Other Botanicals Growers' Stories References

High Falls Gardens. highfallsgardens.net

Home Garden References

Davis, D. D., L. J. Kuhns, and T. L. Harpster. 2005. Use of mushroom compost to suppress artillery fungi. *Journal of Environmental Horticulture* 23(4):212–215.

Mountain Moss. Annie Martin's website. mountainmoss.com

Murray, T. Voles. Garden friends and foes. Whatcom County Extension, Washington State University. whatcom.wsu.edu/ag/homehort/pest/voles.htm

Salmon, T. P., and W. P. Gorenzal. 2010. Voles (Meadow mice) Integrated pest management for home gardeners and landscape professionals. University of California, Integrated Pest Management Program, Publication 7439. ipm.ucdavis.edu/PDF/PESTNOTES/pnvoles.pdf

Wild-harvesting References

Good Stewardship Harvesting of Wild American Ginseng Brochures. 2013. American Herbal Products Association. ahpa.org/default.aspx?tabid=154

Ginseng. 2013. West Virginia Division of Forestry. wvforestry.com/ginseng.cfm?menucall=ginseng

Wild Ginseng: Regulations and guidelines for sustainable harvest. 2012. Wisconsin Department of Natural Resources. PUB-LE-005. dnr.wi.gov/files/PDF/pubs/LE/LE0005.pdf

Good Agricultural, Collection, and Manufacturing Practices References

Alkemists Labs, Costa Mesa, CA. alkemists.com

Bionetwork Natural Products Laboratory, Candler, NC. abtech.edu/category/department/bionetwork

Food Safety and Modernization Act. fda.gov/Food/GuidanceRegulation/FSMA/default.htm

Good agricultural and collection practice for herbal raw materials. 2006. American Herbal Products Association. ahpa.org/portals/0/pdfs/06_1208_AHPA-AHP_GACP.pdf

Good manufacturing practices, U.S. Food and Drug Administration. fda.gov/Food/GuidanceRegulation/CGMP/ucm079496.htm

Guidelines on good agricultural and collection practices for medicinal plants. 2003. World Health Organization. apps.who.int/medicinedocs/en/d/Js4928e/

Natural health products regulations, Health Canada. hc-sc.gc.ca/dhp-mps/prodnatur/index-eng.php

Joe-Ann McCoy's Disease List References

Abad, G., J. Abad, and A. Thomas. 2005. Black root and crown rot of black cohosh (*Actaea racemosa*) is associated to *Phytophthora* and *Pythium* species. Southwest Center Agricultural Experiment Station Workshop Series. aes.missouri.edu/swcenter/fieldday/2005/page41.stm

Case, F., and R. Case. 1997. *Trilliums*. Timber Press, Portland, OR.

Dimitre, I., and M. Bernards. 2012. Ginsenosides and the pathogenicity of *Pythium irregular. Phytochemistry*. 78:44–53.

Elmer, W. H., and R. E Marra. 2012. First report of crown rot of bloodroot (*Sanguinaria canadensis*) caused by *Fusarium oxysporum* in the United States. *Plant Disease*. 96(10):1577.

Farr, D., G. Bills, G. Chamuris, and A. Rossman. 1989. *Fungi on plants and plant products in the United States*. American Phytopathological Society Press. St. Paul, MN.

Farr, D. F., A. Y. Rossman, M. E. Palm, and E. B. McCray. (n.d.). Fungal databases, systematic botany and mycology laboratory, ARS, USDA. nt.ars-grin.gov/fungaldatabases

Henricott, B. 2010. New diseases of *Trillium* in the UK caused by *Colletotrichum lineola* and *Urocystis trillii. New Disease Reports*. 22, 32. dx.doi.org/10.5197/j.2044-0588.2010.022.032

Hill, S. N., O. P. Hurtado-Gonzales, K. H. Lamour, and M. K. Hausbeck. 2008. First report of mefenoxam sensitivity and pathogenicity of *Phytophthora citricola* isolated from American ginseng (*Panax quinquefolium*). *Plant Disease* 92(12).

Kirk, P., P. Cannon, J. David, and J. Stalpers. 2001. *Dictionary of the fungi*, 9th ed. CAB International. Wallingford, UK.

Pirone, P. 1978. *Diseases and Pests of Ornamental Plants*. John Wiley and Sons, New York, NY.

Planer, F. R. 1972. Aboveground stem infection caused by *Ditylenchus destructor* B nematode. *Nematologica* 18(1): 417.

Putnam, M. L., and L. J. Du Toit. 2002. First report of Alternaria blight caused by *Alternaria panax* on ginseng (*Panax quinque-*

folius L) in Oregon and Washington, USA. *New Disease Reports* 6(5).

Reeleder, R. D. 2003. The ginseng root pathogens *Cylindrocarpon destructans* and *Phytophthora cactorum* are not pathogenic to the medicinal herbs *Hydrastis canadensis* and *Actaea racemosa*. *Canadian Journal of Plant Pathology* 25(2): 218–221.

Westcott, C. 1971. *Plant Disease Handbook*, 3rd ed. Van Nostrand Reinhold Company, New York, NY.

Zhang, G. Z., and H. W. Zhang. 2007. First report of root rot of American ginseng (*Panax quinquefolium*) caused by *Ditylenchus destructor* in China. *Plant Disease* 91(4).

Helpful Resources, Websites, and Organizations

Goldenseal Resources

Davis, J. M., and A. Hamilton. 2008. Dryers for Commercial Herb Growers: A Construction Guide. North Carolina State University. ces.ncsu.edu/fletcher/programs/herbs/pdf /herb_dryer_leaflet.pdf

Diver, S. 2002. Notes on compost tea. attra .ncat.org/attra-pub/summaries/summary .php?pub=125

Environment Canada for CITES and export information. ec.gc.ca

Foot, D. K. with D. Stoffman. 2001. *Boom, Bust & Echo: Profiting from the Demographic Shift in the 21st Century*. Stoddart, Toronto, Canada.

Greenbook Data Solutions. A searchable database for plant protection product labels. greenbook.net

National Organic Program. United States Department of Agriculture. ams.usda.gov /AMSv1.0/nop

Pennsylvania Department of Environmental Protection. Compost Tea, As Easy as 1-2-3. dep.state.pa.us/dep/deputate/airwaste/wm /recycle/tea/tea1.htm

Plant Hardiness Zone Maps. Agriculture and Agri-Food Canada. sis.agr.gc.ca/cansis/ns db/climate/hardiness/index.html

Plant Hardiness Zone Map. United States Department of Agriculture. planthardiness.ars .usda.gov/PHZMWeb

Sustainable Agriculture Research and Education Program. SARE. sare.org

United Plant Savers. unitedplantsavers.org.

United States Fish and Wildlife Service. fws.gov

Ramps Resources

Canada Species at Risk Public Registry. Containing general information about the plants and animals of concern in Canada. This is where the federal species at risk list is. sararegistry.gc.ca

Government of Quebec Species at Risk. This is where the provincial species at risk list is. It is in French. mddefp.gouv.qc.ca/biodivers ite/especes/index.htm

Ontario Ministry of Natural Resources Species at Risk. This is where the provincial species at risk list is. mnr.gov.on.ca/en/Business /Species/index.html

Smoky Mountain Native Plant Association. Ramp meal and stone-ground corn meal are available for order by phone, mail, and email. smnpa.org

United States Department of Agriculture. The website contains a hyperlinked version of the USDA Plant Hardiness Zone Map. planthardiness.ars.usda.gov/PHZMWeb/

Rootstock, Plants, and Seed Sources

Below is a list of a few of the nurseries I am familiar with that sell rootstock, plants, and seeds for the species included in Parts 3 through 5. Others can be found through an online search.

There are not many nurseries that supply planting material in large quantities. That must often be purchased through a raw materials buyer who will source from a wild-harvester.

Nursery		Goldenseal	Ramps	Other Botanicals	Home Garden
Catoctin Mountain Botanicals	catoctinginseng.com	X		X	X
Companion Plants	companionplants.com	X		X	X
Elk-Mountain Nursery	elk-mountain.com			X	X
Garden Medicinals and Culinaries	gardenmedicinals.com			X	X
Gardens of the Blue Ridge	gardensoftheblueridge.com	X		X	X
Horizon Herbs	horizonherbs.com	X		X	X
J.L. Hudson Seedsman	jlhudsonseeds.net			X	X
Johnny's Selected Seeds	johnnyseeds.com	X		X	X
MoonBranch Botanicals	moonbranch.com			X	X
Mountain Gardens	mountaingardensherbs.com	X	X	X	X
Mountain Moss	mountainmoss.com				X
Nature's Cathedral	naturescathedral.com	X			X
North Carolina Ginseng and Goldenseal Co.	ncgoldenseal.com	X			X
Prairie Nursery	prairienursery.com			X	X
Prairie Moon Nursery	prairiemoon.com		X	X	X
Ramp Farm Specialties	rampfarm.com		X		X
Red Root Natives	redrootnatives.com	X	X	X	X
Richters Herbs	richters.com	X		X	X
Sandy Mush Herb Nursery	sandymushherbs.com	X	X	X	X
Southern Exposure Seed Exchange	southernexposure.com	X			X
The Thyme Garden Herb Company	thymegarden.com			X	X
United Plant Savers	unitedplantsavers.org	X			X
Well-Sweep Herb Farm	wellsweep.com		X	X	X

Comprehensive Resource Directory

A&L Analytical Laboratories, Inc., 2790 Whitten Road, Memphis, TN 38133, Phone: 901-213-2400, Fax: 901-213-2440, Website: allabs.com

A&L Heartland Laboratories, Inc., 111 Linn Street, PO Box 455, Atlantic, IA 50022, Phone: 712-243-6933, Fax: 712-243-5213, Website: alheartland.com

A&L Eastern Laboratories, 7621 Whitepine Road, Richmond, VA 23237, Phone: 804-743-9401, Fax: 804-271-6446, Website: al-labs-eastern.com

Alkemists Labs, 1260 Logan Avenue, #B2, Costa Mesa, CA 92626, Phone: 714-754-4372, Fax: 714-754-4372, Email: sales@alkemist.com, Website: alkemists.com

American Botanical Council, (Mark Blumenthal), Mailing Address: P.O. Box 144345, Austin, TX 78714-4345, Physical Address: 6200 Manor Road, Austin, TX 78723, Phone: 512-926-4900, Fax: 512-926-2345, Email: abc@herbalgram.org, Website: abc.herbalgram.org

American Botanicals, (Allen Lockard), PO Box 158, 24750 Hwy FF, Eolia, MO 63344, Phone: 573-485-2300, Fax: 573-485-3801, Email: info@americanbotanicals.com, Website: americanbotanicals.com

American Herbal Products Association, (Michael McGuffin), 8630 Fenton Street, Suite 918, Silver Spring, MD 20910, Phone: 301-588-1171, Fax: 301-588-1174, Email: ahpa@ahpa.org, Website: ahpa.org

ARBICO Organics, P.O. Box 8910, Tucson, AZ 85738-0910, Phone: 520-825-9785, Fax: 520-825-2038, Website: arbico-organics.com

The Association of Ginseng Growers of British Columbia (TAGG), (Doug Murdock), 1274 McGill Rd., Kamloops, BC, Canada V1B 3E8, Phone: 250-851-2880, Website: bcginseng.com

ATTRA, National Sustainable Agriculture Information Service, PO Box 3838, Butte, MT 59702, Website: attra.ncat.org

Beyfuss, Robert L., 136 Schuessler Lane, Preston Hollow, NY 12469.

BioNetwork Natural Products Laboratory, 1463 Sand Hill Road, Candler, NC 28715, Phone: 828-398-7945, Email: SarahASchober@abtech.edu, Website: abtech.edu/category/department/bionetwork

Bob's Goldenseal & Ginseng, (Robert Tipp), 35538 North State Highway 47, Warrenton, MO 63383, Phone: 636-456-2829, Email: sang_seal@hotmail

Boom, Bust & Echo books, (David Foot), Footwork Consulting, Phone: 416-484-9973, Fax: 416-484-8720, Website: footwork.com

British Columbia Ministry of Agriculture Ginseng Production Guide: agf.gov.bc.ca/speccrop/ginseng/ginseng_production_guide.htm

Buetsch Implement Co., Inc., 2895 County Road South, Marathon, WI 54448, Phone: 715-443-2276.

Canada Species at Risk Public Registry, Environment Canada, Inquiry Centre, 351 St. Joseph Boulevard, Gatineau, QC Canada K1A 0H3, Phone: 819-997-2800, Fax: 819-994-1412, Email: enviroinfo@ec.gc.ca, Website: sararegistry.gc.ca

Catoctin Mountain Botanicals, (Steve Galloway), PO Box 454, Jefferson, MD 21755, Phone: 301-473-4351, Email: info@catoctinginseng.com, Website: catoctinginseng.com

Companion Plants, 7247 N. Coolville Ridge Road, Athens, OH 45701, Phone: 740-592-4643, Fax: 740-593-3092, Email: sales@companionplants.com, Website: companionplants.com

Convention on International Trade in Endangered Species of Wild Fauna and Flora (CITES), CITES Secretariat, International Environment House, 11 Chemin des Anémones, CH-1219 Châtelaine, Geneva, Switzerland, Phone: +41-(0)22-917-81-39/40, Fax: +41-(0)22-797-34-17, Email: info@cites.org, Website: cites.org

Davis, Jeanine, North Carolina State University, Mountain Horticultural Crops Research and Extension Center, 455 Research Drive, Mills River, NC 28759, Phone: 828-684-3562, Email: Jeanine_Davis@ncsu.edu, Websites: ncherb.org, ncorganic.org

Eclectic Institute, 755 NE 6th Street, Gresham, OR 97030, Phone: 800-332-4372.

Elk Mountain Nursery, (Craig Mailloux), PO Box 3, Alexander, NC 28701, Phone: 828-683-9330, Email: craig@elk-mountain.com, Website: elk-mountain.com

Emerald Castle Farms, PO Box 324, Ashland, OH 44805, Phone: 419-651-8158, Email: brad@emeraldcastlefarms.com, Website: emeraldcastlefarms.com

Environment Canada, CITES and export information, Inquiry Centre, 10 Wellington, 23rd Floor, Gatineau QC, Canada K1A 0H3, Phone: 819-997-2800, Fax: 819-994-1412, Email: enviroinfo@ec.gc.ca, Website:ec.gc.ca

Food Safety Modernization Act, United States Food and Drug Administration, 5100 Paint Branch Pkwy, Wiley Building, HFS-009, Attn: FSMA Outreach, College Park, MD 20740, Email: FSMA@fda.hhs.gov, Website: fda.gov/Food/GuidanceRegulation/FSMA/default.htm

Gaia Herbs, Inc., (Jackie Greenfield), 101 Gaia Herbs Drive, Brevard, NC 28712, Phone: 800-831-7780, Fax: 800-717-1722, Email: jtg@gaiaherbs.com, Website: gaiaherbs.com

Gardens of the Blue Ridge, (Robyn Fletcher), PO Box 10, Pineola, NC 28662, Phone: 828-733-2417, Fax: 828-733-8894, Email: contact@gardensoftheblueridge.com, Website: gardensoftheblueridge.com

Garden Medicinals and Culinaries, PO Box 460, Mineral, VA 23117, Phone: 540-872-8351, Email: gardenmedicinals@gmail.com, Website: gardenmedicinals.com

Ginseng, West Virginia Division of Forestry, Website: wvforestry.com/ginseng.cfm?menucall=ginseng

Ginseng & Herb Co-op, A Wisconsin Ginseng Grower Cooperative, P.O. Box 581, 3899 County Rd. B, Marathon, WI 54448, Phone: 715-443-3355, Email: ginsengherbcoop@frontier.com, Website: ginsengherbco-op.com

Gintec Shade Technologies, (Mark Lucas), 266 Elmwood Avenue, Buffalo, NY 14222, Phone: 877-443-4743.

Good agricultural and collection practice for herbal raw materials, American Herbal Products Association, Website: ahpa.org/portals/0/pdfs/06_1208_AHPA-AHP_GACP.pdf

Good stewardship harvesting of wild American ginseng brochures, American Herbal Products Association. Website: ahpa.org/default.aspx?tabid=154

Government of Quebec Species at Risk (in French), Website: mddefp.gouv.qc.ca/bio diversite/especes/index.htm

Green Gold Enterprises, Inc., (W. Scott Persons, President), PO Box 236, Tuckasegee, NC 28783, Phone and fax: 828-293-5189, Email: wasp3@frontier.com

Greenbook Data Solutions, greenbook.net, searchable database for plant protection product labels.

Harding's Ginseng Farm, (Larry Harding), PO Box 267, Friendsville, MD 21531, Phone: 301-746-5380, Fax: 301-746-4516, Email: ginseng@hardingsginsengfarm.com, Website: hardingsginsengfarm.com

Health Canada, Natural Health Products, Health Canada, Address Locator 0900C2, Ottawa, ON, Canada K1A 0K9, Phone: 613-957-2991, Fax: 613-941-5366, Email: Info@ hc-sc.gc.ca, Website: hc-sc.gc.ca/dhp-mps /prodnatur/nhp-new-nouvelle-psn-eng.php

Herb Growing & Marketing Network, (Maureen Rogers), PO Box 245, Silver Spring, PA 17575, Phone: 717-393-3295, Fax: 717-393-9261, Website: herbnet.com

Herb Pharm, (Ed Smith), PO Box 116, Williams, OR 97544, Phone: 541-846-6262, Fax: 800-454-7392, Email: info@ herb-pharm.com, Website: herb-pharm.com

Herbalist & Alchemist, Inc., 51 South Wandling Avenue, Washington, NJ 07882, Phone: 908-689-9020/800-611-8235, Fax: 908-689-9071, Email: herbalist@nac.net, Website: herbalist -alchemist.com

Hershey's International, Inc., (James Aiello) 8210 Carlisle Pike, York Springs, PA 17372, Phone: 717-528-4495, Fax: 717-528-8503, Email: hershey@hersheyintl.com, Website: hersheysintl.com

High Falls Gardens, P.O. Box 125, Philmont, NY 12565, Phone: 518-672-7365, Website: highfallsgardens.net

Horizon Herbs, LLC, (Richo Cech), PO Box 69, Williams, OR 97544, Phone: 541-846-6704, Fax: 541-846-6233, Website: horizon herbs.com

Hsu's Ginseng Enterprises, Inc., (Paul Hsu), PO Box 509, Wausau, WI 54402-0509, Phone: 715-675-2325/800-826-1577, Fax: 715-675-3175, Email: info@hsu.ginseng.com, Website: hsuginseng.com

Ison, Tony, 24 Destiny Lane, Viper, KY 41774, Phone: 606-439-1494.

J. L. Hudson Seedsman, Box 337, La Honda, CA 94020-9733, Website: jlhudsonseeds.net

Johnny's Selected Seeds, PO Box 299, Waterville, ME 04903, Phone: 207-861-3900, Fax: 1-800-738-6314, Website: johnnyseeds.com

Lowe Fur and Herb, Inc., (Arthur C. Lowe, Jr.), 108 5th Street, North Wilkesboro, NC 28659, Phone: 336-838-3881, Fax: 336-667-3736, Email: alowe72849@aol.com

Marathon Feed Inc., (Jason), 3901 South Highway 107, Marathon, WI 54448, Phone: 715-443-2424, Email: marathonfeed@aol.com

Mike Dammen, 83552 270th St, Hollandale, MN 56045, Phone: 507-889-3303, Email: weirdtornadoe@yahoo.com

MoonBranch Botanicals, (Robin Suggs), 5294 Yellow Creek Road, Robbinsville, NC 28771, Phone: 828-479-2788, Email: moon branch@earthlink.net, Website: moon branch.com

Mountain Gardens, (Joe Hollis), 546 Shuford Creek Road, Burnsville, NC 28714, Phone: 828-675-5664, Email: joehollis@gmail.com, Website: mountaingardensherbs.com

Mountain Moss, 40 Holly Ridge Road, Pisgah Forest, NC 28768, Phone: 828-577-1321, Email: mossinannie@gmail.com, Website: mountainmoss.com

National Organic Program, Miles McEvoy, Deputy Administrator, Phone: 202-720-3252, Fax: 202-205-7808, Email: miles .mcevoy@ams.usda.gov, Website: ams.usda .gov/AMSv1.0/nop

Nature's Cathedral, (Leroy Ballard), 1995 78th Street, Blairstown, IA 52209, Phone: 319-454-6959, Fax: 319-454-9049, Email: natcath@netins.net, Website: naturescathedral.com

NatureServe Explorer, 4600 N. Fairfax Drive, 7th Floor, Arlington, VA 22203, Phone: 703-908-1800, Fax: 703-229-1670, Website: natureserve.org/explorer

North Carolina Arboretum Germplasm Repository, (Joe-Ann McCoy, Director), 100 Frederick Law Olmsted Way, Asheville, NC 28806-9315, Phone: 828-665-5178 ext. 268, Email: jmmcoy@ncarboretum.org, Website: ncarboretum.org

North Carolina Consortium on Natural Medicines, (Jeanine Davis and Susan Gaylord), University of North Carolina-Chapel Hill and North Carolina State University, Website: ncmedicinalherbs.org

North Carolina Department of Agriculture and Consumer Services, Plant Conservation Program/Plant Protection Section/Plant Industry Division, 1060 Mail Service Center, Raleigh, NC 27699-1060, Phone: 919-707-3753, Fax: 919-733-1041, Email: Phil.Wilson@ncagr.gov, Website: ncagr.gov/plantindustry/plant/plantconserve/index.htm

North Carolina Ginseng and Goldenseal Co., (Robert Eidus), 300 Indigo Bunting Lane, Marshall, NC 28753, Phone: 828-649-3536, Email: robert@ncgoldenseal.com, Website: ncgoldenseal.com

North Carolina State University, Website for ginseng guide: ces.ncsu.edu/depts/hort/hil/pdf/ ag-323.pdf

Ohio River Ginseng & Fur, Inc., (Mitchell McCullough), PO Box 2347, East Liverpool, OH 43920, Phone: 330-385-1832, Fax: 330-385-1842, Email: info@ohioriverginseng.net, Website: ohioriverginseng.net

Ontario Ginseng Growers Association, Box 587, 1283 Blueline Rd, Simcoe, ON, Canada N3Y 4N5, Phone: 519-426-7046, Email: info@ginsengontario.com, Website: ginsengontario.com

Ontario Ministry of Agriculture, Food and Rural Affairs (OMAFRA), 1 Stone Road West, Guelph, ON, Canada N1G 4Y2, Phone: 888-466-2372, Website for ginseng publication: omafra.gov.on.ca/english/crops/pub610

Ontario Ministry of Natural Resources Species at Risk, Species at Risk List, Phone: 800-667-1940, Fax: 705-755-1677, Email: mnr.nric.mnr@ontario.ca, Website: wmnr.gov.on.ca/en/Business/Species/index.html

Organic Materials Research Institute (OMRI), Box 11558, Eugene, OR 97440, Phone: 541-343-7600, Fax: 541-343-8971, Website: omri.org

Ozark Mountain Ginseng, (Dennis Lindberg), Route 1, Box 1223, Thayer, MO 65791, Phone and Fax: 417-264-2448, Email: lindy08@socket.net, Website: ozarkmountainginseng.com

Peaceful Valley Farm Supply, PO Box 2209, Grass Valley, CA 95945, Phone: 530-272-4769, Email: helpdesk@groworganic.com, Website: groworganic.com

Pennsylvania Department of Environmental Protection, Website for compost tea: dep.state.pa.us/dep/deputate/ airwaste/wm/recycle/tea/tea1.htm

Plant Hardiness Zone Maps, Agriculture and Agri-Food Canada, Website: sis.agr.gc.ca/cansis/nsdb/climate/hardiness/index.html

Plant Hardiness Zone Map, United States Department of Agriculture, Agricultural Research Service, Website: planthardiness.ars.usda.gov/PHZMWeb

Prairie Nursery, PO Box 306, Westfield, WI 53964, Phone: 800-476-9453, Fax: 608-296-2741, Website: prairienursery.com

Prairie Moon Nursery, 32115 Prairie Lane, Winona, MN 55987, Phone: 507-452-1362,

Fax: 507-454-5238, Email: info@prairie moon.com, Website: prairiemoon.com

Purdue University, Center for New Crops & Plant Products, Department of Horticulture, 625 Agriculture Mall Drive, West Lafayette, IN 47907, Fax: 765-494-0391, Website: hort.purdue.edu/newcrop

Quality Northern Ginseng Co., Inc., (Steve Korshak), 554 Cummings Road, Barre, VT 05641, Email: korshak@wildginseng.com

RRAWR, Inc., (Bruce Phetteplace), 701 Lyon Brook Road, Norwich, NY 13815, Phone: 607-334-4942, Email: BPhetteplace@Road runner.com

Ramp Farm Specialties, (Glen Facemire, Jr.), PO Box 48, Richwood, WV 26261, Phone: 304-846-4235, Email: rampfarm@wirefire .com, Website: rampfarm.com

Red Moon Herbs, PO Box 8023, Asheville NC 28814, Phone: 888-929-0777, Website: red moonherbs.com

Red Root Natives, 48 Wolfe Cove Road, Asheville, NC 28804, Phone: 828-545-0565, Email: redrootnatives@gmail.com, Website: redrootnatives.com

Richters Herbs, 357 Hwy 47, Goodwood, ON, Canada L0C 1A0, Phone: +1–905-640-6677, Fax: +1-905-640-6641, Website: richters .com

Ridge Runner Trading Company, Inc., (Tony Hayes), PO Box 391, Boone, NC 28607, Phone: 828-264-3615, Fax: 828-262-3605, Email: herbalogic@yahoo.com

Rural Action, (Tanner Filyaw), 9030 Hocking Hills Drive, The Plains, OH 45780, Phone: 740-677-4047 ext. 24, Fax: 740-767-4957, Website: ruralaction.org/forestry

Sandy Mush Herb Nursery, (Fairman and Kate Jayne), 316 Surrett Cove Road, Leicester, NC 28748-5517, Phone: 828-683-2014, Email: info@sandymushherbs.com, Website: sandymushherbs.com

Sego's Herb Farm, (Roger Sego), 4820 NE 306th Circle, Le Center, WA 98629, Phone: 360-263-7757, Fax: 360-263-7749, Email: info@segoherbfarm.com, Website: segoherb farm.com

Shelter Shade, (Glen Robinson), 132 Carruth Dr., Marietta, GA 30060, Phone: 800-997-4233.

Smoky Mountain Native Plant Association, PO Box 761, Robbinsville, NC 28771, Phone: 828-479-8788, Email: smnpa.info@gmail .com, Website: smnpa.org

Soil Food Web, Natural Resources and Conservation Services, Soil Primer, Soil Food Web, Website: soils.usda.gov/sqi/concepts /soil_biology/soil_food_web.html

Southern Exposure Seed Exchange, PO Box 460, Mineral, VA 23117, Phone: 540-894-9480, Fax: 540-894-9481, Email: gardens@ southernexposure.com, Website: southern exposure.com

Springland Trading Inc., (Jeff Boehner), 125-127 West 29th Street, New York, NY 10001, Phone: 212-594-9595, Email: jeffrey.boehner @gmail.com

Strategic Sourcing, Inc., (Edward Fletcher), 115 Snow Ridge Road, Banner Elk, NC 28604, Phone: 828-898-7642, Fax; 828-898-7647, Email: efletcher@strategicsourcinginc.net, Website: strategicsourcinginc.net

Sustainable Agriculture Research and Education Program (SARE), USDA-CSREES, Kim Kroll, Associate Director, 1122 Patapsco Building, University of Maryland, College Park, MD 20742-6715, Phone: 301-405-9912, Email: assoc_dir@sare.org, Website: sare .org

The Thyme Garden Herb Company, (Rolfe and Janet Hagen), 20546 Alsea Highway, Alsea, OR 97324, Phone: 541-487-8671, Email: herbs@thymegarden.com, Website: thymegarden.com

United Plant Savers, PO Box 400, East Barre, Vermont 05649, Phone: 802-476-6467, Fax: 802-476-3722, Email: office@UnitedPlantSavers.org, Website: unitedplantsavers.org

United States Department of Agriculture (USDA), Animal and Plant Health Inspection Service (APHIS), PPQ-APHIS-USDA, Permit Unit, 4700 River Road, Riverdale, MD 20737-1236, Phone: 877-770-5990, Website: aphis.usda.gov/ppq/permits/plantproducts

United States Department of Agriculture (USDA), National Agroforestry Center, East Campus—UNL, Lincoln, NE 68583-0822, Phone: 402-437-5178, Fax: 402-437-5712, Website: unl.edu/nac

United States Fish and Wildlife Service, Phone: 1-800-344-WILD, Website: fws.gov, For permits: International Affairs, US Fish & Wildlife Service, 4401 N. Fairfax Drive, Arlington, VA 22203, Phone for permits: 800-358-2104, Email for permits: managementauthority@fws.gov, Website for permits: fws.gov/international

United States Food and Drug Administration, Current Good Manufacturing Practices for Dietary Supplements, 10903 New Hampshire Avenue, Silver Spring, MD 20993, Phone: 888-463-6332, Website: fda.gov/Food/GuidanceRegulation/CGMP/

Vermont Ginseng Association, (Steve Norse), PO Box 92, Manchester Center, VT 05225, Phone: 802-867-0390.

Voles, Garden Friends and Foes, T. Murray, Whatcom County Extension, WA, Website: whatcom.wsu.edu/ag/homehort/pest/voles.htm

Voles (Meadow mice) Integrated pest management for home gardeners and landscape professionals, T. P. Salmon and W. P. Gorenzal, University of California, Website: ipm.ucdavis.edu/PDF/PESTNOTES/pnvoles.pdf

Well-Sweep Herb Farm, (Louise and David Hyde), 205 Mt. Bethel Road, Port Murray, NJ 07865, Phone: 908-852-5390, Website: wellsweep.com

West Virginia Trappers Association, Email: info@wvtrappers.com, Website: wvtrappers.com

Wild Ginseng: Regulations and guidelines for sustainable harvest, Wisconsin Department of Natural Resources., Website: dnr.wi.gov/files/PDF/pubs/LE/LE0005.pdf

Wild Grown.com, P.O. Box 103, Montgomeryville, PA 18936, Phone: 215-540-5890 or 888-675-7264, Email: info@wildgrown.com, Website: wildgrown.com

Wise Woman Herbals, (Jamie Nelson and Sue Sierralupe), PO Box 279, Creswell, OR 97426, Phone: 541-895-5152, Fax: 541-895-5174, Email: wise1@wisewomanherbals.com, Website: wisewomanherbals.com

World Health Organization, Guidelines on good agricultural and collection practices for medicinal plants, Avenue Appia 20, 1211 Geneva 27, Switzerland, Phone: + 41 22 791 21 11, Fax: + 41 22 791 31 11, Website: apps.who.int/medicinedocs/en/d/Js4928e/

Yat Chau (U.S.A.) Inc., (Benson), 40–20 Main Street, Flushing, NY 11354, Phone: 718-886-0300, Email: yatchauusa@hotmail.com

Index

A

acetic acid (vinegar), 68
Actaea racemosa. See black cohosh
Affolter, Jim, 308
Agriculture Canada, 33
agroforestry, 274
Alaskan ginseng, 13
Allium tricoccum. See ramps
Alternaria blight, 29–30, 106–108, 112
American ginseng. *See* ginseng
American Ginseng (Nash), 28
American Herbal Products Association, 434
Animal and Plant Health Inspection Service (APHIS), 16
Appalachia, 26–27
Apsley, Dave, 195
Aralia racemosa (spikenard), 345–348
Arbuckle, Bonnie, 409–420
artificial shade cultivation
  advantages of, 51–52
  disadvantages of, 52–56
  of goldenseal, 240–248
  harvesting, 139
  history of, 27–28, 34–35
  polypropylene shade cloth, 47–48, 50–51
  preparation and planting, 46–47
  profitability, 56–57
  wood lath shade structures, 47–50
artillery fungus, 396
Asarum canadense (wild ginger), 349–351
Asian ginseng (*Panax ginseng*), 12, 21, 23
Asian mayapple (*Podopyllum emodi*), 337
The Associated Ginseng Growers (TAGG), 34–35
Astor, John Jacob, 24–25

B

balloon flower (*Platycodon grandiflorum*), 362–363
bamboo ginseng (*Panax japonicum*), 12, 13, 23
Baptisia tinctoria (wild indigo), 353–356
Baskin, Carol, 327
Becker, H., 336
berberine, 190
berries and seeds, ginseng
  depulping, 129–132
  development of, 8, 10, 127–128
  diseases, 137–138
  germination requirements, 136–137
  harvesting, 128–129, 168–169
  producers, 86

production of, 153–156
berry plants, 389
bethroot (*Trillium erectum*), 285–290
Beyfuss, Robert, 46, 56, 62–64, 133–134, 238
biological controls, 113–114
Bir, Dick, 332
black cohosh (*Actaea racemosa*), 291–301
  in home garden, 385
  for interplanting, 114
  profitability, 368
black walnut (*Juglans nigra*), 61, 64
blight, 29–30, 106–108, 112
bloodroot (*Sanguinaria canadensis*), 303–313
  dye recipe, 408
  for interplanting, 114
  ornamental varieties, 385
  planting stock, 366
  propagation, 368–369
blossom snipping, 121
blue cohosh (*Caulophyllum thalictroides*), 315–321
  profitability, 361
Boone, Daniel, 25
Bordeaux mix, 112, 118
British Columbia, ginseng history, 34–35
Brunn, Randy, 76
Bullington, Bob, 410
Bullington Gardens, 409–414
Burkhart, Eric, 38
buyers
  ginseng, 16, 17–18, 156–157
  goldenseal, 222–223
  locating, 434

C

calcium carbonate (lime), 65
calcium requirements
  ginseng, 61–65
  goldenseal, 199–200
  ramps, 259
calcium sulfate (gypsum), 64–65
Canada
  export regulations, 15–16
  ginseng history, 23–24, 33–35
Canadian ginseng. *See* ginseng
Canadian Scientific Authority, 15
Carolina silver bells (*Halesia carolina*), 61
Carroll, Chip, 195
Caulophyllum thalictroides. *See* blue cohosh
Cech, Richo, 205, 206, 243, 287, 289, 311, 319, 320, 327, 350
Cecil, Chad, 191
Certificates of Origin, 14–15
Chai-Na-Ta Corporation, 34–35
Chamaelirium luteum. *See* false unicorn
Chan, Dennis, 39–40
China
  ginseng history, 19–20, 35–38
  market for ginseng, 159
CITES (Convention on International Trade in Endangered Species of Wild Fauna and Flora)
  bethroot, 290
  black cohosh, 292
  ginseng, 14–18, 31, 80
  goldenseal, 186, 187–188, 193, 202
  wild-harvesters and, 439–440
Civil War, 26
Classic Herbal (Shen Nung), 20
companion plants, 60–61, 231
Conner, Joe, 332
"Control of Diseases, Pests, and Weeds in Cultivated Ginseng" (Hausbeck), 105, 112
copper sulfate and lime, 112

Corbin, Jim, 64, 79–81
cultivation, ginseng
companion plants, 60–61, 231
history of, 27–31
initial trial, 151–152
options, 43, 147–148
outside native habitat, 11–12
See also artificial shade cultivation; wild-simulated planting; woods cultivation
Curtis, I.C., 29–30
Cushman, Kent, 337

D

damping-off, 106, 112, 113, 114
Davis, Donald, 396
dealers. *See* buyers
deer, 117, 390–391, 399–400
Department of Agriculture (USDA), 16
DeValue, J., 263
devil's club (*Echinopanax horridum*), 13
Deyton, Claude, 332
Dickman, Marvin, 30
direct marketing, 157–159, 223
disease
control, 77–78, 114–115
ginseng berries, 137–138
goldenseal, 211–214, 244
identification, 105–110
list, 477–481
in monocultures, 53–56
prevention, 110–114
ramps, 262–263
Division of Management Authority (DMA), 14, 31
Division of Scientific Authority (DSA), 14–15
documentation, 5
dogs, 116–117
Douglas, J.A., 206
downy mildew, 114
drying rooms, 141–142, 220–221
dwarf ginseng (*Panax trifoius*), 12, 13
dye, 189, 303, 353, 408

E

elderberry (*Sambucus pubens*), 61
elderberry syrup, 407
Ellis, Peggy, 422
Environmental Protection Agency (EPA), 55, 68
Evans, Brian, 23–24
export regulations, 14–16

F

Facemire, Glen, 272–282
false unicorn (*Chamaelirium luteum*), 323–328
potential of, 366
propagation, 369–370
ferns, 388
fertilizer
artificial shade cultivation, 46–47
bethroot, 286
black cohosh, 294
bloodroot, 305
effect on ginseng roots, 31, 41–42, 43
goldenseal, 198–201, 242
ramps, 258–259
soil test results and, 393–395
wild-simulated planting, 78
woods cultivation, 98–99
field cultivation. *See* artificial shade cultivation
Fish and Wildlife Service (USFWS), 5, 14
florist industry, 329–330
Foley, Ceara, 420–426
Follett, J.M., 206

Food and Drug Administration, 38
Forest Service, 330
Fromm brothers, 30
fungal diseases. *See* disease
fungicides. *See* pesticides

G

Gagnon, Daniel, 255, 257
galax (*Galax urceolata*), 329–333
giant blue cohosh (*Caulophyllum giganteum*), 315
Giblette, Jean, 360
ginseng
  cultivated vs. wild, 31–33
  future markets, 159–163
  history, 19–31, 33–38
  in home garden, 403–406
  life cycle, 7–10
  obtaining planting stock, 148–150
  range, 10–12
  regulations, 14–18
  related species, 12–14
  studies of, 4–5
  tea recipe, 407–408
Ginseng Act, 26
ginseng berries and seeds
  depulping, 129–132
  development of, 8, 10, 127–128
  diseases, 137–138
  germination requirements, 136–137
  harvesting, 128–129, 168–169
  producers, 86
  production of, 153–156
  stratification, 132–136
Ginseng Board of Wisconsin, 54–55
ginseng cultivation
  companion plants, 60–61, 231
  history of, 27–31
  initial trial, 151–152
  options, 43, 147–148
  outside native habitat, 11–12
  See also artificial shade cultivation; wild-simulated planting; woods cultivation
Ginseng Growers Association of Canada (GGAC), 33
ginseng harvest. *See* ginseng roots
"Ginseng Production Guide for Commercial Growers" (BC Ministry of Agriculture), 45, 138
ginseng roots
  development of, 10
  digging rate, 88
  grading, 39
  harvesting and drying, 138–145, 169–171
  storage, 145–146
  when to harvest, 156
ginseng seedlings, development of, 7
ginseng seeds. *See* ginseng berries and seeds
ginseng, selling
  cultivated vs. wild, 31–33
  domestic market, 38–39
  green crops, 233–234
  Hong Kong auction market, 39–40
  options, 156–159, 223
  recent prices, 40–42
ginseng, wild
  vs. cultivated, 31–33
  regulations, 16–17, 439–440
  suppliers of, 38
  threat to, 5
Gladstar, Rosemary, 237
goldenseal (*Hydrastis canadensis* L.)
  artificial shade cultivation, 240–248
  description, 188–189
  disease, 211–214, 244
  foliage loss, 305
  harvesting and drying, 216–222

history, 185–187
in home garden, 386
for interplanting, 114
irrigation, 210
medicinal uses, 189–191
mulching, 208–210
pest control, 210–211
planting, 207–208, 231, 238–239
profitability, 223–227
propagation, 201–207
regulations, 187–188
seed collection, 214–216
selling, 222–223, 232–233
site selection and preparation, 193–201
soap recipe, 408
Goldenseal Sanctuary, 325
Good Manufacturing Practices for Dietary Food Supplements, 298
government regulation, ginseng, 14–18
grading ginseng, 39
Greenfield, Jackie, 227, 259, 264, 272, 301, 313, 331
The "Green" Gold Rush (Oliver), 34
groundhogs, 400
Growing At-Risk Medicinal Herbs (Cech), 243, 287, 311, 319
gypsum (calcium sulfate), 64–65, 199–200, 259

H

Hankins, Andy, 84, 152
Hankins method, 84–88
Hardy, Karen, 307
harvesting ginseng. *See* ginseng roots
Hausbeck, M.K., 55, 105, 112
Having Your Ramps and Eating Them Too (Facemire), 280–281
Helicobacter pylori, 303
Hellyer, Clarence, 33
Hellyer Brothers, 33
herbicides. *See* pesticides
Hollis, Joe, 357–366
home garden
design, 379–384
examples of, 407–430
ginseng, 403–406
harvesting, 401–402
irrigation, 397
mulching, 396–397
plant choices, 385–391
planting, 395–396
site assessment, 373–379
weeding, 397–398
Hong Kong, ginseng consumption, 159
Hong Kong auction market, 39–40
Hus, Henri, 205
hydrastine, 190
Hydrastis canadensis L. *See* goldenseal

I

immune system, 5
insecticides. *See* pesticides
insects, 117–119, 399
irrigation
bloodroot, 305
ginseng, 120–121
goldenseal, 210
home garden, 397
ramps, 257

J

Japan, ginseng history, 23
Japanese ginseng (*Panax japonicum*), 12, 13, 23
Jartoux, Father, 23–24
Jenkins, Donan, 147
Joe-Pye weed (*Eupatorium purpureum*), 385, 411
Jones, Frances, 409–414, 418, 420

K

Kentucky, ginseng production, 38
Kershaw, John, 240–248, 298, 366–370
Konsler, Tom, 120, 136, 183, 259
Korea, 20–22, 38
Korean Tobacco and Ginseng Corporation, 22
Korean Tobacco and Ginseng Research Institute, 20–21

L

Lafitau, Joseph Francis, 24
Lambert, Juanita, 409–414, 418, 420
Lambert, Larason, 409–414, 418, 420
Lapointe, L., 287
Lass, William E., 25–26
Latta, John, 34–35
Lee, C.T., 337
leek moth, 263, 282
lime (calcium carbonate), 65
Littler, R.A., 206
Lobstein, M.B., 309

M

Manget, Luke, 26–27
Maqbool, Muhammad, 337
Marco Polo, 23
Marino, P.C., 305
mayapple (*Podophyllum peltatum*), 335–339
  foliage loss, 305
McCoy, Joe-Ann, 293, 294, 298–299
meadow mice, 400–401
medicinal plants
  cultivation, 283
  ornamental varieties, 386
  recipes, 407–408
  See also specific plants
mesh bags, 308
microchips, 79–81
Millspaugh, C.F., 324, 353
Minnesota, 25–26
Moraes, Rita, 337
mosses, 389
Mountain Gardens, 357–366
mulches
  ginseng, 47, 73–74, 86–87, 103–105
  goldenseal, 208–210, 215–216
  home garden, 396–397
  ramps, 257, 262
  retaining, 67
  straw, 118, 132
  tree sources, 98
Murphy, John, 409
Murphy, Laura, 4–5
mushrooms, 389–390
mycorrhizae, 54

N

Nash, George V., 28
National Center for Natural Products Research, 316
Neem oil, 112–113
nematodes, 211–212
New Jersey tea (*Ceanothus americanus*), 387
Nixon, Richard, 30
North Carolina, ginseng production, 38

O

oak trees, 94, 98
Oliver, Al, 34
Onofrietti, Maria, 303, 307
Ontario
  export regulations, 15
  ginseng history, 33–34
  ginseng production, 38
Ontario Ginseng Grower's Association, 15
Ontario Ministry of Agriculture, Food

and Rural Affairs (OMAFRA), 55, 105, 136
organic certification, 68–69, 158
Organic Materials Research Institute (OMRI), 68
Oriental ginseng (*Panax ginseng*), 12, 21, 23

P

Panax ginseng (Asian ginseng), 12, 21, 23
Panax japonicum Nees (Japanese ginseng), 12, 13, 23
Panax notoginseng Burkill (Sanchi ginseng), 12–13
Panax quinquefolius (American ginseng). *See* ginseng
Panax trifolius L. (dwarf ginseng), 12, 13
Park, Hoon, 22
partridge berry (*Mitchella repens*), 386
passionflower (*Passiflora incarnata*), 386
pesticides
  ginseng, 53, 55–56, 68–69
  goldenseal site preparation, 197–198
  for pests, 118–119
  preventative application, 111–113
"Pest Management in the Future: A Strategic Plan for the Michigan and Wisconsin Ginseng Industry" (Hausbeck), 55
pests
  ginseng, 115–119
  goldenseal, 210–211, 244
  in home garden, 399–400
  ramps, 263, 282
pH
  ginseng, 61–65
  goldenseal, 199–200
  ramps, 259
Philadelphia, 25
phosphorus, 62, 63
Phytobiotics, 311–312
pinkroot (*Spigelia marilandica*), 341–343
  in home garden, 386
plant inventories, 422
plant marking program, 79–81
poachers, 78–81
podophyllotoxin, 336–337
Podophyllum peltatum. *See* mayapple
polypropylene shade cloth, 47–48, 50–51
potassium, 62, 63
powdery mildew, 114
The Practical Guide to Growing Ginseng (Beyfuss), 46
Proctor, John, 120, 136
"Producing and Marketing Wild Simulated Ginseng in Forest and Agroforestry Systems, Publication 354-312," 85
"Production Recommendations for Ginseng" (OMAFRA), 55
protected status, 439

Q

Quebec, export regulations, 15
queen of the meadow (*Eupatorium purpureum*), 385, 411
quiche recipe, 408

R

Ramp Farm Specialties, 277–279
ramps (*Allium tricoccum*)
  description, 251–253
  disease, 262–263
  harvesting, 265–267, 270, 275
  mulching, 262
  pests, 263, 282
  profitability, 361

propagation, 259–262, 270, 280
protected status, 250–251
quiche recipe, 408
rebuilding populations, 273–274
reproduction, 275–277
seed stratification, 263–265
selling, 267–268
site selection and preparation, 255–259, 279, 280
traditions, 249
uses, 253
value-added products, 271–272, 277–279
range, American ginseng, 10–12
record keeping, 105
red ginseng, 13–14, 22
Reeleder, Rick, 109–110, 212
regulations, 14–18
See also CITES; United Plant Savers
Rock, Janet, 266
Rockwood, L.L., 309
Rocky Mountain ginseng, 13
rodents, 115–116, 211
root knot nematodes, 211–212
root rot, 108–110, 112, 113
roots, ginseng
development of, 10
digging rate, 88
harvesting and drying, 138–145, 169–171
storage, 145–146
when to harvest, 156
root washers, 219
Routhier, M.C, 287
rusty root, 108–109

S

Saint-Onge, Steve, 236–240
Sanchi ginseng (*Panax notoginseng*), 12–13
Sanguinaria canadensis. *See* bloodroot
sanguinarine, 303, 305, 311–312
security, 78–81
sedum, 368
seedlings, ginseng, 7
seeds and berries, ginseng
depulping, 129–132
development of, 8, 10, 127–128
diseases, 137–138
germination requirements, 136–137
harvesting, 128–129, 168–169
producers, 86
production of, 153–156
selling ginseng
domestic market, 38–39
green crops, 233–234
Hong Kong auction market, 39–40
options, 156–159
recent prices, 40–42
regulations, 14–18
wild vs. cultivated, 31–33
selling goldenseal, 222–223, 232–233
Septoria leaf spot, 262–263
shade, 91, 376–377
Shiitake mushrooms (*Lentinula edodes*), 389–390
Shipman, Vestal, 229–236
shrubs, in home garden, 387
Siberian ginseng (*Eleutherococcus senticosis*), 13
Sinclair, Adrianne, 15–16
slugs, 117–119, 210–211, 398–399
Smithsonian, 38
Smoky Mountain Native Plant Association, 268, 269–272
soap, 408
soil testing, 61–65, 393–395
Souther, Wallace, 426–430
southern ginseng (*Gynostemma pentaphyllum*), 361, 363–364

South Korea, ginseng history, 20–22
Special Crops Magazine, 29–30
spicebush (*Lindera benzoin*), 61
Spigelia marilandica. See pinkroot
spikenard (*Aralia racemosa*), 345–348
Stanton, George, 27–28
State Certificate of Legal Take, 16–17
State Certificate of Origin, 17
Stoltz, L.P., 62
sugar maples (*Acer saccharum*), 61, 63–64
Suggs, Robin, 305

T

TAGG (The Associated Ginseng Growers), 34–35
Taiwan, 159
Takeda, Fumiomi, 211
tea, 407–408
Tennessee, ginseng production, 38
trees
  calcium, 63–64
  hardwood canopy, 61, 195, 255, 376
  in home garden, 379, 387
  medicinal, 388
  shade provided by, 91
  site assessment, 93
  site clearing, 94
trenches
  in artificial shade cultivation, 48
  isolating root rot, 111
  in woods cultivation, 96–97, 100, 105, 110
Trichoderma, 113
Trillium erectum (bethroot), 285–290
triterpene glycosides, 300–301
tulip poplar trees (*Liriodendron tulipifera*), 61, 64
turkeys, 117, 400

U

United Plant Savers
  bethroot, 290
  black cohosh, 292
  bloodroot, 312
  blue cohosh, 320
  false unicorn, 323
  goldenseal, 186–187, 193
  mayapple, 336
  pinkroot, 341
  spikenard, 345
United States
  ginseng history, 24–33
  ginseng production, 38
United States Department of Agriculture (USDA), 16
United States Environmental Protection Agency (EPA), 55, 68
United States Fish and Wildlife Service (USFWS), 5, 14
United States Food and Drug Administration, 38
United States Forest Service, 330
US Good Manufacturing Practices for Dietary Food Supplements, 298

V

Vasseur, Liette, 255, 257
Viadent toothpaste, 303
vinegar (acetic acid), 68
vines, 386
Virginia Cooperative Extension, 85
voles, 400–401

W

wahoo (*Euonymus atropurpureus*), 386
wasabi, 364
weeds
  artificial shade cultivation, 53
  goldenseal, 195, 196–198, 211

mulch and, 104
wild-simulated planting, 78, 86
woods cultivation, 94–95, 105, 110, 114, 119–120
See also pesticides
Western Carolina Botanical Club, 410
Whanger, P., 253
Whisman, Abraham, 27
white ginseng, 14
Whitehead, Beverly, 269
wild ginger (*Asarum canadense*), 349–351
wild ginseng
cultivated vs., 31–33
regulations, 16–17, 439–440
suppliers of, 38
threat to, 5
wild-harvesting
about, 433–435, 437–438
goldenseal, 230
regulations, 439–440
wild indigo (*Baptisia tinctoria*), 353–356
wild-simulated planting
alternative planting methods, 75–77, 84–88
maintenance, 77–78
preparation and planting, 65–75
profitability, 81–84
security, 78–81
seed production, 154
site choice, 59–61
soil testing, 61–65
wild yam (*Dioscorea villosa*), 386
Wisconsin
family farms, 34
ginseng production, 30, 38
pesticides registrations, 56
woman's herbs, 291–292, 316
Wood, Oscar, 165–171
woodland medicinals
cultivation, 283
ornamental varieties, 386
recipes, 407–408
See also specific plants
wood lath shade structures, 47–50
woods cultivation
export from Canada, 15–16
maintenance, 119–121, 167–168
planting, 99–104
profitability, 121–126, 152–153
record keeping, 105
seed production, 154
site preparation, 91–99, 166–167
work involved, 89–91

Y

York, I. E., 33

New Society Publishers

ENVIRONMENTAL BENEFITS STATEMENT

New Society Publishers has chosen to produce this book on recycled paper made with 100% post consumer waste, processed chlorine free, and old growth free.

For every 5,000 books printed, New Society saves the following resources:[1]

66	Trees
5,948	Pounds of Solid Waste
6,544	Gallons of Water
8,536	Kilowatt Hours of Electricity
10,812	Pounds of Greenhouse Gases
47	Pounds of HAPs, VOCs, and AOX Combined
16	Cubic Yards of Landfill Space

[1] Environmental benefits are calculated based on research done by the Environmental Defense Fund and other members of the Paper Task Force who study the environmental impacts of the paper industry.